Covid-19 Infection and Pregnancy

Covid-19 Infection and Pregnancy

Edited by

AHMED M. MAGED EL-GOLY
Obstetrics and Gynecology
Kasr Alainy Medical School
Cairo University, Egypt

ELSEVIER

ACADEMIC PRESS
An imprint of Elsevier

Library of Congress Cataloging-in-Publication Data
A catalog record for this book is available from the Library of Congress

British Library Cataloguing-in-Publication Data
A catalogue record for this book is available from the British Library

ISBN: 978-0-323-90595-4

For information on all Academic Press publications visit our website at https://www.elsevier.com/books-and-journals

Publisher: Stacy Masucci
Acquisitions Editor: Ana Claudia A Garcia
Editorial Project Manager: Sam W. Young
Production Project Manager: Niranjan Bhaskaran
Cover Designer: Alan Studholme

Typeset by TNQ Technologies

Working together
to grow libraries in
developing countries
www.elsevier.com • www.bookaid.org

To my mam who supported me all through my life.
To my beautiful wonderful girls, Nour and Salma.

List of Contributors

Shimaa Mostafa Abd-El-Fatah, MD
Assistant Professor Obstetrics and Gynecology
Kasr Al-Ainy School of Medicine
Cairo University
Cairo, Egypt

Amr Sherif Hamza, MD
Senior Consultant of Obstetrics and Gynecology
Department of Obstetrics and Prenatal Medicine
Kantonsspital Baden
Baden, Switzerland

Yosra Mohamed Hassan, MD
Lecturer of Clinical and Chemical Pathology
Kasr Al-Ainy School of Medicine
Cairo University
Cairo, Egypt

Ahmed Mohamed Maged El-Goly, MD
Professor of Obstetrics and Gynecology
Kasr Al-Ainy School of Medicine
Cairo University
Cairo, Egypt

Ahmed Abdelmeguid Metwally, MD
Assistant Professor of Obstetrics and Gynecology
Kasr Al-Ainy School of Medicine
Cairo University
Cairo, Egypt

Adel Mohamed Nada, MD
Professor of Obstetrics and Gynecology
Kasr Al-Ainy School of Medicine
Cairo University
Cairo, Egypt

Noha Salah Soliman, MD
Lecturer of Clinical and Chemical Pathology
Kasr Al-Ainy School of Medicine
Cairo University
Cairo, Egypt

Ahmed Alaa-el-din Wali, MD
Assistant Professor of Obstetrics and Gynecology
Kasr Al-Ainy School of Medicine
Cairo University
Cairo, Egypt

Preface

By the end of 2019, the Chinese government informed the World Health Organization (WHO) about the appearance of many cases of unfamiliar pneumonia subsequently causing outbreak in Wuhan city. At the present time, the whole world is facing a battle with the pandemic caused by the seventh member of human coronaviruses (SARS-CoV-2). After its initial emergence in China, SARS-CoV-2 massively has spread to many countries all over the world and has reached almost all continents imposing a real global threat. The WHO raised a public health emergency of international concern in January 2020. In March, 2020, it was finally declared by the WHO that COVID-19 can be defined as a pandemic. There is no specific treatment for COVID-19 infection in the general population till the present time. Treatment of such infection is more challenging during pregnancy as successful drugs that can be used in nonpregnant women may have a hazardous effect on the growing fetus. Most clinical trials for a particular therapy do not include pregnant women for safety reasons. Data about the pathology, manifestations, prevention, and treatment are changing each day with the progress of the clinical trials and publishing their results. We extensively searched all the available evidence regarding the infection to reach the most appropriate management to reach the best prognosis for both the mother and her child. Data actually changing daily and all currently available lines of treatment were discussed in detail with special considerations to recommendations of their use.

Ahmed M. Maged El-Goly
Professor of Obstetrics and Gynecology
Kasr Alainy Medical School
Cairo University
10/02/2021

Contents

1 **Epidemiology, virology, and history of Covid-19 infection**, *1*
   Noha S. Soliman, MD, Yosra M. Hassan, MD and Adel M. Nada, MD

2 **Immunological Changes in Pregnancy and Its Relation to COVID-19 Infection**, *23*
   Amr S. Hamza, MD

3 **Diagnosis of COVID-19 Infection in Pregnancy**, *39*
   Ahmed M. Maged El-Goly, MD and Ahmed A. Metwally, MD

4 **Management of COVID-19 Infection During Pregnancy, Labor, and Puerperium**, *63*
   Ahmed M. Maged El-Goly, MD

5 **Lines of Treatment of COVID-19 Infection**, *91*
   Ahmed M. Maged El-Goly, MD

6 **Prognosis and Outcomes of COVID-19 infection During Pregnancy**, *145*
   Ahmed A. Wali, MD and Shimaa M. Abd-El-Fatah, MD

AUTHOR INDEX, *167*
SUBJECT INDEX, *175*

Epidemiology, virology, and history of Covid-19 infection

NOHA S. SOLIMAN, MD • YOSRA M. HASSAN, MD • ADEL M. NADA, MD

(Dr. Prof. M Fadel Shaltout. Prof. of Obstetrics and Gynecology, Cairo University, Faculty of Medicine)

Covid-19 Infection and Pregnancy. https://doi.org/10.1016/B978-0-323-90595-4.00005-4

INFECTION HISTORY

COVID-19 disease is a threatening infection that has appeared in December 2019 and has spread widely all over the globe to form a pandemic. According to the World Health Organization (WHO), it is considered the fifth pandemic since the Spanish flu 1918 pandemic (Liu et al., 2020a,b).

By the end of 2019, the Chinese government informed the WHO about the appearance of many cases of unfamiliar pneumonia subsequently causing outbreak in Wuhan city (Shereen et al., 2020). The events started in a seafood market located in Wuhan city, China, which frequently sells live animals such as frogs, bats, birds, snakes, and rabbits (Wang et al., 2020). It was reported that more than 50 people were rapidly infected and suffered from fever, dry cough, malaise, and dyspnea suggesting viral pneumonia as reported by the National Health Commission of China in January 2020. On the basis of genome sequencing of the virus, it was found that it is a novel coronavirus that belongs to group b of coronaviruses (Cui et al., 2019; Lai et al., 2020).

The newly discovered virus is considered the seventh among the coronaviruses that infect humans (Wu et al., 2020). The new virus was termed by the WHO as 2019 novel coronavirus (2019nCoV) in January 2020, and the infectious disease was officially named as COVID-19 in February 2020. The genomic characterization of the virus provided full analysis, and according to the International Committee on Taxonomy of Viruses (ICTV), it was named as SARS-CoV-2 (Coronaviridae Study Group of the International Committee on Taxonomy of Viruses).

Back to history, in the 1930s, several variants of coronaviruses were discovered, and since 1965, four human coronaviruses (HCoVs) have been identified to cause minor respiratory infection (Chauhan, 2020; McIntosh, 1974). The first two discovered human coronaviruses were HCoV-OC43 and HCoV-229E. The HCoV-NL63 and HCoV-HKU1 were other two human coronaviruses identified in the 1960s (Yesudhas et al., 2020). In 2003, a new virus that belonged to beta subgroup of coronaviruses infected Chinese population in Guandong province. The infected patients suffered from symptoms of severe pneumonia with expanded alveolar injury, which caused acute respiratory distress syndrome (ARDS), and the whole world witnessed the first appearance of severe acute respiratory syndrome (SARS) (Pyrc et al., 2007).

Initially, SARS coronavirus appeared in Guandong, China, and then disseminated all over the world rapidly to infect >8000 and kill 776 persons. Ten years later in 2012, another highly pathogenic member of coronaviruses emerged in Saudi Arabia (Shereen et al., 2020). The virus was considered the sixth among the emerged human coronaviruses that belongs to beta subgroup; however, it differs in phylogenetics from other human coronaviruses and was named as Middle East respiratory syndrome coronavirus (MERS-CoV). It emphasized the capability of coronaviruses to unexpectedly be transmitted from animals to humans (Zaki et al., 2012).

As reported by the WHO, about 2482 persons were infected with MERS-CoV with 838 deaths (Rahman and Sarkar, 2019). The infection by MERS-CoV varies from mild respiratory injury up to severe respiratory distress. Like SARS coronavirus, MERS-CoV exhibits symptoms of pneumonia that may reach up to acute respiratory distress and kidney failure (Memish et al., 2013). MERS-CoV started in Saudi Arabia and then spread to many Middle East countries (Shereen et al., 2020). Although MERS-CoV was relatively considered slowly spreading coronavirus, it recorded fatality rates of 36% (Yesudhas et al., 2020).

At current, the whole world is facing a battle with the pandemic caused by the seventh member of human coronaviruses (SARS-CoV-2). After its initial emergence in China, SARS-CoV-2 massively has spread to many countries all over the world and has reached almost all continents imposing a real global threat. The WHO raised a Public health emergency of international concern in January 2020 (Chan et al., 2020; Li et al., 2020a,b). In March 2020, it was finally declared by the WHO that COVID-19 can be defined as a pandemic (Liu et al., 2020a,b).

Prior to the emerged outbreaks by coronaviruses, the whole world previously suffered from various outbreaks by different strains of influenza viruses that reaped millions of human deaths as H1N1 (Spanish flu, 1918), H2N2 (Asian flu, 1957), H3N2 (Hong Kong flu, 1968), and H1N1 pandemic flu (Liu et al., 2020a,b).

The emergence of the latest SARS-CoV-2 pandemic has put the world in a state of extreme confusion and has shed the light on the countless flaws in the health systems in modern societies and the unpreparedness of many governments to face this scenario of extensive spread of the virus specially with the exponential rise of infections beyond the capacity of public hospitals. Generally, the outcome of pandemics crucially depends on world cooperation that is considered imperative in containing infection and facing the devastating consequences of the pandemic (Häfner, 2020).

EPIDEMIOLOGY

Reservoirs and Hosts of Coronaviruses

Studying the origin and transmission of infections is of utmost importance to help break the chain of transmission and develop strategies to contain infections and prevent their spread (Shereen et al., 2020).

All human coronaviruses originally stemmed from animals that act as natural hosts. Mainly bats were considered the natural hosts of HCoV-NL63, HCoV-229E, SARS-CoV, and MERS-CoV. However, rodents were probably the animal origin for HKU1 and HCoV-OC43. Generally, bats are considered the key reservoirs for alpha and beta coronaviruses (Woo et al., 2012). Rhinolophus bats are claimed to be the natural hosts for SARS CoV, while recent researches detected MERS-CoV in *Perimyotis* and *Pipistrellus* bats (Annan et al., 2013).

The transmission of coronaviruses from natural hosts to human requires the presence of intermediate hosts. Mostly, domestic animals act as intermediate hosts, as they get diseased by the virus and then transmitted it to humans. For example, palm civets and camels played a key role in transmitting SARS-CoV and MERS-CoV to humans, respectively, by being intermediate hosts for the viruses (Haagmans et al., 2014). The whole-genome sequencing of these viruses showed 96.2% similarity at the full-length genome level to a coronavirus (Bat-CoVRaTG13) whose natural host was a bat named as *Rhinolophus affinis* that lives in Yunnan Province at a distance of 1500 km from Wuhan (Zhou et al., 2020).

The presence of an intermediate host facilitates the transmission of viruses from their natural hosts to humans. Similar to other coronaviruses, bats are claimed to be the natural hosts for SARS-CoV-2, which has been transmitted to human either through direct contact or via an intermediate host. The role of intermediate hosts in transmitting SARS-CoV-2 remains inconclusive and has no solid evidence, as no enough samples were taken from suspected intermediate hosts to be tested by the scientists in the beginning when infections appeared in wild life and sea food markets in Wuhan, as wild animals could be the source of zoonotic infections. However, many researchers of phylogenetic analysis still work on tracing sources of COVID-19 infection assuming that the infection had multiple sources in the beginning of its spread (Liu et al., 2020a,b).

It was highly suggested that pangolins act as intermediate hosts for SARS-CoV-2 that might have carried the virus from bats to humans, as the whole-genome sequencing showed that the coronavirus detected in samples taken from Malayan pangolins (*Manis javanica*) in Guandong, China, was highly identical to SARS-CoV-2 (Lam et al., 2020). Also, researchers suspected that raccoon dogs and palm civets had possibly transmitted SARS-CoV-2 infection to humans (Liu et al., 2020a,b). However, molecular tests previously showed positive results for corona-like viral RNA in samples taken from civets at the food market, suggesting that civet palm might act as intermediate hosts (Shereen et al., 2020).

Surprisingly, molecular analysis was performed in a study on samples taken from healthy individuals in Hong Kong in 2001 and showed a frequency rate of 2.5% for SARS-CoV, which suggests the circulation of SARS coronaviruses in humans before the first outbreak appearance in 2003 (Zheng et al., 2020)

Modes of Transmission

As agreed by the majority of researchers, SARS-CoV-2 outbreak initially started by transmission of the virus from a natural host to human either directly or through an intermediate host, and then subsequently, human-to-human transmission had been reported. The human-to-human transmission of SARS-CoV-2 virus can occur directly through exposure to respiratory droplets from infected patients generated by coughing, sneezing, or even talking at a distance of less than 1 m. Moreover, indirect transmission may occur through touching surfaces, clothes, or personal belongings contaminated by the virus. SARS-CoV-2 mainly spreads through big respiratory droplets. However, the accelerated exponential rise in the rates of SARS-CoV-2 raised the suspicions about the possibility of viral transmission through aerosols in air, yet no clear data is available to prove or disprove the theory of airborne transmission (Yesudhas et al., 2020).

The rate of infection (R0) is defined as the number of people acquiring microbial infection by an infected individual. For SARS-CoV-2, the R0 value was estimated in the range of 1.5−3.5, which was found to be close to the R0 value (2.75) of SARS pandemic in 2003. However, the R0 value of MERS-CoV-2 in 2012 was estimated to be around 1, and for H1N1 influenza in 2009 was 1.46−1.48. The difference in R0 values between various coronaviruses appears to be minimal (Phelan et al., 2020; Somsen et al., 2020).

The hardships faced with SARS-CoV-2 in controlling the high rates of infection are owed to various reasons as follows: (1) overlapping symptoms with other noncorona respiratory viruses, (2) inconsistency of clinical course of infection among various patients with uncertain incubation period, (3) many infected individuals may not show symptoms; however, they are capable of transmitting the infection, and (4) variable risk predisposition to acquiring infection among different population. All of these factors need further researches to uncover more facts about the virus that may help in overcoming the challenges faced in controlling its spread (Yesudhas et al., 2020).

At-Risk Populations

People with underlying health problems such as diabetes, hypertension, cardiovascular, chronic respiratory, chronic renal diseases, cancer, and immune suppression are liable to acquire COVID-19 infection and most probably develop severe course of illness with poor or fatal outcome (European Center for Disease Prevention and Control, 2020).

In a European multicenter study, the most common risk factors for severe illness and intensive care unit (ICU) admission in adolescents and children were the presence of underlying health problems such as chronic lung disease, congenital heart disease, malignancy, and chronic kidney disease (European Center for Disease Prevention and Control, 2020).

People residing long-term care facilities specially those of old age (more than 60 years) with underlying medical problems are vulnerable to infection with high likelihood to adverse consequences and unfavorable outcomes. Other settings with medically vulnerable people include long-term care hospital wards, daycare centers, hostels, and home-based centers (European Center for Disease Prevention and Control, 2020)

The category of healthcare workers is considered of the highest risk of COVID-19 infection due to the high chance of exposure to infected patients. According to a study done in the United Kingdom, the risk of infection among the frontline healthcare workers is 3.4-fold higher than people living in the community. In china, it was recorded that 3.8% of SARS-CoV-2-infected cases were healthcare workers, and 14.8% of them had severe disease (European Center for Disease Prevention and Control, 2020).

In May 2020, it has been reported by the International Council of Nurses that about 90,000 healthcare workers have been infected with SARS-CoV-2 with more than 260 deaths during the pandemic. In June 2020, the Unites states reported 600 deaths due to COVID-19 among the frontline healthcare workers. Poor compliance and malpractice in dealing with personal protective equipment was considered the key factor for the elevated rates of infection among healthcare workers (European Centre for Disease Prevention and Control, 2020).

Asymptomatic patients infected with the virus are considered hidden sources that mediate transmission to healthy individuals. It was estimated in a systematic review that asymptomatic cases account for 6%—41% of SARS-CoV-2 positive patients. In asymptomatic cases, symptoms may start to appear later than in usual symptomatic cases, or they may remain without appearance of any symptoms or signs. However, in these cases, the SARS-CoV-2 viral shedding continues in gastrointestinal and respiratory tract samples carrying the risk of transmitting the virus to healthy individuals. Non- or late appearance of clinical symptoms and signs makes it difficult to trace asymptomatic transmission, which in turn may hinder the ability to estimate or quantify the actual number of infected cases (Byambasuren et al., 2020; Koh et al., 2020).

In terms of age, the available data showed that the chance of developing infection in children was 0.26 time slower than in old people (Jing et al., 2020). Children may contract SARS-CoV-2 infection from any gathering places such as schools, daycare centers, and sport clubs or through exposure to an infected family member at home. Publication data from Italy showed that 55% of infected children acquired the infection from a source outside family (Parri et al., 2020). However, a study in Italy reported that the majority of infected children (67%) acquired infection due to contact with at least one infected parent (Garazzino et al., 2020).

A matter of concern in children who go out for school or daycare centers is that many of children who get infected may be asymptomatic or exhibit mild nonspecific symptoms, which may not predict their infection with SARS-CoV-2. Nonapparent infection with SARS-CoV-2 makes it hardly suspected. Even the symptomatic children may continuously shed the virus in the early phase of acute illness before appearance of symptoms or before being confirmed of having SARS-CoV-2 by laboratory testing. The danger lies in the potential risk of transmitting infection to their parents or elderly family members who may have underlying medical problems ending into adverse consequences up to death (European Center for Disease Prevention and Control, 2020). It was reported from publication data in Germany that viral loads of SARS-CoV-2 in symptomatic children are comparable with middle-age and old-age persons (Wolfel et al., 2020; Jones et al., 2020). However, in another study, higher viral load was detected in symptomatic children (under and above 5 years of age), as well as adults (Heald-Sargent et al., 2020).

Occupational settings and work places with unfavorable environmental health conditions are considered epicenters for emergence of multiple outbreaks with COVID-19 infection. This is worsened by defective implementation of infection control measures and malpractices of workers inside these settings. Since the emergence of SARS-CoV-2, multiple outbreaks were reported in many occupational settings (Waltenburg et al., 2020). Several contributing factors are involved in occurrence of outbreaks in work settings such as (1) small working indoor spaces, (2) sharing same work tools and facilities in office and accommodation spaces, (3) inadequate compliance with the recommended social distancing, (4) shortage of personal protective equipment or improper use in donning and doffing, and (5) fear of losing job that may force some infected individuals with impaired awareness to ignore their illness, deny reporting, and continue going to their work exposing other employees to the risk of acquiring infection (Park et al., 2020; Baker, 2020).

In April 2020, the confirmed COVID-19 cases were estimated at 2,114,269 worldwide with about 60% of cases occurred mainly in Spain, Italy, Germany, France, and the United States as the distribution shown in Fig. 1.1. According to the WHO, as of December 8, 2020, the total number of reported cases all over

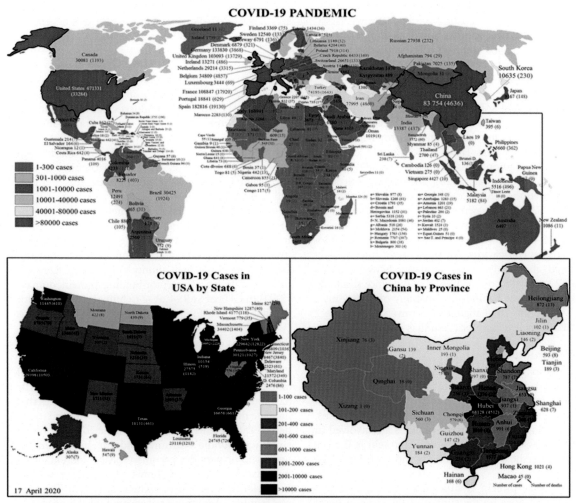

FIG. 1.1 The geographical distribution of SARS-CoV-2-infected cases according to the European CDC in April 2020 (https://www.ecdc.europa.eu/en/geographical-distribution-2019- ncov-cases). *CDC*, Centers for Disease Control and Prevention; *SARS*, severe acute respiratory syndrome coronavirus-2.

the world reached 65.8 million with 1.5 million deaths in 220 countries all over the globe (da Costa et al., 2020).

In respect to previous outbreaks by coronaviruses, Fig. 1.2 demonstrates the count of cases for SARS and MERS coronaviruses and their geographical distribution worldwide. In 2003, the SARS outbreak caused 8096 cases with774 deaths. In April 2012, the laboratory-confirmed cases for MERS-CoV were 2519 with a fatality rate of 34.3% (da Costa et al., 2020).

VIROLOGY

Viral Taxonomy

The Coronaviridae family belongs to Nidovirales order. The name of "corona" refers to crown-like spikes on the surface of the virus. The Coronaviridae family is classified into alpha, beta, gamma, and delta groups of coronaviruses. The alpha and beta coronaviruses can infect animals. On genetic basis, human coronaviruses mostly belong to beta coronavirus genus (B-CoV). The B-CoVs are further classified into four different lineages: A, B, C, and D. SARS and SARS-CoV-2 belong to the lineage B, whereas MERS-CoV is grouped in lineage C (Letko et al., 2020). As for other human coronaviruses, the alpha group includes HCoV-NL63 and HCoV-229E, whereas beta coronavirus includes HCoV-OC43 and HCoV-HK1 (Wan et al., 2020). The phylogenetic analysis showed that SARS-CoV-2 has 80% similarity with that of SARS coronavirus and is 96% identical to BatCoV-RaTG3 coronavirus (Zhou et al., 2020). However, MERS-CoV showed 54% relatedness to HKU4 Tylonycteris bat coronavirus. The sequence of spike protein showed about 76%−78% similarity between SARS-CoV-2 and SARS coronavirus. This sequence similarity is the root cause behind the capability of both viruses to bind to the same angiotensin-converting enzyme 2 (ACE2) host cell receptor (Yesudhas et al., 2020).

Viral Structure

The coronaviruses are enveloped spherical particles about 120 nm in diameter containing genetic material of single-stranded RNA. The outer surface is made of membrane (M), envelope (E), and spike (S) proteins. The envelope and membrane proteins are involved in the virus assembly, while the spike protein is the key element for host cell recognition and virus entry (Li, 2016).

The spike protein is structured as peplomers that form protrusions on the surface of the virus giving it the shape as if the coronavirus carries a crown, hence the name "corona," which is a Latin word that means

a crown. The spike protein is divided into three segments: (1) ectodomain that consists of S1 and S2 receptor-binding subunit, (2) transmembrane domain, and (3) intracellular domain (Yesudhas et al., 2020).

The SARS-CoV-2 RNA genome contains 29,903 nucleotides with a 50-methyl-guanosine cap and poly(A)-tail (Wu et al., 2020). The SARS-CoV-2 has nine transcribed subgenomic RNAs, and its genome contains a 50-untranslated region that includes a 50 leader sequence, an opening reading frame (ORF)1a/ab that encodes nonstructural proteins (nsp) needed for replication: four structural proteins (spike, membrane, envelope, and nucleocapsid); accessory proteins (ORF3a6,7a/b and 8); and a 30-untranslated region. The polyprotein pp1a/b is broken down into 16 nonstructural proteins including nsp3 and nsp5 (proteases), nsp13 (helicase), and nsp12 (RNA-dependent RNA polymerase) (Liu et al., 2020a,b).

Pathogenesis and Viral Life Cycle

There are two pathways for coronaviral entry into host cells: endocytic and nonendosomal pathways (Zumla et al., 2016). The endocytic pathway (clathrin-dependent endocytosis) was demonstrated through various studies for MERS-CoV and SARS-CoV viral entry. It has been reported that the same mechanism is used for SARS-CoV-2. The exact mechanism of viral entry is dependent on the type of both virus and host cell (Yesudhas et al., 2020).

In viral infection, the spike protein is cleaved by the proteases of host cell into S1 receptor binding and S2 membrane fusion subunits. The S1 subunit is divided into N-terminal (NTD) and C-terminal (CTD) domains. The CTD of S1 has high affinity to human ACE2 receptor. The affinity of receptor-binding protein domain (RBD) in CTD of SARS-CoV-2 to human ACE2 receptor is higher 10−20 folds than RBD of SARS coronavirus (Wrapp et al., 2020). The putative cycle of SARS-CoV-2 inside host cells starts with binding of viral spike protein and human ACE2 receptors (Liu et al., 2020a,b). The S1 subunit of the viral spike protein binds to sugar and ACE2 receptors on the surface of host cell, while the S2 subunit is subjected to conformational changes that mediate the fusion of the viral envelope with cell membrane. During this state, the trimeric S2 conforms a six-helical bundle structure, and the hidden fusion hydrophobic peptides become exposed and entangled into the host cell membrane facilitating viral and host cell membrane fusion (Gui et al., 2017). This process requires large amount of energy, which is needed to accelerate the membrane fusion. Proteases such as elastases and transmembrane protease serine 2

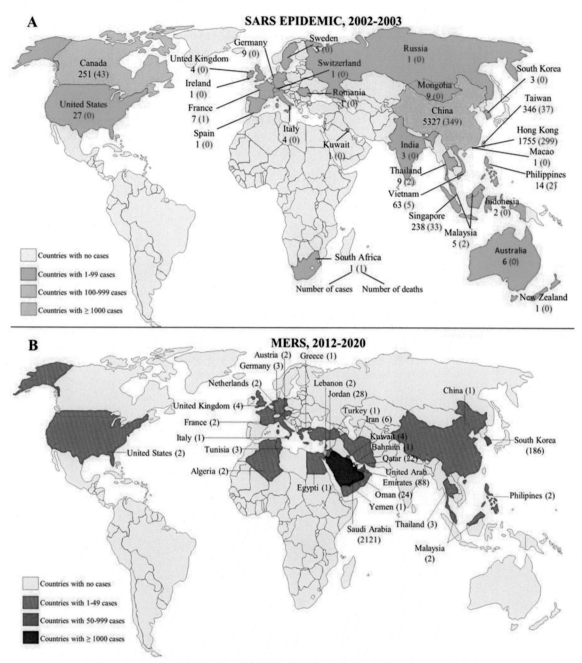

FIG. 1.2 The geographical distribution of SARS-CoV (**A**) and MERS-CoV worldwide (**B**) according to the World Health Organization (https://www.who.int/csr/sars/country/table2004_04_21/en/), (http://www.emro.who.int/health- topics/mers-cov/mers-outbreaks.html). *MERS-CoV*, Middle East respiratory syndrome coronavirus; *SARS*, severe acute respiratory syndrome coronavirus.

(TMPRSS2) on the surface of respiratory tract cells and lung cells play an important role in spike protein priming to activate membrane fusion (Fig. 1.3) (Yesudhas et al., 2020).

After membrane fusion, the viral genome is released into the cytoplasm and becomes translated into pp1a and 1ab replicase viral polyproteins. A group of subgenomic mRNAs undergoes transcription by polymerase enzyme; then membrane proteins enter the Golgi apparatus and endoplasmic reticulum, while the N protein binds to the genomic RNA forming nucleoprotein complex. The nucleoprotein complex, structural proteins, and viral envelope are assembled to form new viral particles, which are released from infected cells to enter new host cells repeating the same cycle (Liu et al., 2020a,b).

Viral Spike Protein Active and Inactive States

SARS-CoV S1 subunit of the spike protein is composed of beta strand structures, formed of three C-terminal

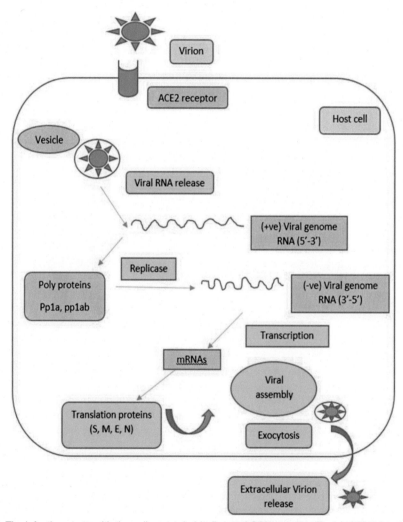

FIG. 1.3 The infection starts with the spike protein binding to ACE2 receptor. The viral RNA is released into the host cell and is translated into pp1a and pp1ab viral replicase polyproteins. The replicase enzyme uses the RNA genome positive strand as a template to produce copies of negative-strand viral RNA genome. In transcription phase, RNA polymerase enzyme produces subgenomic RNAs that become translated into viral proteins spike (S), membrane (M), envelope (E), and nucleocapsid (N). The viral RNA genome and proteins are assembled to be released into vesicles out of the cell.

domains (CTD1, CTD2, and CTD3) and N-terminal domain (NTD). The NTD is bound to CTD1 through 295—319 residues linker, where CTD1 acts as receptor-binding domain (RBD) for SARS-CoV-2 and binds with the ACE2 receptor (Yesudhas et al., 2020). All the conformations that occur to the spike glycoprotein depend on the position of CTD1. The three-monomer spike glycoprotein interlaces with each other and forms homotrimer. The head of this trimer is taken place by CTD1 and NTD of S1 subunits, where the CTD1s are placed in the center, while the NTDs are outside of this head. The S2 subunits constitute the stem of this trimer, which is then surrounded by CTD2 and CTD3 of the S1 subunit. When the spike protein is in the inactive state, the S2 subunit becomes covered by CTD1 (head portion), which takes "down" position causing steric clashes for binding between ACE2 and spike protein. In the active state, one CTD1 turns outward "up confirmation," which uncovers S2 subunit and thus allows the interaction between ACE2 receptor and spike protein (Fig. 1.4) (Yesudhas et al., 2020).

SARS-CoV-2 and SARS-CoV Key Residues

The similarity in the sequence of SARS-CoV and SARS-CoV-2 spike protein may support the theory of having the same ACE2 receptor in host cell (Wan et al., 2020). The sequence studies identified the key residues that are involved in spike—ACE2 receptor interactions. The key residue at 493 position in RBD of SARS-CoV-2 is Gln, whereas the corresponding 479 residues in SARS-CoV of humans and civets are Asn and Lys, respectively. As the 479 residue of RBD is near to the virus binding Lys31 residue of human ACE2 receptor, the Lys residue in civet leads to stearic clashes weakening binding to human ACE2 receptor. However, the Lys 479-Asn mutation showed that the Asn lying in 479 position of human SARS-CoV strengthens the binding of virus with human ACE2 receptor. The Gln493 residue in SARS-CoV-2 RBD is matching with hot spot Lys31 residue of human ACE2 receptor explaining the target cell identification (Wan et al., 2020). Likely, the 501 residue in the RBD of SARS-CoV-2 is Asn, while the corresponding residues at 487 position in civet RBD are Ser and Thr in humans.

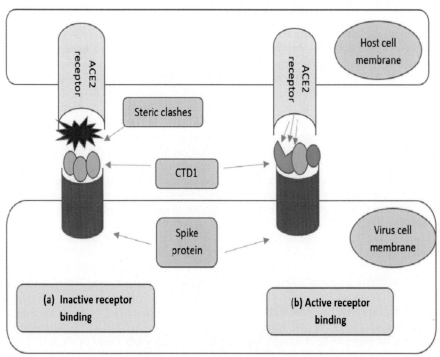

FIG. 1.4 Conformations of spike protein in active and inactive states. **(A)** The conformations in S2 subunit in the inactive state where CTD1s completely cover the S2 subunits "down position." This hinders the binding of the spike protein with ACE2 receptor. **(B)** Spike protein conformation in the active state in which one of the CTD1s (showed in green) is in an open state facilitating the binding of the spike protein with ACE2 receptor. *ACE2*, angiotensin-converting enzyme 2; *CTD*, C-terminal domain.

This residue plays a key role in mediating the interaction with Lys353 hot spot residue of ACE2 receptor. The analysis of Sr487-Thr mutation shows a significant role in human ACE2 receptor binding and in human-to-human transmission. Thus, interacting with Lys353 will be more favorable for human SARS-CoV threonine and SARS-CoV-2 Asn than civet serine (Wan et al., 2020).

The Lue455, Ser494, and Phe486 residues in SARS-CoV-2 RBD and Tyr442, Asp480, and Leu472 residues in human and civet SARS-CoV are also considered important for binding with the human ACE2 receptor. The interaction of Tyr 442 residue of civet and human SARS-CoV RBDs with Lys31 hot spot residue of ACE-2 receptor is considered unfavorable; however, Lue455 residue of SARS-CoV-2 shows favorable interactions. The Phe486 residue of SARS-CoV-2 RBD interacts more favorably with Lys31 residue of human ACE2 receptor than Leu472 residue of human–civet SARS-CoV. The Ser49 residue shows positive affinity for the hot spot residue Lys353 of human ACE2 receptors. Also, the Asp480 residue of SARS-CoV shows favorable interaction with the Lys353 residue (Yesudhas et al., 2020).

In describing the viral entry process, the Lys353 and Lys 31 are named as "hot spot" residues of human ACE2 receptors, which are made of salt bridge in a crypt of hydrophobic medium and contribute vitally for binding of virus and host cell receptor. This key residue comparison of human and civet SARS-CoV with SARS-CoV-2 emphasizes how SARS-CoV-2 can actively choose and bind with human ACE2 receptors, which subsequently can cause human-to-human transmission (Wan et al., 2020). The heterogenic amino acids in ACE2 receptors contribute for wavering the virus- and host-binding affinities. However, the variations in ACE2 host cell receptors can resist binding and entry of the invading pathogen. As reported by Hussein et al., S19P and E329G mutations in ACE2 receptor hinder interaction with viral spike protein. Therefore, resistance against infection by SARS-COV-2 can be attributed to variations in ACE2 receptor or viral spike protein (Hussain et al., 2020).

Intrinsically disorder regions in coronaviruses

The intrinsically disordered regions (IDRs) have a chief role in many biological functions such as protein binding and DNA/RNA binding, as they allow easy access to the corresponding sites. The protein–RNA recognition needs structural changes in both protein and RNA, which is mediated by flexibility in the structure of disordered residues. The IDRs have functional role in proteins transcription, translation, and cell. Studying the role of IDPs can help in identifying the viral life cycle and its pathophysiology (Yesudhas et al., 2020).

Evolution of SARS-CoV-2 virus

Coronaviruses being RNA viruses have high likelihood for genomic mutations that allow these viruses to accommodate new environments rendering them as long-term persistent threats. Mutations could be generated during replication of RNA viruses due to the low RNA-dependent RNA polymerase (RdRP) proofreading ability. These genomic mutations by viral RdRP allow an emerging virus to adapt to a new environment. The rate of synonymous substitution for coronaviruses is lower than that of some RNA viruses. During coronavirus replication, the mutation rate could be controlled by viral exoribonuclease (nsp14) (Denison et al., 2011). During the SARS-CoV-2 pandemic, viral evolution has been continuously occurring worldwide. The SRAS-CoV-2 was first collected from Wuhan city, China, in December 2019, and it was shown that the submitted sequences in databases varied from the latest sequence submitted in North America in April 2020. In the light of the continuous variations in viral genomic sequence, establishing a phylogenetic network is considered vital to investigate the capability of the virus to adapt in different environments and populations (Liu et al., 2020a,b).

It was claimed by a recent study that there are three genetic types of SARS-CoV-2 globally (Forster et al., 2020). The study demonstrated the genotypes in correlation to geographical locations; however, still the sample size and the analytical method are argued in the field of research (Mavian et al., 2020). Therefore, it will be still early to confirm if the evolution of SARS-CoV-2 is correlated to varied genetic and immunological factors in different human populations. Investigating SARS-CoV-2 variation among different geographical areas can provide useful information to develop vaccines for different populations. Moreover, regular sequencing is considered crucial to update information about viral evolution.

Clinical Features of SARS-CoV-2 Viral Infection

Generally, coronaviruses can affect different body systems (gastrointestinal, respiratory, and central nervous system). The clinical features are overlapping between different coronaviruses; however, they vary in intensity from mild to severe and in the most dominating symptoms and signs (da Costa et al., 2020). SARS-CoV-2 is characterized by variable wide range of incubation period from 1 to 12 days, with a median of 4 days (Guan et al., 2020). What is remarkably noted about the SARS-CoV-2 is the extreme variation in clinical presentation, severity, course of disease, prognosis, and clinical outcome among exposed individuals, a matter that renders consequences of infection unpredictable. Some of SARS-CoV-2-exposed individuals experience no symptoms (asymptomatic) or very mild symptoms (subclinical infection), which may be mistaken for other respiratory viral infections. Mostly, these cases are accidentally discovered by screen testing. Even the symptomatic patients vary in their clinical presentation between gastrointestinal, respiratory, and neurological symptoms according to the dominant affected body system (da Costa et al., 2020). SARS-CoV-2 infection most commonly presents with fever (85%), cough (58%), fatigue (29%), dyspnea (17%), pharyngitis (13%), diarrhea (7%), headache (7%), nausea and vomiting (4%), and abdominal pain (2%) (da Costa et al., 2020).

Globally, the cases and deaths records for COVID-19 pandemic had exceeded those of SARS and MERS coronaviruses (European Center for Disease Prevention and Control, 2020). According to the recent reports from the European Centre for Disease Prevention and Control, there were estimates of 145,144 deaths. The distribution for most of deaths was recorded in the country order of the United States, Italy, Spain, France, and United Kingdom (European Center for Disease Prevention and Control, 2020). Usually the clinical outcome is related to the age and the presence or absence of underlying health conditions, which makes fatality rate is expected to be high among old-age patients with chronic illnesses. Unexpectedly, there were unexplainable deaths records among middle age and SARS-CoV-2 young individuals who had no history of underlying medical problems. It is noted that most of SARS-CoV-2-infected patients have clinical manifestations of moderate severity and well prognosis. However, some patients can progress to complicated cases who die of organ system failure as acute cardiac injury, acute renal injury, and shock (da Costa et al., 2020). So far, the pathophysiology behind the rapid deterioration and loss of SARS-CoV-2-infected patients with severe clinical course is not fully understood. However, recent studies primarily introduced a suggested theory of "cytokine storm," which may associate the severe cases of COVID-19. In cytokine storm, immunological proinflammatory cytokines, e.g., INFγ, IL1B, MCP, and IP10, are produced in large amounts and poured in circulation resulting into shock and organ failure. The role of cytokine storm in severe cases was evidenced by reporting higher levels of cytokines in COVID-19 patients who required ICU admission than other patients with mild or moderate severity of disease. In addition to the reported increase in the secretion of cytokines from T-helper1 cells, recently a similar increase was observed in production of T-helper2 cytokines (e.g., IL10 and IL4) (da Costa et al., 2020). In this perspective, more studies are needed in the research field to introduce detailed emphasis on the different patterns of immune response in SARS-CoV-2-infected patients, which in turn can help to trace the dynamics of virus pathophysiology (da Costa et al., 2020).

SARS-CoV-2 Viral Shedding

The shedding of viral RNA is the highest at the onset of symptoms and then reduces after days and weeks from the beginning of symptoms (Lavezzo et al., 2020). Nevertheless, the viral RNA can be detected in respiratory specimens during the incubation period (1−2 days before the appearance of symptoms) reaching the peak few days after, and it can remain positive for up to 8 days in mild patients and can extend for longer periods in severe cases (Wolfel et al., 2020). It has been reported that viral RNA shedding may extend for up to 67 days in nasopharyngeal specimens among adult cases and for >1 month in feces of pediatric cases (Perera et al., 2020).

The late clearance of viral RNA (≥15 days after disease onset) is more reported with cases with old age, severe course of illness, hypertension, or delayed hospital admission (Xu et al., 2020a,b,c). Not mere polymerase chain reaction (PCR) detection of viral RNA will necessarily confirm infectivity unless the infectious particles are recovered through the isolation of virus from cultured samples. Moreover, the viral infectivity correlates proportionally with viral load, which is also considered an index for the disease course and prognosis. It has been reported that the viral loads are 60 times higher in severe than in mild cases (Wolfel et al., 2020). The viral load profile of SARS-CoV-2 is found to be similar to that of influenza virus, as in both, the viral load reaches the peak at around the onset of symptoms (Cheung et al., 2020). In contrast, SARS-CoV viral load reaches the peak about 10 days after the onset of symptoms, whereas MERS-CoV viral load peaks at the second week after the onset of symptoms (Cheung et al., 2020). Reaching high viral load around the time of the onset of symptoms can predict high infectivity and easy transmission of SARS-CoV-2 in early stage of illness (Ackerman et al., 2020).

Respiratory infection with non-SARS-CoV-2 pathogen does not exclude having SARS-CoV-2. A study conducted in North Carolina that 20.7% of COVID-19 patients had coinfection with additional viral pathogens (Kim et al., 2020). The most common copathogens were rhinovirus, respiratory syncytial virus, and seasonal coronaviruses.

Pneumonia in pregnancy

Pneumonia was considered the third common cause for indirect maternal deaths in one study done in the United States between 1960 and 1968 (Visscher and Visscher, 1971).

It is the most common cause of nonobstetric infections in pregnancy, and for that, pneumonia of infectious origin is considered very important cause for maternal morbidity and mortality (Benedetti et al., 1982; Berkowitz and LaSala, 1990; Madinger et al., 1989).

It is estimated that 25% of pregnant females who developed pneumonia will need ICU admission and ventilatory support (Madinger et al., 1989).

As a rule viral pneumonia is more serious than bacterial pneumonia as regard maternal morbidity and mortality although bacterial pneumonia may be serious even with bacterial agents sensitive to antibiotics (Rigby and Pastorek, 1996).

Many etiologies stand behind increase risk of pneumonia in pregnancy as alteration in cell-mediated immunity and changes in pulmonary functions (Jamieson et al., 2006; Sargent and Redman, 1992; Nyhan et al., 1983; Weinberger et al., 1980).

There is historic evidence proved that pregnancy adversely affects the course of pneumonia especially in late pregnancy (third trimester); during influenza pandemic in 1918–19, the case fatality rate (CFR) for pregnant female was 27% and reaches 50% if pneumonia occurred in third trimester(Harris, 1919).

There was other proof during pandemic of Asian flu between 1957 and 1958, as the CFR was doubled in pregnant women than nonpregnant women, and pregnant deaths constituted 10% of all deaths (Eickhoff et al., 1961).

There were many reported adverse complications of pneumonia on pregnancy; they include premature labor or delivery, premature rupture of membranes (PROM), intrauterine fetal death (IUFD), and fetal growth restriction (Visscher and Visscher, 1971; Benedetti et al., 1982; Berkowitz and LaSala, 1990).

Severe acute respiratory syndrome and pregnancy

There were several case reports and small studies that described effects of SARS on pregnant females; those effects included severe maternal illness or death and spontaneous abortion or preterm labor (PTL) (Jamieson et al., 2006; Wong et al., 2003, 2004; Ng et al., 2003, 2004).

As usual, the complications were more, and clinical outcome was worse in pregnant female with SARS than nonpregnant female in Hong Kong (Maxwell et al., 2017).

During the period from February 1, 2003, and July 31, 2003, Wong et al. observed cohort of pregnant female infected by SARS, and they reported 60% abortion rate in those infected early in pregnancy and 80% PTL in those infected in third trimester (Wong et al., 2004).

Another case control study was done in Princess Margarit Hospital in Hong Kong; this study compared pregnant (10 patients) and nonpregnant females (40 patients), who all had SARS infection. The investigators reported three deaths in pregnant females (30%) versus no deaths in nonpregnant females ($P = .006$). The risk of renal failure was also more in pregnant females ($P = .006$), and also risk of disseminated intravascular coagulopathy was more in pregnant group ($P = .006$). Not only that ICU admission was more in pregnant females, but they also needed more intubation and ventilatory support (Lam et al., 2004).

In 2017, Maxwell and his colleagues studied seven pregnant females infected by SARS-CoV in SARS unit, and they reported two deaths out of seven (CFR = 28%) and that four out of seven needed ICU admission and mechanical ventilation (57%). They also reported better outcome on nonpregnant females with CFR about 10%, and 20% only needed ICU admission. For the remaining 5 recovered pregnant females, two of them had delivered babies with low birth weight due to intrauterine growth restriction, but none of those babies were infected by SARS-CoV, and their mothers were advised to cancel breast feeding to avoid vertical transmission of infection (Maxwell et al., 2017).

During the peak of SARS epidemic in 2003, Zhang and his colleagues, reported their clinical analysis for five primigravida females infected by SARS-CoV, and they found two out of five primigravida were infected in second trimester, while three were infected in the third trimester, and also two out of five had hospital-acquired infection, while three were community-acquired infection. All primigravidas developed fever and abnormal radiograph in chest X-ray or CT. Four of them developed cough, four of them developed hypoproteinemia, three developed elevated liver enzymes, namely, alanine transferase, three developed rigors, two out of five developed lymphopenia, and two developed thrombocytopenia. They reported also one ICU admission, but fortunately all pregnant females were recovered without any maternal deaths and also their babies were proved to be SARS free (Zhang et al., 2003).

Robertson et al. had reported detailed case report of American pregnant lady. She was 36 years old and presented by intermittent cough and no fever. She had traveled to Hong Kong during SARS epidemic in 2003, and during her stay in Hong Kong, she was exposed to a person who was proved to be SARS-CoV positive, When she reached 19 weeks of her pregnancy, she developed fever, loss of appetite, continuous cough, headache, weakness, and dyspnea. When she had come back to the United States, she was hospitalized for pneumonia and tested for SARS-CoV and proved to be positive. Ultrasound was normal except of placenta previa; fortunately, she recovered and was discharged from hospital. Then after she was readmitted at 38 weeks for elective cesarean delivery due to placenta previa, she delivered healthy girl; after delivery, whole maternal blood, maternal serum, nasopharyngeal swab, rectal swab, placenta, cord blood sample, amniotic fluid sample, and breast milk sample were collected and tested by PCR, but no viral RNA was detected in all samples.

However, antibodies to SARS-CoV were detected by enzyme immune assay and indirect immunofluorescence assay in maternal serum, breast milk, and cord blood (Robertson et al., 2004; Schneider et al., 2004).

Another American pregnant female 38 years old was reported to have SARS; she traveled to Hong Kong at 7 weeks' gestation and had exposed to SARS-CoV-positive person at her stay in Hong Kong. When she had come back to the United States, her husband developed manifestations of SARS, and then after 6 days, she developed fever, muscle pain, headache, cough, expectoration, wheezy chest, and dyspnea. She was hospitalized for SARS, and later on, she recovered. Serum samples were taken 4 and 9 weeks postonset of her illness; both samples were proven to be positive for SARS-CoV by enzyme immunoassay and immunofluorescence assay. She had continued her pregnancy without any manifestations except high blood glucose. At 36 weeks' gestation, cesarean section was done due to preterm premature rupture of membranes and reported fetal distress, and she had delivered healthy boy baby. At time of delivery, serum sample was taken and reported to be positive for SARS-CoV antibodies, but samples from umbilical cord and placenta were negative. Breast milk was also negative for SARS-CoV antibodies, and it was tested twice: 12 and 30 days postdelivery. Samples from maternal blood, nasopharyngeal swab, stool, cord blood, infant blood, and infant stool were tested for SARS-CoV virus by reverse transcriptase—polymerase chain reaction (RT-PCR), and all were negative (Stockman et al., 2004).

Yudin and his colleagues observed a 33-year-old pregnant woman in Canada who was hospitalized at 31 weeks' gestation with fever, dry cough, and abnormal chest X-ray with patchy infiltrations; she had infected by SARS via contact to a person from her family. She had stayed in hospital for 3 weeks, but she did not need ICU admission or ventilatory support. After convalescence, she was positive for antibodies for coronavirus. Her pregnancy continued uneventful, and she delivered healthy normal girl baby with no evidence of infection (Yudin et al., 2005).

Shek and his colleagues studied that five neonates delivered from infected pregnant women had been proved to be positive for SARS-CoV during Hong Kong epidemic in 2003. Fortunately, serial testing by RT-PCR, viral culture, and antibodies assay for the infants were negative, and none of them had any respiratory manifestations so they were clinically and microbiologically free. From this study and other studies, it is proven that there is no vertical transmission from pregnant mother infected by SARS CoV to their infants during Asian epidemic in 2003 (Shek et al., 2003).

Pathological changes in placenta of SARS patients

Ng et al. studied the placentae of seven pregnant females infected by SARS-CoV; two out of seven were infected and recovered in the first trimester, and they found no pathological abnormalities in their placentae, other three placentae were delivered from mothers of acute SARS-CoV, and they found fibrin deposition subchorionic and intervillous. These findings are common with impaired placental blood flow, and the two placenta examined from pregnant women who recovered from SARS-CoV in third trimester were highly abnormal; they found extensive fetal vascular thrombosis, fetal hypo- or malperfusion, and avascular villi. These two pregnancies were also complicated by fetal growth restriction, oligo hydramnios, and poor fetal outcome, but it was strange that no signs of inflammation as villitis were found in those placentae (Ng et al., 2006).

Middle East respiratory syndrome with pregnancy

Unfortunately, we have few data about MERS and pregnancy as regards clinical course, fetal outcome, and postpartum period. It seems that alteration in cell-mediated immunity may increase risk of MERS and its severity in pregnant women (Malik et al., 2016).

As clinical outcome was much worse in pregnant females with SARS-CoV than nonpregnant females, so it is expected that related coronavirus causing MERS will have similar course and effect and from the previous study of pregnant females with MERS 10 out of 11 had adverse complications (Malik et al., 2016).

From November 2012 to February 2016, the Ministry of Health in Saudi Arabia (MOH) had reported 1038 cases with MERS; only five pregnant females were reported to have MERS, and unfortunately, all of them had adverse complications. All five females were admitted to ICU, with two maternal deaths; although their ages were young from 27 to 34 years old, there are two perinatal deaths: one of them was IUFD and the other was early neonatal death shortly after cesarean delivery. Two of those females were nurses, and this confirmed that the healthcare workers are at risk for MERS (Assiri et al., 2016).

Then, Alserihi et al. described another case report of MERS with pregnancy; they observed a 33-year-old pregnant nurse that was working in critical care unit who got infected in her third trimester of pregnancy during hospital major outbreak. After admission to the hospital, the nurse's conditions deteriorated, and she was admitted to ICU and put on intermittent positive pressure ventilator. She received dexamethasone for lung maturity of her fetus; then emergency cesarean section at 32 weeks' gestation was done, and then she was readmitted to ICU. Fortunately, the mother had recovered, and her baby was admitted to neonatal ICU for observation, but he was discharged also after several days in good conditions (Alserehi et al., 2016).

It was supposed that her young age, immune response early delivery, and steroids therapy may contribute to the good maternal outcome.

Then; Alfaraj et al. reported two pregnant cases infected by MERS-CoV. The cases were recorded in PMBAH (Prince Mohammed bin Abdulaziz Hospital). The infection was confirmed by RT-PCR from nasopharyngeal swab samples. One of them was relatively young 29 years old, had no history of comorbidity, and was 6 weeks gestation, but the other one was 39 years old, with history of hypertension, renal failure, and repeated hemodialysis. Both patients were tested negative for MERS-CoV later on and then were discharged after several days. The young patient delivered a healthy baby but no report about the other one after discharge (Alfaraj et al., 2019).

Another report from Jordan in 2012 described a case of second trimester IUFD due to MERS CoV infection during outbreak in Al Zarqa region. The mother was presented by fever, abdominal pain, mild vaginal bleeding, fatigue, cough, and with history of exposure to MERS CoV. One week after onset of symptoms,

ultrasound was repeated and confirmed IUFD. The infection was confirmed by antibody testing and history of contact to a known family member with MERS CoV. This was the first report of IUFD with MERS CoV (Payne et al., 2014).

Another report came from the United Arab of Emirates (UAE) in 2013. When Malik et al. reported that a pregnant female developed adult respiratory distress syndrome (ARDS), she got infected via community-acquired pneumonia. She was admitted to ICU and developed respiratory failure and hypotension. The mother was tested by RT-PCR for MERS CoV and proved to be positive. Emergency cesarean section was done to ameliorate her condition with delivery of healthy baby with good Apgar score. She received ribavirin and peginterferon-α. However, her condition deteriorated into septic shock and died despite vigorous treatment and ventilatory support. It was strange that viral shedding and chest radiograph improved on treatment despite the patient died (Malik et al., 2016).

MERS CoV was also reported outside Middle East in South Korea. Jeong et al., in 2015 described a 39-year-old pregnant female who was exposed to MERS CoV contact in during third trimester of her pregnancy. She was confirmed to be MERS-CoV by RT-PCR testing. This lady delivered a healthy baby by emergency cesarean Section despite placental abruption and sudden antepartum hemorrhage and premature rupture of membranes. Antibody testing of her baby was negative as regard MERS-CoV IgG, IgM, and IgA (Jeong et al., 2017).

Other coronaviruses with pregnancy
The alpha coronaviruses include HCoV 229E and NL63.

There are beta coronaviruses, and they include HKU1 and OC43. They can infect human and present by common cold symptoms. Gagneur et al. investigated the vertical transmission (from mother to infant) of all four mentioned viruses. They collected maternal samples from vagina and respiratory tract during labor and collected gastric aspirate from newly born babies. All the samples were evaluated by RT-PCR for HCoV 229E and NL63, HKU1, and OC43. From July 2003 to August 2005, the collected samples are from 159 pregnant mothers and their infants. The human coronavirus was detected in 12 samples; all were from only seven maternal—infant pairs.

Three pairs were positive respiratory samples, two were positive for respiratory and vaginal (total four), two were positive for vaginal-only, and the other remaining three samples were positive from neonatal gastric samples. Fortunately, all babies were free of symptoms (Gagneur et al., 2007, 2008).

Vertical transmission of COVID-19 (SARS-CoV-2) from mother to infant

On January 13, 2020; the Wuhan Children's Hospital in Hubei Province reported delivery of a healthy baby. Then later on, his nursemaid was confirmed to be COVID-19 positive, and the mother also was tested few days later and was also COVID-19 positive. 16 days after delivery, the baby started to develop symptoms. It is not clear who infected who, but it supposed that the nursemaid had transmitted infection to the mother through direct contact, and then the mother transmitted it to her baby, but other scenario may be true.

On February 5, the hospital reported a delivery of a baby who was tested positive for COVID-19 after 30 h of his delivery; the mother was also known to be COVID-19.

The bay had no fever or cough, but he developed tachypnea and abnormal chest radiograph and elevated liver enzymes.

So vertical transmission from mother to infant should not be excluded (Steinbuch, 2020; Woodward, 2020; Gillespie, 2020).

Immunological response to COVID-19 (SARS-CoV-2), SARS-CoV, and MERS-CoV

The most constant immune response to all mentioned coronaviruses infections is severe lymphopenia (Booth et al., 2003; Lee et al., 2003; Panesar, 2003; Ajlan et al., 2014; Arabi et al., 2015; Ko et al., 2016; Xu et al., 2020a,b,c; Zhang et al., 2020a,b).

Also decreases in CD4 and CD8 lymphocytes were noted in the early course of COVID-19 (SARS-CoV-2), SARS-CoV, and MERS-CoV and were associated with adverse complications. In SARS-CoV infection, lymphocyte involvement was detected by expression of CD25, CD28, and CD69. In MERS-CoV infection, the virus negatively regulates MHC-I, MHC-II, and CD80/86 in antigen-presenting cells, which results in inhibition of the T lymphocyte response. These events can further impair the function of B and T cells through negative regulation of DPP4 receptors by MERS-CoV (Yu et al., 2003; Cai et al., 2004; Josset et al., 2013; Chu et al., 2014; Ying et al., 2016).

MERS-CoV infection may induce T lymphocytes apoptosis, and this leads to the obvious immune suppression seen during infection. DPP4 (which is negatively affected during MERS-CoV) plays an important role in activating T lymphocyte during infection by MERS-CoV (Ishii et al., 2001; Manni et al., 2014).

Ying et al. reported that T CD4 helper cells are more affected during MERS-CoV infection through apoptosis by extrinsic and intrinsic pathways (Ying et al., 2016).

A significant increase in cytokines IL-17 was observed during MERS-CoV infection; thus, it is thought that MERS-CoV infection induces production of Th 17 cytokines. These Th17 cytokines recruit monocytes and neutrophils to the site of infection and inflammation and lead to production of many cytokines in serial reactions such as IL-1, IL-6, TNF-α, TGF-β, IL-8, and MCP-1 (Jin and Dong, 2013; Mahallawi et al., 2018).

Normally, the immune system can produce neutralizing antibodies to any viral infection, which block the viruses and prevent their entry to host cells. These antibodies appear in patient serum and thus are used as diagnostic method to confirm any viral infection or MERS-CoV infection. We can detect serum antibodies to MERS-CoV infection 14–21 days after infection, and this may continue up to 18 months. The antibodies for SARS-CoV may even last more up to 24 months and may take 6 years to disappear (Park et al., 2015; Corman et al., 2016; Alshukairi et al., 2016; Liu et al., 2006; Tang et al., 2011).

In patients recovered from SARS, specific memory CD4[+] T lymphocytes for HLA-DR08 and HLA-DR15 restricted epitopes of the SARS-CoV S protein have been identified. Virus-specific CD4[+] T lymphocytes mainly exhibited a central memory phenotype (CD45RA[−] CCR7[+] CD62L[−]), while CD8[+] memory T lymphocytes were identified as memory effector cells (CD45RA[+] CCR7[−] CD62L[−]).

These findings suggest that there is an activity of specific memory T lymphocytes in long-term protection against SARS-CoV infections. In addition, the response of memory B lymphocytes to SARS-CoV decreases significantly after 1–2 years of infection (Openshaw, 2004; Yang et al., 2006, 2007; Libraty et al., 2007; Martin et al., 2008).

The immune response to SARS-CoV-2 or COVID-19 needs time to be fully investigated. Some reported a decrease in CD4 T cells and CD8 T cells in the peripheral blood of infected patients (Xu et al., 2020a,b,c), a reduction in total lymphocytes (35%), and an increase of serum IL-6 (52%) and C-reactive protein (84%) (Peng et al., 2020).

Cytokines Storm

Many reports point to the pathology beyond the severe cases of SARS-CoV-2 or COVID-19 that need ICU admission and ventilatory support. Hence the expression of cytokines storm went to force. There is an increase in humoral immunity in patients admitted to ICU, such as IL-2, IL-7, IL-10, G-CSF, IP-10, MCP-1, MIP-1A, and TNF-α.

As mentioned before, there is a decrease in CD4 and CD8 T cells in peripheral blood of infected patients but with increased activation of remaining cells with marked expression of HLA-DR (CD4 3.47%) and CD38 (CD8 39.4%) double-positive fractions (Huang et al., 2020; Fang, 2016).

Clinical course of COVID-19 and other coronaviruses in pregnancy

Contact to MERS-CoV or SARS-CoV or SARS-CoV-2 (COVID-19) can trigger asymptomatic carrier state or common cold symptoms with or without fever. Unfortunately, symptoms may progress into severe pneumonia, adult respiratory distress, gastroenteritis, liver damage, septic shock, renal failure, multiple organ failure, and death. As regard severe cases of COVID-19, there we found multifocal nodules and ground glass opacity in lungs, for pregnant women. In addition to all mention manifestations, there is risk of vertical transmission and obstetric complications such as PTL and IUFD. As few reports about COVID-19 in pregnancy are available, but in general severe cases with pregnancy usually need ventilatory support and induction of PTL (WHO, 2020; Hui et al., 2018; ECDC, 2018; Kong and Agarwal, 2020; Santos, 2003; Alserehi et al., 2016; Assiri et al., 2016; CDC, 2013).

Cytokines produced by T-helper (Th) lymphocytes regulate immunity and inflammation. Th1-type cytokines are antimicrobial and proinflammatory and mainly include interferon-γ (IFN-γ), interleukin (IL)−1α, IL-1β, IL-6, and IL-12, but Th2-type cytokines are antiinflammatory and comprise IL-4, IL-10, IL-13, and transforming growth factor-β (TGF-β).

In pregnancy, the attenuation in cell-mediated immunity by Th1 cells due to the physiological shift to a from Th1 to Th2 dominant environment contributes to overall infectious morbidity by increasing maternal susceptibility to intracellular pathogens such as viruses (Nelson-Piercy, 2015; Berger, 2000).

The CFR in pregnant females with COVID-19 (SARS-CoV-2), SARS-CoV, and MERS-CoV are 0%, 18%, and 25%, respectively. The disease looks milder in pregnant COVID-19 than in SARS and MERS pregnant. The explanation is not fully understood. However, in COVID-19, a range of immune responses has been described, and early adaptive immune responses may be predictive of milder disease severity. We postulate that changes in the hormonal milieu in pregnancy, which influence immunological responses to viral pathogens together with the physiological transition to a Th2 environment favoring the expression of antiinflammatory cytokines (IL-4 and IL-10) and other unidentified immune adaptations, may serve as the predominant immune response to SARS-CoV-2, resulting in the lesser severity of COVID-19 compared with that in nonpregnant individuals. These immune responses should be further characterized in gravidas and nongravidas with COVID-19 of different disease severities (Wong et al., 2003; Assiri et al., 2016; Thevarajan et al., 2020).

Fetal complications of COVID-19 include miscarriage (2%), intrauterine growth restriction (IUGR; 10%), and preterm birth (39%). Fever, with a median temperature of 38.1−39.0°C, is the prevailing symptom in COVID-19. Cohort studies in patients with other infections have not shown increased risks of congenital anomalies from maternal pyrexia in the first trimester, although childhood inattention disorders are more common, possibly related to hyperthermic injury to fetal neurons (Sass et al., 2017).

There is a theoretical risk of vertical transmission, similar to that seen in SARS, as the ACE2 receptor is widely expressed in the placenta, with a similar RBD structure between SARS-CoV and SARS-CoV-2.

Most recently, two neonates from COVID-19-infected mothers are said to have tested positive for SARS-CoV-2 shortly following delivery, casting concerns about the possibility of vertical transmission (Levy et al., 2008; Woodward 2020; Murphy, 2020; Li et al., 2020a,b; Liu et al., 2020a,b).

However, there have been no confirmed instances of vertical transmission among the 46 other neonates born to COVID-19-infected mothers reported thus far, supported in turn by evidence demonstrating an absence of viral isolates in the amniotic fluid, cord blood, breast milk, and neonatal throat swabs in a subset of these patients.

It is notable, however, that the overwhelming majority of these women acquired COVID-19 in the third trimester; there are currently no data on perinatal outcomes when the infection is acquired in early pregnancy. Regardless of the risk, it is reassuring that COVID-19 appears to manifest as a mild respiratory disease in the pediatric population (Zhu et al., 2020; Zhang et al., 2020a,b; Chen et al., 2020a,b,c; Xu et al., 2020a,b,c; Cai et al., 2020).

REFERENCES

Ackermann, M., Verleden, S.E., Kuehnel, M., Haverich, A., Welte, T., Laenger, F., et al., 2020. Pulmonary vascular endothelialitis, thrombosis, and angiogenesis in Covid-19. N. Engl. J. Med.

Ajlan, A.M., Ahyad, R.A., Jamjoom, L.G., Alharthy, A., Madani, T.A.(, 2014. Middle East respiratory syndrome coronavirus (MERS-CoV) infection: chest CT findings. Am. J. Roentgenol. 203 (4), 782−787. https://doi.org/10.2214/AJR.14.13021 [PubMed] [CrossRef] [Google Scholar].

Alfaraj, S.H., Al-Tawfiq, J.A., Memish, Z.A., 2019. Middle East respiratory syndrome coronavirus (MERS-CoV) infection during pregnancy: report of two cases & review of literature. J. Microbiol. Immunol. Infect. 52, 501−503 [PMC free article] [PubMed] [Google Scholar].

Alserehi, H., Wali, G., Alshukairi, A., Alraddadi, B., 2016. Impact of Middle East respiratory syndrome coronavirus (MERS-CoV) on pregnancy and perinatal outcome. BMC Infect. Dis. 16, 105. https://doi.org/10.1186/s12879-016-1437-y [PMC free article] [PubMed] [CrossRef] [Google Scholar].

Alshukairi, A.N., Khalid, I.K., Ahmed, W.A., Dada, A.M., Bayumi, D.T., Málica, L.S., 2016. Antibody response and disease severity in healthcare worker MERS survivors. Emerg. Infect. Dis. 22 (6), 1113. https://doi.org/10.3201/eid2206.160010 [PMC free article] [PubMed] [CrossRef] [Google Scholar].

Annan, A., Baldwin, H.J., Corman, V.M., Klose, S.M., Owusu, M., Nkrumah, E.E., et al., 2013. Human betacoronavirus 2c EMC/2012−related viruses in bats, Ghana and Europe. Emerg. Infect. Dis. 19 (3), 456.

Arabi, Y.M., Harthi, A., Hussein, J., Bouchama, A., Johani, S., Hajeer, A.H., 2015. Severe neurologic syndrome associated with Middle East respiratory syndrome corona virus (MERS-CoV). Infection 43 (4), 495−501. https://doi.org/10.1007/s15010-015-0720-y [PMC free article] [PubMed] [CrossRef] [Google Scholar].

Assiri, A., Abedi, G.R., Almasry, M., Bin Saeed, A., Gerber, S.I., Watson, J.T.(, 2016. Middle East respiratory syndrome coronavirus infection during pregnancy: a report of 5 cases from Saudi Arabia. Clin. Infect. Dis. 63, 951−953. https://doi.org/10.1093/cid/ciw412 [PMC free article] [PubMed] [CrossRef] [Google Scholar].

Baker, M.G., 2020. Nonrelocatable occupations at increased risk during pandemics: United States, 2018. Am. J. Publ. Health e1−e7.

Benedetti, T.J., Valle, R., Ledger, W.J., 1982. Antepartum pneumonia in pregnancy. Obstet. Gynecol. 144, 413−417. https://doi.org/10.1016/0002-9378(82)90246-0 [PubMed] [CrossRef] [Google Scholar].

Berger, A., 2000. Th1 and Th2 responses: what are they? BMJ 321, 424 [PMC free article] [PubMed] [Google Scholar].

Berkowitz, K., LaSala, A., 1990. Risk factors associated with the increasing prevalence of pneumonia during pregnancy. Am. J. Obstet. Gynecol. 163, 981−985. https://doi.org/10.1016/0002-9378(90)91109-P [PubMed] [CrossRef] [Google Scholar].

Booth, C.M., Matukas, L.M., Tomlinson, G.A., Rachlis, A.R., Rose, D.B., Dwosh, H.A.(, 2003. Clinical features and short-term outcomes of 144 patients with SARS in the greater Toronto area. J. Am. Med. Assoc. 289, 2801−2809. https://doi.org/10.1001/jama.289.21.JOC30885 [PubMed] [CrossRef] [Google Scholar].

Byambasuren, O., Cardona, M., Bell, K., Clark, J., McLaws, M.-L., Glasziou, P., 2020. Estimating the extent of true asymptomatic COVID-19 and its potential for community transmission: systematic review and meta-analysis. medRxiv.

Cai, C., Zeng, X., Ou, A.H., Huang, Y., Zhang, X., 2004. Study on T cell subsets and their activated molecules from the convalescent SARS patients during two follow-up surveys. Chin. J. Cell. Mol. Immunol. 20 (3), 322−324 [PubMed] [Google Scholar].

Cai, J., Xu, J., Lin, D., 2020. A case series of children with 2019 novel coronavirus infection: clinical and epidemiological features. Clin. Infect. Dis. [Epub ahead of print] [PMC free article] [PubMed] [Google Scholar].

CDC, Centers for Disease Control and Prevention Update, 2013. Severe respiratory illness associated with Middle East Respiratory Syndrome Coronavirus (MERS-CoV) worldwide, 2012−2013. Morb. Mortal. Wkly. Rep. 62 (23), 480−483 [PMC free article] [PubMed] [Google Scholar].

Chan, J.F.W., Yuan, S., Kok, K.H., To, K.K.W., Chu, H., Yang, J., Xing, F., Liu, J., Yip, C.C.Y., Poon, R.W.S., Tsoi, H.W., 2020. A familial cluster of pneumonia associated with the 2019 novel coronavirus indicating person-to-person transmission: a study of a family cluster. Lancet 395, 514−523.

Chauhan, S., 2020. Comprehensive review of coronavirus disease 2019 (COVID-19). Biomed. J. 43, 334e40.

Chen, H., Guo, J.M.S., Chen, W., 2020a. Clinical characteristics and intrauterine vertical transmission potential of COVID-19 infection in nine pregnant women: a retrospective review of medical records. Lancet 395 (10226), 809−815 [PMC free article] [PubMed] [Google Scholar].

Chen, S., Chen, S., Huang, B., 2020b. Pregnant women with new coronavirus infection: a clinical characteristics and placental pathological analysis of three cases. Zhonghua Bing Li Xue Za Zhi 49, E005 [PubMed] [Google Scholar].

Chen, Y., Peng, H., Wang, L., 2020c. Infants born to mothers with a new coronavirus (COVID-19). Front. Pediatr. 8 [Epub ahead of print] [PMC free article] [PubMed] [Google Scholar].

Cheung, K.S., Hung, I.F.N., Chan, P.P.Y., Lung, K.C., Tso, E., Liu, R., et al., 2020. Gastrointestinal manifestations of SARSCoV-2 infection and virus load in fecal samples from the Hong Kong cohort and systematic review and metaanalysis. Gastroenterology 159.

Chu, H., Zhou, J., Wong, B.H.Y., Li, C., Cheng, Z.S., Lin, X., 2014. Productive replication of Middle East respiratory syndrome coronavirus in monocyte-derived dendritic cells modulates innate immune response. Virology 454, 197−205. https://doi.org/10.1016/j.virol.2014.02.018 [PMC free article] [PubMed] [CrossRef] [Google Scholar].

Corman, V.M., Albarrak, A.M., Omrani, A.S., Albarrak, A.M., Farah, M.E., Almasri, M., 2016. Viral shedding and antibody response in 37 patients with Middle East respiratory syndrome coronavirus infection. Clin. Infect. Dis. 62 (4), 477–483. https://doi.org/10.1093/cid/civ951 [PMC free article] [PubMed] [CrossRef] [Google Scholar].

da Costa, V.G., Moreli, M.L., Saivish, M.V., 2020. The emergence of SARS, MERS and novel SARS-2 coronaviruses in the 21st century. Arch. Virol. 165, 1517–1526.

Cui, J., Li, F., Shi, Z.-L., 2019. Origin and evolution of pathogenic coronaviruses. Nat. Rev. Microbiol. 17 (3), 181–192.

Denison, M.R., Graham, R.L., Donaldson, E.F., Eckerle, L.D., Baric, R.S., 2011. Coronaviruses: an RNA proofreading machine regulates replication fidelity and diversity. RNA Biol. 8, 270e9.

Eickhoff, T.C., Sherman, I.L., Serfling, R.E., 1961. Observations on excess mortality associated with epidemic influenza. J. Am. Med. Assoc. 176, 776–782. https://doi.org/10.1001/jama.1961.03040220024005 [PubMed] [CrossRef] [Google Scholar].

European Centre for Disease Prevention and Control, November 2020. Threats and Outbreaks, COVID-19, Latest Evidence. https://www.ecdc.europa.eu/en/covid-19/latest-evidence/epidemiology.

European Centre for Disease Prevention and Control (ECDC), 2018. Rapid Risk Assessment: Severe Respiratory Disease Associated With Middle East Respiratory Syndrome Coronavirus (MERS-CoV) 22nd Update [Google Scholar].

Fang, L., 2016. Structure, function, and evolution of coronavirus spike proteins. Annu. Rev. Virol. 3, 237–261. https://doi.org/10.1146/annurev-virology-110615-042301 [PMC free article] [PubMed] [CrossRef] [Google Scholar].

Forster, P., Forster, L., Renfrew, C., Forster, M., 2020. Phylogenetic network analysis of SARS-CoV-2 genomes. Proc. Natl. Acad. Sci. U.S.A. 117, 9241e3.

Gagneur, A., Dirson, E., Audebert, S., Vallet, S., Quillien, M.C., Baron, R., Laurent, Y., Collet, M., Sizun, J., Oger, E., et al., 2007. Vertical transmission of human coronavirus. Prospective pilot study. Pathol. Biol. 55, 525–530. https://doi.org/10.1016/j.patbio.2007.07.013 [PMC free article] [PubMed] [CrossRef] [Google Scholar].

Gagneur, A., Dirson, E., Audebert, S., Vallet, S., Legrand-Quillien, M.C., Laurent, Y., Collet, M., Sizun, J., Oger, E., Payan, C., 2008. Materno-fetal transmission of human coronaviruses: a prospective pilot study. Eur. J. Clin. Microbiol. Infect. Dis. 27, 863–866. https://doi.org/10.1007/s10096-008-0505-7 [PMC free article] [PubMed] [CrossRef] [Google Scholar].

Garazzino, S., Montagnani, C., Dona, D., Meini, A., Felici, E., Vergine, G., et al., 2020. Multicentre Italian study of SARSCoV-2 infection in children and adolescents, preliminary data as at 10 April. Euro Surveill. 25 (18), 2000600.

Gillespie, T., 2020. Coronavirus: Doctors Fear Pregnant Women Can Pass on Illness After Newborn Baby Is Diagnosed. Available online: https://news.sky.com/story/coronavirus-doctors-fear-pregnant-women-can-pass-on-illness-after-newborn-baby-is-diagnosed-11926968. (Accessed 8 February 2020).

Guan, W.J., Ni, Z.Y., Hu, Y., et al., 2020. Clinical characteristics of coronavirus disease in 2019 in China. N. Engl. J. Med. 382, 1708–1720.

Gui, M., Song, W., Zhou, H., Xu, J., Chen, S., Xiang, Y., Wang, X., 2017. Cryoelectron microscopy structures of the SARS-CoV spike glycoprotein reveal a prerequisite conformational state for receptor binding. Cell Res. 27, 119–129.

Haagmans, B.L., Al Dhahiry, S.H., Reusken, C.B., Raj, V.S., Galiano, M., Myers, R., et al., 2014. Middle East respiratory syndrome coronavirus in dromedary camels: an outbreak investigation. Lancet Infect. Dis. 14, 140e5.

Häfner, S.J., 2020. Pandemic number five - latest insights into the COVID-19 crisis. Biomed. J. 43, 305–410.

Harris, J.W., 1919. Influenza occurring in pregnant women; a statistical study of thirteen hundred and fifty cases. J. Am. Med. Assoc. 72, 978–980. https://doi.org/10.1001/jama.1919.02610140008002 [CrossRef] [Google Scholar].

Heald-Sargent, T., Muller, W.J., Zheng, X., Rippe, J., Patel, A.B., Kociolek, L.K., 2020. Age-related differences in nasopharyngeal severe acute respiratory syndrome coronavirus 2 (SARS-CoV-2) levels in patients with mild to moderate coronavirus disease 2019 (COVID-19). JAMA Pediatr.

Huang, C., Wang, Y., Li, X., Ren, L., Zhao, J., Hu, Y., 2020. Clinical features of patients infected with 2019 novel coronavirus in Wuhan, China. Lancet 395 (10223), 497–506. https://doi.org/10.1016/S0140-6736(20)30183-5 [PMC free article] [PubMed] [CrossRef] [Google Scholar].

Hui, D.S., Azhar, E.I., Kim, Y.J., Memish, Z.A., Oh, M.D., Zumla, A., 2018. Middle East respiratory syndrome coronavirus: risk factors and determinants of primary, household, and nosocomial transmission. Lancet Infect. Dis. 18 (8), e217–e227. https://doi.org/10.1016/S1473-3099(18)30127-0 [PMC free article] [PubMed] [CrossRef] [Google Scholar].

Hussain, M., Jabeen, N., Raza, F., Shabbir, S., Baig, A.A., Amanullah, A., Aziz, B., 2020. Structural variations in human ACE2 may influence its binding with SARS-CoV-2 spike protein. J. Med. Virol. https://doi.org/10.1002/jmv.25832.

Ishii, T., Ohnuma, K., Murakami, A., Takasawa, N., Kobayashi, S., Dan, N.H., 2001. CD26-mediated signaling for T cell activation occurs in lipid rafts through its association with CD45RO. Proc. Natl. Acad. Sci. U. S. A. 98 (21), 12138–12143. https://doi.org/10.1073/pnas.211439098 [PMC free article] [PubMed] [CrossRef] [Google Scholar].

Jamieson, D.J., Theiler, R.N., Rasmussen, S.A., 2006. Emerging infections and pregnancy. Emerg. Infect. Dis. 12, 1638–1643.

Jeong, S.Y., Sung, S.I., Sung, J.H., Ahn, S.Y., Kang, E.S., Chang, Y.S., Park, W.S., Kim, J.H., 2017. MERS-CoV infection in a pregnant woman in Korea. J. Kor. Med. Sci. 32, 1717–1720.

Jin, W., Dong, C., 2013. IL-17 cytokines in immunity and inflammation. Emerg. Microb. Infect. 2 (1), 1–5. https://doi.org/10.1038/emi.2013.58 [PMC free article] [PubMed] [CrossRef] [Google Scholar].

Jing, Q.-L., Liu, M.-J., Yuan, J., Zhang, Z.-B., Zhang, A.-R., Dean, N.E., et al., 2020. Household secondary attack rate of COVID-19 and associated determinants. medRxiv. https://doi.org/10.1101/2020.04.11.20056010.

Jones, T.C., Mühlemann, B., Veith, T., Zuchowski, M., Hofmann, J., Stein, A., et al., 2020. An analysis of SARS-CoV-2 viral load by patient age. medRXiv. Available from: https://zoonosen.charite.de/fileadmin/user_upload/microsites/m_cc05/virologieccm/dateien_upload/Weitere_Dateien/analysis-of-SARS-CoV-2-viral-load-by-patient-age.pdf.

Josset, L., Menachery, V.D., Grajinski, L.E., Agnihothram, S., Sova, P., Carter, V.S., 2013. Cell host response to infection with novel human coronavirus EMC predicts potential antivirals and important differences with SARS coronavirus. mBio 4 (3). https://doi.org/10.1128/mBio.00165-13 [PMC free article] [PubMed] [CrossRef] [Google Scholar].

Kim, D., Quinn, J., Pinsky, B., Shah, N.H., Brown, I., 2020. Rates of co-infection between SARS-CoV-2 and other respiratory pathogens. JAMA 323.

Ko, J.H., Park, G.E., Lee, J.Y., Lee, J.Y., Cho, S.Y., Ha, Y.E., 2016. Predictive factors for pneumonia development and progression to respiratory failure in MERS-CoV infected patients. J. Infect. 73 (5), 468–475. https://doi.org/10.1016/j.jinf.2016.08.005 [PMC free article] [PubMed] [CrossRef] [Google Scholar].

Koh, W.C., Naing, L., Rosledzana, M.A., Alikhan, M.F., Chaw, L., Griffith, M., et al., 2020. What do we know about SARS CoV-2 transmission? A systematic review and meta-analysis of the secondary attack rate, serial interval, and asymptomatic infection. medRxiv. https://doi.org/10.1101/2020.05.21.20108746.

Kong, W., Agarwal, P.P., 2020. Chest imaging appearance of COVID-19 infection. Radiol. Cardiothorac. Imag. 2 (1), e200028. https://doi.org/10.1148/ryct.2020200028 [CrossRef] [Google Scholar].

Lai, C.-C., Shih, T.-P., Ko, W.-C., Tang, H.-J., Hsueh, P.-R., 2020. Severe acute respiratory syndrome coronavirus 2 (SARS-CoV-2) and corona virus disease-2019 (COVID19): the epidemic and the challenges. Int. J. Antimicrob. Agents 105924.

Lam, C.M., Wong, S.F., Leung, T.N., Chow, K.M., Yu, W.C., Wong, T.Y., Lai, S.T., Ho, L.C., 2004. A case-controlled study comparing clinical course and outcomes of pregnant and non-pregnant women with severe acute respiratory syndrome. BJOG 111, 771–774. https://doi.org/10.1111/j.1471-0528.2004.00199.x [PMC free article] [PubMed] [CrossRef] [Google Scholar].

Lam, T.T., Shum, M.H., Zhu, H.C., Tong, Y.G., Ni, X.B., Liao, Y.S., et al., 2020. Identifying SARS-CoV-2 related coronaviruses in Malayan pangolins. Nature (in press).

Lavezzo, E., Franchin, E., Ciavarella, C., Cuomo-Dannenburg, G., Barzon, L., Del Vecchio, C., et al., 2020. Suppression of a SARS-CoV-2 outbreak in the Italian municipality of Vo'. Nature 584.

Lee, N., Hui, D., Wu, A., Chan, P., Cameron, P., Joynt, G., 2003. A major outbreak of severe acute respiratory syndrome in Hong Kong. N. Engl. J. Med. 348, 1986–1994 [PubMed] [Google Scholar].

Letko, M., Marzi, A., Munster, V., 2020. Functional assessment of cell entry and receptor usage for SARS-CoV-2 and other lineage B betacoronaviruses. Nat. Microbiol. 5, 562–569.

Levy, A., Yagil, Y., Bursztyn, M., Barkalifa, R., Scharf, S., Yagil, C., 2008. ACE2 expression and activity are enhanced during pregnancy. Am. J. Physiol. Regul. Integr. Comp. Physiol. 295, 1953–1961 [PubMed] [Google Scholar].

Li, F., 2016. Structure, function, and evolution of coronavirus spike proteins. Annu. Rev. Virol. 3, 237–261.

Li, Q., Guan, X., Wu, P., Wang, X., Zhou, L., Tong, Y., Ren, R., Leung, K.S., Lau, E.H., Wong, J.Y., Xing, X., 2020a. Early transmission dynamics in Wuhan, China, of novel coronavirus-infected pneumonia. N. Engl. J. Med. 382, 1199–1207.

Li, Y., Zhao, R., Zheng, S., 2020b. Lack of vertical transmission of severe acute respiratory syndrome coronavirus 2, China. Emerg. Infect. Dis. 26 [PMC free article] [PubMed] [Google Scholar].

Libraty, D.H., O'Neil, K.M., Baker, L.M., Acosta, L.P., Olveda, R.M., 2007. Human CD4+ memory T-lymphocyte responses to SARS coronavirus infection. Virology 368 (2), 317–321. https://doi.org/10.1016/j.virol.2007.07.015 [PMC free article] [PubMed] [CrossRef] [Google Scholar].

Liu, W., Fontanet, A., Zhang, P.H., Zhan, L., Xin, Z.T., Baril, L., 2006. Two-year prospective study of the humoral immune response of patients with severe acute respiratory syndrome. J. Infect. Dis. 193 (6), 792–795. https://doi.org/10.1086/500469 [PMC free article] [PubMed] [CrossRef] [Google Scholar].

Liu, Y., Chen, H., Tang, K., Guo, Y., 2020a. Clinical manifestations and outcome of SARS-CoV-2 infection during pregnancy. J. Infect. [Epub ahead of print] [PMC free article] [PubMed] [Google Scholar].

Liu, Y.-C., Kuo, R.-L., Shih, S.-R., 2020b. COVID-19: the first documented coronavirus pandemic history. Biomed. J. 43 (4), 328–333. http://www.sciencedirect.com/science/article/pii/S2319417020300445.

Madinger, N.E., Greenspoon, J.S., Eilrodt, A.G., 1989. Pneumonia during pregnancy: has modern technology improved maternal and fetal outcome? Am. J. Obstet. Gynecol. 161, 657–662. https://doi.org/10.1016/0002-9378(89)90373-6 [PubMed] [CrossRef] [Google Scholar).

Mahallawi, W.H., Khabour, O.F., Zhang, Q., Makhdoum, H.M., Suliman, B.A., 2018. MERS-CoV infection in humans is associated with a pro-inflammatory Th1 and Th17 cytokine profile. Cytokine 104, 8–13. https://doi.org/10.1016/j.cyto.2018.01.025 [PMC free article] [PubMed] [CrossRef] [Google Scholar].

Malik, A., El Masry, K.M., Ravi, M., Sayed, F., 2016. Middle East respiratory syndrome coronavirus during pregnancy, Abu Dhabi, United Arab Emirates 2013. Emerg. Infect. Dis. 22, 515–517. https://doi.org/10.3201/eid2203.151049 [PMC free article] [PubMed] [CrossRef] [Google Scholar].

Manni, M.L., Robinson, K.M., Alcorn, J.F., 2014. A tale of two cytokines: IL-17 and IL-22 in asthma and infection. Expet

Rev. Respir. Med. 8 (1), 25–42. https://doi.org/10.1586/17476348.2014.854167 [PMC free article] [PubMed] [CrossRef] [Google Scholar].

Martin, J.E., Louder, M.K., Holman, L.A., Gordon, I.J., Enama, M.E., Larkin, B.D., 2008. A SARS DNA vaccine induces neutralizing antibody and cellular immune responses in healthy adults in a phase I clinical trial. Vaccine 26 (50), 6338–6343. https://doi.org/10.1016/j.vaccine.2008.09.026 [PMC free article] [PubMed] [CrossRef] [Google Scholar].

Mavian, C., Pond, S.K., Marini, S., Magalis, B.R., Vandamme, A.M., Dellicour, S., et al., 2020. Sampling bias and incorrect rooting make phylogenetic network tracing of SARS-COV-2 infections unreliable. Proc. Natl. Acad. Sci. U. S. A. (in press).

Maxwell, C., McGeer, A., Tai, K.F.Y., Sermer, M., 2017. No. 225-management guidelines for obstetric patients and neonates born to mothers with suspected or probable severe acute respiratory syndrome (SARS). J. Obstet. Gynaecol. Can. 39, e130–e137. https://doi.org/10.1016/j.jogc.2017.04.024 [PMC free article] [PubMed] [CrossRef] [Google Scholar].

McIntosh, K., 1974. Coronaviruses: a comparative review. In: Current topics in Microbiology and Immunology/Ergebnisse der mikrobiologie und immunitatsforschung, vol. 25. Springer, Berlin, € Heidelberg, pp. 85–129.

Memish, Z.A., Zumla, A.I., Al-Hakeem, R.F., Al-Rabeeah, A.A., Stephens, G.M., 2013. Family cluster of Middle East respiratory syndrome coronavirus infections. N. Engl. J. Med. 368 (26), 2487–2494.

Murphy, S., 2020. Newborn Baby Tests Positive for Coronavirus in London. https://www.theguardian.com/world/2020/mar/14/newborn-baby-tests-positive-for-coronavirus-in-london.

Nelson-Piercy, C., 2015. Handbook of Obstetric Medicine. CRC Press, Boca Raton, FL, pp. 63–84. Respiratory Disease; [Google Scholar].

Ng, P.C., So, K.W., Leung, T.F., Cheng, F.W., Lyon, D.J., Wong, W., Cheung, K.L., Fung, K.S., Lee, C.H., Li, A.M., et al., 2003. Infection control for SARS in a tertiary neonatal centre. Arch. Dis. Child. Fetal Neonatal Ed. 88, F405–F409. https://doi.org/10.1136/fn.88.5.F405 [PMC free article] [PubMed] [CrossRef] [Google Scholar].

Ng, P.C., Leung, C.W., Chiu, W.K., Wong, S.F., Hon, E.K., 2004. SARS in newborns and children. Biol. Neonate 85, 293–298. https://doi.org/10.1159/000078174 [PubMed] [CrossRef] [Google Scholar].

Ng, W.F., Wong, S.F., Lam, A., Mak, Y.F., Yao, H., Lee, K.C., Chow, K.M., Yu, W.C., Ho, L.C., 2006. The placentas of patients with severe acute respiratory syndrome: a pathophysiological evaluation. Pathology 38, 210–218. https://doi.org/10.1080/00313020600696280 [PMC free article] [PubMed] [CrossRef] [Google Scholar].

Nyhan, D., Bredin, C., Quigley, C., 1983. Acute respiratory failure in pregnancy due to staphylococcal pneumonia. Ir. Med. J. 76, 320–321 [PubMed] [Google Scholar].

Openshaw, P.J.M., 2004. What does the peripheral blood tell you in SARS? Clin. Exp. Immunol. 136 (1), 11–12. https://doi.org/10.1111/j.1365-2249.2004.02448.x [PMC free article] [PubMed] [CrossRef] [Google Scholar].

Panesar, N.S., 2003. Lymphopenia in SARS. Lancet 361, 1985. https://doi.org/10.1016/S0140-6736(03)13557-X [PMC free article] [PubMed] [CrossRef] [Google Scholar].

Park, W.B., Perera, R.A.P.M., Choe, P.G., Lau, E.H.Y., Choi, S.J., Chun, J.Y., 2015. Kinetics of serologic responses to MERS coronavirus infection in humans, South Korea. Emerg. Infect. Dis. 21 (12), 2186. https://doi.org/10.3201/eid2112.151421 [PMC free article] [PubMed] [CrossRef] [Google Scholar].

Park, S.Y., Kim, Y.M., Yi, S., Lee, S., Na, B.J., Kim, C.B., et al., 2020. Coronavirus disease outbreak in call center, South Korea. Emerg. Infect. Dis. 26 (8), 1666–1670.

Parri, N., Lenge, M., Buonsenso, D., 2020. Children with Covid-19 in pediatric emergency departments in Italy. New Times.

Payne, D.C., Iblan, I., Alqasrawi, S., Al Nsour, M., Rha, B., Tohme, R.A., Abedi, G.R., Farag, N.H., Haddadin, A., Al Sanhouri, T., et al., 2014. Stillbirth during infection with Middle East respiratory syndrome coronavirus. J. Infect. Dis. 209, 1870–1872. https://doi.org/10.1093/infdis/jiu068 [PMC free article] [PubMed] [CrossRef] [Google Scholar].

Peng, Z., Yang, K.L., Wang, Z.G., Hu, B., Zhang, L., Zhang, W., 2020. Discovery of a novel coronavirus associated with the recent pneumonia outbreak in humans and its potential bat origin. BioRxiv 01 (22), 914952. https://doi.org/10.1101/2020.01.22.914952 [CrossRef] [Google Scholar].

Perera, R., Tso, E., Tsang, O.T.Y., Tsang, D.N.C., Fung, K., Leung, Y.W.Y., et al., 2020. SARS-CoV-2 virus culture and subgenomic RNA for respiratory specimens from patients with mild coronavirus disease. Emerg. Infect. Dis. 26 (11).

Phelan, A.L., Katz, R., Gostin, L.O., 2020. The novel coronavirus originating in Wuhan, China: challenges for global health governance. J. Am. Med. Assoc. 323, 709–710.

Pyrc, K., Berkhout, B., Van Der Hoek, L., 2007. Identification of new human coronaviruses. Expert Rev. Anti Infect. Ther. 5 (2), 245–253.

Rahman, A., Sarkar, A., 2019. Risk factors for fatal Middle East respiratory syndrome coronavirus infections in Saudi Arabia: analysis of the WHO Line List, 2013– 2018. Am. J. Publ. Health 109 (9), 1288–1293.

Rigby, F.B., Pastorek, J.G., 1996. Pneumonia during pregnancy. Clin. Obstet. Gynecol. 39, 107–119. https://doi.org/10.1097/00003081-199603000-00011 [PubMed] [CrossRef] [Google Scholar].

Robertson, C.A., Lowther, S.A., Birch, T., Tan, C., Sorhage, F., Stockman, L., McDonald, C., Lingappa, J.R., Bresnitz, E., 2004. SARS and pregnancy: a case report. Emerg. Infect. Dis. 10, 345–348. https://doi.org/10.3201/eid1002.030736 [PMC free article] [PubMed] [CrossRef] [Google Scholar].

Santos, M.L.O., 2003. Ground-glass opacity in diffuse lung diseases: high-resolution computed tomography-pathology correlation. Radiol. Bras. 36 (6) https://doi.org/10.1590/S0100-39842003000600003 [CrossRef] [Google Scholar].

Sargent, I.L., Redman, C., 1992. Immunobiologic adaptations of pregnancy. In: Reece, E.A., Hobbins, J.C., Mahoney, M.J., Petrie, R.H. (Eds.), Medicine of the Fetus and Mother. JB Lippincott Company, Philadelphia, PA, USA, pp. 317–327 [Google Scholar].

Sass, L., Urhoj, S.K., Kjærgaard, J., 2017. Fever in pregnancy and the risk of congenital malformations: a cohort study. BMC Pregnancy Childbirth 17, 413 [PMC free article] [PubMed] [Google Scholar].

Schneider, E., Duncan, D., Reiken, M., Perry, R., Messick, J., Sheedy, C., Haase, J., Gorab, J., 2004. SARS in pregnancy. AWHONN Lifelines 8, 122–128. https://doi.org/10.1177/1091592304265557 [PMC free article] [PubMed] [CrossRef] [Google Scholar].

Shek, C.C., Ng, P.C., Fung, G.P., Cheng, F.W., Chan, P.K., Peiris, M.J., Lee, K.H., Wong, S.F., Cheung, H.M., Li, A.M., et al., 2003. Infants born to mothers with severe acute respiratory syndrome. Pediatrics 112, e254. https://doi.org/10.1542/peds.112.4.e254 [PubMed] [CrossRef] [Google Scholar].

Shereen, M.A., Khan, S., Kazmi, A., Bashir, N., Siddique, R., 2020. COVID-19 infection: origin, transmission, and characteristics of human coronaviruses. J. Advert. Res. 24, 91–98.

Somsen, G.A., van Rijn, C., Kooij, S., Bem, R.A., Bonn, D., 2020. Small droplet aerosols in poorly ventilated spaces and SARS-CoV-2 transmission. Lancet Respir. Med. 8, 658–659. https://www.ecdc.europa.eu/en/geographical-distribution2019-ncov-cases. (Accessed December 2020).

Steinbuch, Y., 2020. Chinese Baby Tests Positive for Coronavirus 30 Hours After Birth. Available online: https://nypost.com/2020/02/05/chinese-baby-tests-positive-for-coronavirus-30-hours-after-birth/. (Accessed 9 February 2020).

Stockman, L.J., Lowther, S.A., Coy, K., Saw, J., Parashar, U.D., 2004. SARS during pregnancy, United States. Emerg. Infect. Dis. 10, 1689–1690. https://doi.org/10.3201/eid1009.040244 [PMC free article] [PubMed] [CrossRef] [Google Scholar].

Tang, F., Quan, Y., Xin, Z.T., Wrammert, J., Ma, M.J., Lv, H., 2011. Lack of peripheral memory B cell responses in recovered patients with severe acute respiratory syndrome: a six-year follow-up study. J. Immunol. 186 (12), 7264–7268. https://doi.org/10.4049/jimmunol.0903490 [PubMed] [CrossRef] [Google Scholar].

Thevarajan, I., Nguyen, T.H.O., Koutsakos, M., 2020. Breadth of concomitant immune responses prior to patient recovery: a case report of non-severe COVID-19. Nat. Med. 26, 453–455 [PMC free article] [PubMed] [Google Scholar].

Visscher, H.C., Visscher, R.D., 1971. Indirect obstetric deaths in the state of Michigan 1960–1968. Am. J. Obstet. Gynecol. 109, 1187–1196. https://doi.org/10.1016/0002-9378(71)90664.

Waltenburg, M.A., Victoroff, T., Rose, C.E., Butterfield, M., Jervis, R.H., Fedak, K.M., et al., 2020. Update: COVID-19 among workers in meat and poultry processing facilities - United States, April-May 2020. Morb. Mortal. Wkly. Rep. 69 (27), 887–892.

Wan, Y., Shang, J., Graham, R., Baric, R.S., Li, F., 2020. Receptor recognition by the novel coronavirus from Wuhan: an analysis based on decade-long structural studies of SARS coronavirus. J. Virol. 94, e00127–e00220.

Wang, C., Horby, P.W., Hayden, F.G., Gao, G.F., 2020. A novel coronavirus outbreak of global health concern. Lancet.

Weinberger, S., Weiss, S., Cohen, W., Weiss, J., Johnson, T.S., 1980. Pregnancy and the lung. Am. Rev. Respir. Dis. 121, 559–581. https://doi.org/10.1164/arrd.1980.121.3.559 [PubMed] [CrossRef] [Google Scholar].

WHO. Emergency Committee Regarding the Outbreak of Novel Coronavirus (2019-nCoV) 2020. Statement on the Meeting of the International Health Regulations. Available from: https://www.who.int/news-room/detail/23-01-2020-statement-on-the-meeting-of-the-international-health-regulations-(2005)-emergency-committee-regarding-the-outbreak-of-novel-coronavirus-(2019-ncov). [Google Scholar].

Wolfel, R., Corman, V.M., Guggemos, W., Seilmaier, M., Zange, S., Muller, M.A., et al., 2020. Virological assessment of hospitalized patients with COVID-2019. Nature 581 (7809), 465–469.

Wong, S.F., Chow, K.M., de Swiet, M., 2003. Severe acute respiratory syndrome and pregnancy. BJOG 110, 641–642. https://doi.org/10.1046/j.1471-0528.2003.03008.x [PMC free article] [PubMed] [CrossRef] [Google Scholar].

Wong, S.F., Chow, K.M., Leung, T.N., Ng, W.F., Ng, T.K., Shek, C.C., Ng, P.C., Lam, P.W., Ho, L.C., To, W.W., et al., 2004. Pregnancy and perinatal outcomes of women with severe acute respiratory syndrome. Am. J. Obstet. Gynecol. 191, 292–297. https://doi.org/10.1016/j.ajog.2003.11.019 [PMC free article] [PubMed] [CrossRef] [Google Scholar].

Woo, P.C., Lau, S.K., Lam, C.S., Lau, C.C., Tsang, A.K., Lau, J.H., et al., 2012. Discovery of seven novel Mammalian and avian coronaviruses in the genus deltacoronavirus supports bat coronaviruses as the gene source of alphacoronavirus and betacoronavirus ad avian coronaviruses as the gene source of gammacoronavirus and deltacoronavirus. J. Virol. 86, 3995e4008.

Woodward, A., 2020. A Pregnant Mother Infected with the Coronavirus Gave Birth, and Her Baby Tested Positive 30 Hours Later. Available online: https://www.businessinsider.com/wuhan-coronavirus-in-infant-born-from-infected-mother-2020-2. (Accessed 8 February 2020).

Wrapp, D., Wang, N., Corbett, K.S., Goldsmith, J.A., Hsieh, C.L., Abiona, O., et al., 2020. Cryo-EM structure of the 2019-nCoV spike in the prefusion conformation. Science 367, 1260e3.

Wu, F., Zhao, S., Yu, B., Chen, Y.M., Wang, W., Song, Z.G., et al., 2020. A new coronavirus associated with human respiratory disease in China. Nature 579, 265e9.

Xu, Y., Li, X., Zhu, B., 2020a. Characteristics of pediatric SARS-CoV-2 infection and potential evidence for persistent fecal viral shedding. Nat. Med. 26, 502–505 [PMC free article] [PubMed] [Google Scholar].

Xu, K., Chen, Y., Yuan, J., Yi, P., Ding, C., Wu, W., et al., 2020b. Factors associated with prolonged viral RNA shedding in patients with coronavirus disease 2019 (COVID-19). Clin. Infect. Dis. 71 (15), 799–806.

Xu, L., Liu, J., Lu, M., Yang, D., Zheng, X., 2020c. Liver damage during highly pathogenic human coronavirus infections. Liver Int. 00, 1–7. https://doi.org/10.1111/liv.14435 [PMC free article] [PubMed] [CrossRef] [Google Scholar].

Yang, L.T., Peng, H., Zhu, Z.L., Li, G., Huang, Z.T., Zhao, Z.X., 2006. Long-lived effector/central memory T-cell responses to severe acute respiratory syndrome coronavirus (SARS-CoV) S antigen in recovered SARS patients. Clin. Immunol. 120 (2), 171–178. https://doi.org/10.1016/j.clim.2006.05.002 [PMC free article] [PubMed] [Cross-Ref] [Google Scholar].

Yang, L., Peng, H., Zhu, Z., Li, G., Huang, Z., Zhao, Z., 2007. Persistent memory CD4+ and CD8+ T-cell responses in recovered severe acute respiratory syndrome (SARS) patients to SARS coronavirus M antigen. J. Gen. Virol. 88 (10), 2740. https://doi.org/10.1099/vir.0.82839-0 [PMC free article] [PubMed] [CrossRef] [Google Scholar].

Yesudhas, D., Srivastava, A., Gromiha, M.M., 2020. COVID-19 outbreak: history, mechanism, transmission, structural studies and therapeutics. Infection 1–15.

Ying, T., Li, W., Dimitrov, D.S., 2016. Discovery of T-cell infection and apoptosis by Middle East respiratory syndrome coronavirus. J. Infect. Dis. 213 (6), 877–879. https://doi.org/10.1093/infdis/jiv381 [PMC free article] [PubMed] [CrossRef] [Google Scholar].

Yu, X.Y., Zhang, Y.C., Han, C.W., Wang, P., Xue, X.J., Cong, Y.L., 2003. Change of T lymphocyte and its activated subsets in SARS patients. Acta Acad. Med. Sin. 25 (5), 542–546 [PubMed] [Google Scholar].

Yudin, M.H., Steele, D.M., Sgro, M.D., Read, S.E., Kopplin, P., Gough, K.A., 2005. Severe acute respiratory syndrome in pregnancy. Obstet. Gynecol. 105, 124–127. https://doi.org/10.1097/01.AOG.0000151598.49129.de [PubMed] [CrossRef] [Google Scholar].

Zaki, A.M., Van Boheemen, S., Bestebroer, T.M., Osterhaus, A.D., Fouchier, R.A., 2012. Isolation of a novel coronavirus from a man with pneumonia in Saudi Arabia. N. Engl. J. Med. 367, 1814–1820.

Zhang, J.P., Wang, Y.H., Chen, L.N., Zhang, R., Xie, Y.F., 2003. Clinical analysis of pregnancy in second and third trimesters complicated severe acute respiratory syndrome. Zhonghua Fu Chan Ke Za Zhi 38, 516–520 [PubMed] [Google Scholar].

Zhang, L., Zhang, L., Jiang, Y., 2020a. Analysis of the pregnancy outcomes in pregnant women with COVID-19 in Hubei Province. Zhonghua Fu Chan Ke Za Zhi 55, E009 [PubMed] [Google Scholar].

Zhang, J.J., Dong, X., Cao, Y.Y., Yuan, Y.D., Yang, Y.B., Yan, Y.Q., 2020b. Clinical characteristics of 140 patients infected by SARS-CoV-2 in Wuhan, China. Allergy 00, 1–12. https://doi.org/10.1111/all.14238 [PubMed] [Cross-Ref] [Google Scholar].

Zheng, B.J., Guan, Y., Wong, K.H., Zhou, J., Wong, K.L., Young, B.W.Y., et al., 2020. SARS-related virus predating SARS outbreak, Hong Kong. Emerg. Infect. Dis. 10 (2), 176, 2004 England journal of medicine May 1.

Zhou, P., Yang, X.L., Wang, X.G., Hu, B., Zhang, L., Zhang, W., Si, H.R., Zhu, Y., Li, B., Huang, C.L., Chen, HD. A l, 2020. A pneumonia outbreak associated with a new coronavirus of probable bat origin. Nature 579, 270–273.

Zhu, H., Zhu, H., Wang, L., Fang, C., Peng, S., 2020. Clinical analysis of 10 neonates born to mothers with 2019-nCoV pneumonia. Transl. Pediatr. 9, 51–60 [PMC free article] [PubMed] [Google Scholar].

Zumla, A., Chan, J.F., Azhar, E.I., Hui, D.S., Yuen, K.Y., 2016. Coronaviruses—drug discovery and therapeutic options. Nat. Rev. Drug Discov. 15, 327–347.

FURTHER READING

Gorbalenya, A., et al., Coronaviridae Study Group of the International Committee on Taxonomy of Viruses, 2020. The species Severe acute respiratory syndrome-related coronavirus: classifying 2019-nCoV and naming it SARS-CoV-2. Nat. Microbiol. 5 (4), 536–544. https://doi.org/10.1038/s41564-020-0695-z.

Li, A.M., Ng, P.C., 2005. Severe acute respiratory syndrome (SARS) in neonates and children. Arch. Dis. Child. Fetal Neonatal Ed. 90, F461–F465. https://doi.org/10.1136/adc.2005.075309 [PMC free article] [PubMed] [CrossRef] [Google Scholar].

World Health Organization (WHO), 2004. The Geographical Distribution of SARS-CoV and MERS-CoV Worldwide. https://www.who.int/csr/sars/country/table2004_04_21/en/, http://www.emro.who.int/health-topics/mers-cov/mers-outbreaks.html.

CHAPTER 2

Immunological Changes in Pregnancy and Its Relation to COVID-19 Infection

AMR S. HAMZA, MD

(Dr. Prof. M Fadel Shaltout. Prof. of Obstetrics and Gynecology, Cairo University, Faculty of Medicine)

Covid-19 Infection and Pregnancy. https://doi.org/10.1016/B978-0-323-90595-4.00006-6

Humans are in constant exposition to microorganisms. Based on their effects on the body, microorganisms are classified into pathogenic and nonpathogenic organisms. This in turn regulates the body response: (1) interaction or (2) defense. The response is carried by an interactive network, which is called the immune system.

With the onset of pregnancy, the immune system has to resume its protective role despite hosting an antigen. In this phase, the aim is to protect the host and the growing antigen, the fetus. Several mechanisms take place to maintain an optimal immunological function without harming the growing intrauterine fetus. An interaction between local uterine components of the innate and adaptive immune response sets a balance to tolerate the "semiallogenic" fetus while securing the host against exogenous infection (Thellin et al., 2000).

OVERVIEW ON THE IMMUNE SYSTEM
Innate Immunity
Leukocytes
In comparison with the nonpregnant state and due to the increased inflammatory response, there is a significant increase in the total white blood cells (Melgert et al., 2012; Efrati et al., 1964). Eosinophil and basophil levels do not increase throughout the whole pregnancy. Yet, an increase in the degranulation of eosinophils has been recorded throughout the whole pregnancy, which decreases with the onset of delivery and drops furthermore until 1 month after labor. Significant higher-end products of eosinophil degranulation were detected after a caesarian section (Matsumoto et al., 2003). The increase was mainly due to the significant rise in neutrophil counts (Belo et al., 2005; Abbassi-Ghanavati et al., 2009). This could be explained by the higher gestational cortisol (Buss et al., 2012) or granulocyte macrophage colony-stimulating factor (GM-CSF) levels (Belo et al., 2005).

Despite the rise in the number of leukocytes due to an increase in neutrophils, a decrease in the phagocytic capacity occurs during pregnancy, as shown in Fig. 2.1 (Lampé et al., 2015).

Monocytes
One major change of the innate immune system is an increase number of monocytes (Luppi et al., 2002; Siegel and Gleicher, 1981). In normal pregnancies, there is a significant rise of the total monocyte count from 0.3 (0.1−0.8) $\times 10^9$ cells/L to 0.6 (0.4−0.9) $\times 10^9$ cells/L (Melgert et al., 2012).

On the contrary, a significant decrease in the phagocytic function of monocytes occurs in healthy pregnancies

when compared with nonpregnant women, as shown in Fig. 2.1. This decrease is part of a maternal immunosuppression, which protects the semiallogenic fetus (Lampé et al., 2015). Parallel to the increase of placental mass associated with the increase in gestational age, a significant upregulation of the activation markers (CD11a, CD54, and CD64 surface antigen) takes place. The upregulation peaks with the onset of labor. In addition, monocytes increasingly produce interleukin-12 during pregnancy (Luppi et al., 2002). Other functional changes include the increased production of oxygen free radicals and different cytokine production. The latter is inconsistently reported in literature. Different studies also are contradictive with regard to the relative change of monocyte subsets (Faas and de Vos, 2017).

Complement system
A balanced activation of the complement system occurs and is protective against complicated pregnancies, e.g., preeclampsia and preterm birth.

While the complement factors C3a, C4a, and C5a are elevated in the second half of the pregnancy and C3, C4d, and C9 and serum complement membrane attack complex throughout the whole pregnancy, it is counterbalanced by an elevation in factor H, decay-accelerating factor (DAF), pregnancy-associated plasma protein A (PAPPA), CD46/CD55 like activities, C1 inhibitor (C1-INH), membrane cofactor protein (MCP), C4-binding protein (C4BP), complement receptor 1 (CR1), mannose-associated serine protease (MASP), and mannose-binding lectin (MBL) (Regal et al., 2015).

A failure to keep this balanced activation can lead into a pathological pregnancy. Many examples exist. A defect in the glycosylphosphatidylinositol (GPI) anchoring of complement regulators CD55 and CD59 in blood cells may result into complement-mediated hemolysis, thrombocytopenia, and thrombosis (Ray et al., 2000). In cases of pregnancy-associated atypical hemolytic−uremic syndrome, the levels of C3, C5, and properdin are low to undetectable, while an upregulation of cell-surface C3b deposition is detected (Zhang et al., 2020). There is also growing evidence of the complement factor dysregulation involvement, e.g., C5a-mediated trophoblasts dysfunction, in the pathogenesis of preeclampsia (Ma et al., 2018). Therefore, keeping a balanced upregulation of the complement system is important in the development of a physiological pregnancy (Regal et al., 2015).

Innate lymphoid cells
Newly, innate lymphoid cells (ILCs) have been identified. These are undifferentiated lymphocytes not expressing

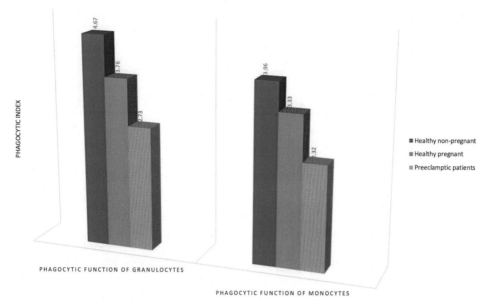

FIG. 2.1 Comparison of the phagocytic index of granulocytes and monocytes in healthy nonpregnant, healthy pregnant, and preeclamptic individuals (Lampé et al., 2015).

the antigen receptors, which are found on T and B cells (Vivier et al., 2018). They are subdivided into three types.

Together with the NK cells, ILC type 1 contributes to the placentation including spiral arteries remodeling and trophoblastic uterine invasion. The necessary tissue remodeling is facilitated by the production of interleukin-8, vascular endothelial growth factor (VEGF), stromal cell—derived factor 1 (SDF-1), and interferon gamma-induced protein 10 (IP-10). NK cells are furthermore subclassified into peripheral (pNK) and decidual NK (dNK) cells, with the decidual subtype being unable to attack trophoblasts (Abu-Raya et al., 2020).

The invasion of the anchoring trophoblasts is promoted by the NK cells and ILC3. Yet, dNK cells have contradicting roles. On one hand, it induces the myometrial invasion by secreting chemokines, e.g., IL-8, IP-10, and XCL1 and GM-CSF. On the other hand, this invasion is inhibited by secreting TGF-β. dNK also promotes spiral artery remodeling by producing angiogenic factors, e.g., VEGF, angiopoietin (Ang) 1 and 2, placental-derived growth factor, and hepatocyte growth factor. To prevent attacking the growing fetus, the gestational hormones (estradiol [E2], progesterone [PRG], and human chorionic gonadotropins [hCGs]), decidual stromal cells (DSCs), and the trophoblasts neutralize the pNKs to induce a functional competence and self-tolerance. E2 and PRG prevent NK degranulation.

hCG and E2 promote furthermore dNK proliferation, which does not attack trophoblasts. DSC suppresses additionally the expression of CD16, rendering the pNK less cytotoxic (Chang et al., 2020; Mendes et al., 2020). Ongoing investigations are directed to whether the cytotoxic pNKs become less in number or function or are increasingly converted to pNKs.

Adaptive Immunity
T lymphocytes
An early comparison of total T cells, T-helper cells (CD4+), and T suppressor cells (CD8+) percentages in the peripheral blood showed no difference on all T cell subsets in the peripheral blood (Coulam et al., 1983). A different behavior of CD3+, CD4+, and CD8+ was also observed in a different population (Mahmoud et al., 2001). More recent observations showed a decrease in the absolute numbers of lymphocytes during pregnancy, which return to the nonpregnant state in puerperium. A decrease in the CD4+ T cells during pregnancy followed by a postpartum increase in CD3+TCRα/β1- and CD3+TCRγ/-1+ may explain the increased susceptibility to viral infection during pregnancy (Watanabe et al., 1997). Additionally a significant increase was observed in HLA-DR+ and CD56+ in the first trimester followed a steady decrease with ongoing pregnancy (Kühnert et al., 1998). Yet, the T-helper cell function remained adequate throughout gestation (Bailey et al., 1985) with a decreased Th1

(IL-2 and interferon γ secretion) and increased Th2 response (IL4 and 10 secretion) throughout the whole pregnancy and postpartum period (Matthiesen et al., 1998).

The Th1 cell response was referred to being destructive toward pregnancy, and vice versa regarding the Th2 cell response (Szekeres-Bartho and Wegmann, 1996). Mostly animal studies highlighted the effect of an immunomodulatory protein known as progesterone-induced blocking factor, which alters the Th1/Th2 balance. In summary, a different cytokine secretion pattern was suspected to decrease the cell-mediated responses during pregnancy (Abu-Raya et al., 2020).

Yet, different conclusions were reported in different studies (Abu-Raya et al., 2020) ranging from a shift toward Th2 cell response only in early gestation (Raghupathy et al., 2000), Th2 cell response domination throughout the whole pregnancy (Wegmann et al., 1993; Wegmann, 1984; Szekeres-Bartho and Wegmann, 1996), a shift toward Th2 response throughout on antigen presentation the whole pregnancy (Marzi et al., 1996), lower Th1 and no Th2 change at the end of third trimester (Saito et al., 1999), rise of the Th1 response after delivery (Aghaeepour et al., 2017), and no change at all (Lissauer et al., 2014).

With regard to T cell subsets, different response among ethnicities is suggested, making populations more prone to infections during pregnancies than others (Mahmoud et al., 2001). Yet, it has to be mentioned that the study populations in all the aforementioned trials is too low to make a conclusion on a whole population.

B lymphocytes and immunoglobulins

B lymphocytes decrease in number in physiological pregnancies. According to a study, this decrease was not statistically significant (Kühnert et al., 1998). In several other studies, the absolute and relative numbers of conventional CD19+ B cells were shown to decrease significantly during pregnancy with a postpartum return to the nonpregnant state (Bhat et al., 1995; Watanabe et al., 1997; Lima et al., 2016; Mahmoud et al., 2001; Zimmer et al., 1998). In contrast, pregnancies ending in preterm labor were associated with an increase in the CD19+ B cells (Busse et al., 2020; Sendag et al., 2002). With increased CD19+ B cells, B-regulatory cells decrease (Busse et al., 2020). This suggests a role of the downregulation of B lymphocytes to maintain a normal pregnancy. In turn, this leads to a decreased production of the IgG, IgA, and IgM. Some studies suggest an initial increase in IgM and IgA at the beginning of the pregnancy followed by a decreased production later in pregnancy (Amino et al., 1978; Miller and Abel, 1984; Yasuhara et al., 1992).

In summary, the immunological response has been controversially described in literature. It is yet to be mentioned that all studies are involving a relatively low number of patients, present often an oversimplified idea of the pregnancy as one unit, carry bias, and do not consider other biological and mechanical changes during pregnancy (Mor and Cardenas, 2010). In addition, populations of different ethnicities seem to respond different to gestation (Mahmoud et al., 2001). This may explain some discrepancies between the aforementioned studies, which needs to be better understood.

Respiratory Adaptations to Pregnancy

In addition to the aforementioned immunological adaptations, which may lead to increased susceptibilities to some viral infection, additional pulmonary adaptations may favor a more severe course in viral respiratory infections (Goodnight and Soper, 2005). With the onset of pregnancy, there is a 30%−40% increase in tidal volume. As the respiratory rate does not change, the minute ventilation is increased by 50%, leading to an increase in minute oxygen uptake. This increase meets the increased maternofetal gestational needs and results in a respiratory alkalosis. The latter is met by an increased renal bicarbonate excretion; thus, the arterial pH remains physiological. No changes in the lung compliance, forced vital capacity, or diffusing capacity was described. As a result of all the aforementioned change, the residual capacity, expiratory reserve volume, and functional residual capacity decrease by 15%−20% at term. Gestational hormones also decrease the total pulmonary resistance. To further improve the oxygenation to encompass the fetal and increased maternal needs, the hemoglobin amount increases. Due to an increase in plasma volume, a physiological anemia develops. The growing uterus elevates the diaphragm. To adapt to that, the transverse chest diameter increases (LoMauro and Aliverti, 2015). All the aforementioned changes result in a decreased ability of the host to contain respiratory infections (Goodnight and Soper, 2005). The details are presented in the lines to come.

Infections and Pregnancy

With onset of pregnancy, the aforementioned changes result in an immunomodulation rather than an immunosuppression, like earlier believed. Different immunological changes throughout the whole pregnancy result in a differential reaction toward an antigen depending on the host parameters, gestational age, and antigen

TABLE 2.1
Different Susceptibility and Clinical Course Based on the Type of Different Infections During Pregnancy in Comparison With Nonpregnant Women (Kourtis et al., 2014).

Infection	Increased Susceptibility	Increased Severity
Influenza	No	Yes
Hepatitis E infection	No	Yes
Herpes simplex infection	No	Yes
Malaria	Yes	Yes
Listeriosis	Yes	No
Measles	No	Yes
Smallpox	No	Yes
HIV type 1	Yes	No
Varicella	No	Yes

characteristics (Mor and Cardenas, 2010). In the coming paragraphs, viral infections during will receive a special focus.

When discussing a viral infection during pregnancy, special aspects become important: the susceptibility during pregnancy, the effect of the pregnancy on the course of the disease, and the maternofetal complications of the disease.

As presented in Table 2.1, there is a variable susceptibility and disease severity depending on the type of infection and other parameters (Kourtis et al., 2014). A possible explanation for the variable body response is summarized in the different and contradictive immunological changes, which is still not completely understood, during pregnancy.

After the emergence of the first cases of human infection with COVID-2 infection, a heavy international debate took place regarding its importance, severity, and implication on daily life. On the night of January 10, 2021, 15 months after the emergence of the first reports in October 2019, we made a PubMed search on the phrase "COVID-19." PubMed reported over 90,000 papers to that date. The gush of papers, with different qualities and contradictive outcomes, fostered the confusion. We need to recall the conflicting studies about immunological changes, especially with regard to a possible ethnical different response, to understand different study outcomes. One of the biggest debates revolves around calling COVID-19, a kind of flu, and its impact similarity with the influenza virus. In the coming lines, we will therefore discuss influenza and COVID-19.

Influenza

Influenza is a highly infectious viral disease responsible for the demise of 0.1−223.5 per 100,000 infected individuals yearly depending on the age, underlying chronic disease and region of residence (Iuliano et al., 2018). Pregnancy may be a moderate but not significant risk factor for the development of severe maternal morbidity (González-Candelas et al., 2012).

In the pregnant population, an influenza infection comes with a higher risk for hospitalization. Yet, no increased mortality has been recorded in a systematic metaanalysis involving 152 studies (OR 1.04; 95% CI 0.81−1.33) (Mertz et al., 2017). A possible explanation is the immunological alternation that occurs during pregnancy and gestational immunomodulatory downregulation of the B cell response (Swieboda et al., 2020). Yet, it remains a subject of debate, whether pregnancy is per se a risk factor or other known or unknown underlying conditions for the development of severe maternal morbidity and mortality. Further research is awaited.

There is no conclusive evidence whether exposure to influenza increases the rate of spontaneous abortion and still births, despite the seasonal variations of spontaneous abortions (Rasmussen et al., 2018). Whether exposure to influenza increases the risk for congenital anomalies is a subject of debate (Luteijn et al., 2014, 2015; Xia et al., 2019).

In a recent study, H1N1 influenza illness was not proven to be a major contributor to preterm delivery in the overall obstetrical population (no pH1N1 diagnosis: adjusted HR [aHR] = 1.0; 95% CI = 0.98, 1.1;

pH1N1 diagnosis: aHR = 1.0; 95% CI = 0.88, 1.2). However, patients with preexisting medical conditions possess a higher risk for preterm delivery after influenza infection (Fell et al., 2018). Furthermore, little and conflicting evidence supports an association between SGA births (pooled odds ratio 1.24; 95% CI 0.96−1.59) and influenza infection. In mild to moderate and severe influenza, the relative risk for fetal demise was 1.9 and 4.2, respectively (Fell et al., 2017).

Based on the aforementioned evidence and interventional studies, several guidelines and studies recommended influenza vaccination during pregnancy to avoid potential adverse maternofetal outcomes (CDC, 2019; Berger et al., 2018) and has been proven to be safe, even if given in the first trimester (Speake et al., 2020).

Comparison With COVID-19

In a French study, the characteristics, morbidity, and mortality of seasonal influenza and COVID-19 infection were compared retrospectively. The study included 89,530 patients with COVID-19 and 45,819 patients with seasonal influenza infections. Almost twice patients were admitted due to COVID-19 in comparison with seasonal influenza. The mortality of hospitalized patients was three times higher, especially in older-age categories. Yet, despite being lower, the mortality rates of children hospitalized due to COVID-19 were four times higher than the mortality rate of adults. With regard to influenza and COVID-19, the mortality rate increased with the presence of comorbidities. The authors suspect that the increased admission, morbidity, and mortality in COVID-19 cases are due to the absence of a vaccine, as is the case in influenza virus (Piroth et al., 2020).

The situation for COVID-19 is quite similar to its impact during World War I, where unknown huge numbers of the international population died in its first waves. Due to the novelty of the virus, hence neither herd immunity nor vaccines are present, the impact of COVID-19 is currently worse than influenza.

COVID-19 AND PREGNANCY ISSUES
Introduction

SARS-COVID-19 (also referred to as SARS-CoV-2/human/Wuhan/X1/2019_XYZ12345) is an enveloped positive-strand RNA virus, which is transmitted via respiratory droplets, i.e., from coughing and sneezing (Wölfel et al., 2020). After being inhaled, it starts replicating in the nasopharynx. The virus is first present in the nasal cavity and induces a limited innate response.

At this stage, the virus can be detected using nasopharyngeal swabs (Corman et al., 2020). Consequently, it propagates to the lower respiratory systems, where it induces a more pronounced innate immune response. It is as this stage, where the viral infections express the known clinical pictures in the host body (Zhu et al., 2020a, 2020b; Coronaviridae Study Group of the International Committee on Taxonomy of Viruses, 2020).

Most of the knowledge about the immune response against SARS-COVID-19, especially with regard to long-term effects, is derived from the previous experience with other type of Coronaviridae, e.g., severe acute respiratory disease syndrome (SARS) of 2003 (Zhao et al., 2010) and Middle East respiratory syndrome (MERS) of 2012 (Alshukairi et al., 2018; Ko et al., 2017).

Initial studies on the immunological response to SARD-COVID-19 showed a leuco- and lymphopenia in 25% and 63% of the patients, respectively. Additionally, prothrombin time and D-dimer levels were higher in patients requiring an intensive care unit (ICU) management. Aspartate aminotransferase and troponin-I were also described to be increased in ICU patients (Huang et al., 2020).

Another study showed significantly lower lymphocyte subsets (CD3$^+$ T cell, CD4$^+$ T cell, CD8$^+$ T cell and B cell (CD19$^+$), and NK cell (CD16$^+$56$^+$)) in patients with severe clinical courses. While T lymphocytes continued to decrease in severe and demised cases, it showed a gradual increase in nonsevere cases with a favorable outcome (Premkumar et al., 2020). The reduction in the T lymphocyte subsets results in a decreased function of antigen-presenting cells, B lymphocytes, other T lymphocytes, secretion of cytokines, and direct killing of target cells (Budd and Fortner, 2013).

A significant increase in the C3 complement activation occurred in severe cases (Premkumar et al., 2020), thus inducing a proinflammatory response responsible for the acute lung injury (Merle et al., 2015). No obvious decrease was detected regarding B and NK cells with when considering the severity of cases (Premkumar et al., 2020).

With regard to the humoral immunity, an initial study showed a seroconversion of total antibodies, IgM, and IgG of 93.1%, 82.7%, and 64.7%, respectively. Within 7 days of the onset of symptoms, <40% of the antibodies are positive. The seroconversion of total antibodies, IgM, and IgG rises quickly after day 15 of the onset of symptoms to 100%, 94.3%, and 79%, respectively (Huang et al., 2020). A metaanalysis of 38 publications from Asia, Europe, and the United States fostered the results of initial trials. Thus, the sensitivity

TABLE 2.2

The Sensitivity and Confidence Interval of the Antibody (Total Antibodies, IgM and IgG) Testing in a Metaanalysis of 38 Studies (Deeks et al., 2020).

Time Interval Between Onset and Testing (Days)	Sensitivity (%)	Confidence Interval
1–7	30.1	21.4–40.7
8–14	72.2	63.5–79.5
15–21	91.4	87.0–94.4
21–35	96	90.6–98.3

for the diagnostic accuracy increased with the increasing interval between the time of onset of the disease and time the antibody testing is done as shown in Table 2.2 (Deeks et al., 2020). Initial studies suggest the presence of a long-term immunity, yet these data need to be verified (Mudd and Remy, 2021).

A current limitation is the lack of studies showing the long-term sensitivity of antibodies after 35 days of the onset of infection and in asymptomatic but infected patients. Due to the short interval between the beginning of the international pandemic and follow-up, these studies are still awaited, especially with regard to the effectiveness of immunized patients. This may give rise to the current scepsis, despite the increasing economic and social need, toward a destined vaccine (Cyranoski, 2020). On the other hand, some evidence suggests that a protracted vaccination program, i.e., not able to vaccinate at least 40% of the population, might not lead normalization of daily life (Moghadas et al., 2020).

Following the Cochrane analysis—specific antibodies against a domain of SARS-CoV-2, the receptor-binding domain (RBD) of the spike protein was identified. The sensitivity of the antibodies was 94%, 77.5%, and 69% for IgG, IgA, and IgM, respectively, after 9 days of the onset of symptoms (Premkumar et al., 2020). In a larger study involving 309 acute and received SARS-COVID-2 cases, mostly with mild disease, IgG antibodies against the RBD alone were sufficient for the serological diagnosis in SARS-COV-2-RNA-positive cases, offering a new option for detection and therapy plans (Indenbaum et al., 2020). Some developers of the awaited vaccine therefore developed a modified mRNA vaccine that encodes the trimerized RBD of the spike glycoprotein of SARS-CoV-2 (Mulligan et al., 2020). The first safety reports revealed similar results to other viral vaccines (Walsh et al., 2020; Polack et al., 2020; Baden et al., 2020).

It is too early to have solid knowledge about the long-term antibody response to SARS-COVID-2 infections and reinfection due to the relatively short duration of the pandemic. Yet, it is known from SARS cases that the long-term antibody response fades away over time with an increased susceptibility to infection after 3 years of the first exposure (Wu et al., 2007). The positive samples (IgG and IgM) over times are shown in Fig. 2.2.

With regard to pregnancy and COVID-19 infection, several questions need to be considered:
1. Are pregnant women more susceptible to an infection?
2. Is the course of the infection in the aforementioned population more severe?
3. Vertical and peripartum infections?

Susceptibility to Infection during Pregnancy

An ideal tool to identify the susceptibility to infection is based on the proper knowledge of the prevalence of infection. Therefore, countries with studies based on active surveillance programs, i.e., screening asymptomatic and symptomatic individuals on a wide scale, are more suitable to reproduce the prevalence of the disease within a society. With regard to viral susceptibility among the pregnant population, the knowledge of the pregnancy status of the screened population is necessary.

In a German prospective trial, all pregnant women presenting between April and May 2020 were tested on SARS-CoV-2. The screening was performed by a nasopharyngeal swab as indicated by the Robert Koch Institute. A total of 234 pregnant women were tested. 96.2% of the patients were tested negative. In a trial to investigate, whether infection has occurred in the past, specific SARS-CoV-2 IgG in serum was tested in around 78% of the patients. In only 0.6% of the study population, antibodies (IgG) were found. In this

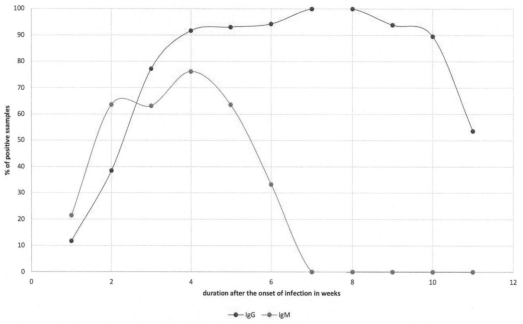

FIG. 2.2 Chart showing the primary rise of IgG and IgM followed by a decline over time after SARS infection (Wu et al., 2007). *SARS*, severe acute respiratory disease syndrome.

population, the prevalence and seroconversion among pregnant patients were below 5% and 1%, respectively. It should be added that the screening occurred during a lockdown, which has lowered the spread of cases (Zöll-kau et al., 2020). Other prevalence studies among pregnant patients showed similar results within a lockdown after the first wave in the beginning of 2020 (Tanacan et al., 2020; Mattern et al., 2020; Trahan et al., 2021). In countries with a similar established and widespread screening programs, the seroprevalence of the SARS-CoV-2 was shown to range from 1.2% to 10.4% (Herr-mann, 2020; Anand et al., 2020; Prabhu et al., 2020; Savirón-Cornudella et al., 2021). Bigger cohorts are awaited from the International COVID-19 and Pregnancy Registry. The project is currently still in the phase of recruitment (Panchaud et al., 2020).

Unlike expected and based on the aforementioned limited data, the infectivity of pregnant patients despite their immunological adaptations and respiratory proinfection gestational changes was shown to be lower than the nonpregnant adult population. Therefore, there is no conclusive evidence for the increased susceptibility based on German observations, which included many asymptomatic infections in comparison with other populations, where testing only occurred with the onset of COVID-19-like symptoms or direct exposure.

Course of the Disease

Initial data from China, the United States, and Italy did not show an increased risk for infection, increased morbidity, or mortality among pregnant patients in comparison with the general population (Chen et al., 2020b; Yan et al., 2020; Ferrazzi et al., 2020; Breslin et al., 2020).

In a Chinese study, 118 pregnant women with COVID-19 infection in Wuhan were identified (Chen et al., 2020b). 42 pregnant women were identified in the Italian collective, and 43 women were identified in an American collective (Breslin et al., 2020; Ferrazzi et al., 2020). In the Chinese collective, 92% had mild disease, and 9.8% had severe disease (hypoxemia), one of whom received noninvasive mechanical ventilation (Chen et al., 2020b). Similar findings were reported in the American and Italian collectives (Breslin et al., 2020; Ferrazzi et al., 2020). The authors of the three trials concluded that the course of illness is similar as in nonpregnant adults. The symptoms and signs are summarized in Fig. 2.3.

Data from a German national survey (CRONOS Registry) included 247 cases from 65 German hospitals. The survey included >20% of the hospitals involved in delivery in Germany. The mortality rate was 0.4% and 9.7% required a more intensive therapy, due to a severe course (Pecks et al., 2020).

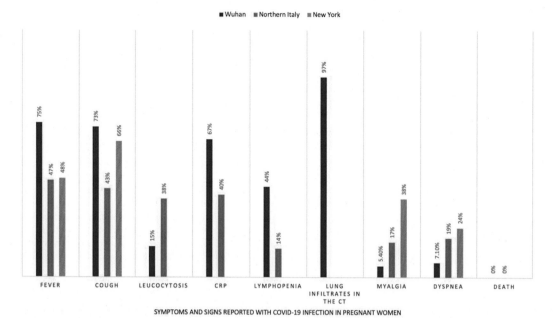

FIG. 2.3 Chart presenting the spectrum of symptoms and signs in the three initial collectives.

On the other hand, the Center for Disease Control in the United States directly compared the characteristics of symptomatic pregnant and nonpregnant women of reproductive age with laboratory-confirmed COVID-19 infection between January 22 and October 3, 2020. Around 400,000 female cases in the reproductive age between 15 and 44 years with known pregnancy status were included. 5.7% of these patients were symptomatic and pregnant. As presented in Table 2.3, pregnant women were at an increased risk for ICU admission, invasive ventilation, and extracorporeal membrane oxygenation. In addition, pregnant COVID-19 patients are at a 70% increased risk of death in comparison with nonpregnant female patients of the same age.

In summary, there is a higher likelihood of a complicated maternal course, including maternal death, if compared with nonpregnant women of the same age. Yet, it remains rare. The patients and caregivers have to be aware of preventive measures. Until vaccination processes prove to be effective and safe, pregnant patients have to be considered a risk group.

Fetal Outcome

A vertical transmission of COVID-19 may occur following a maternal viremia, which leads to placental infection. A high placental viral load might lead to a fetal/neonatal infection (Vivanti et al., 2020). Yet, it is still unknown, which patients and gestations are more prone to fetal infection and which are not.

TABLE 2.3

Comparison of the Rates of Complicated Courses Between Pregnant and Nonpregnant Women Between 15 and 44 Years of Age According to (Zambrano et al., 2020).

	RATE PER 1000 CASES			
	Pregnant	**Nonpregnant**	**aRR**	**95% CI**
Admission to ICU	10.5	3.9	3.0	2.6–3.4
Invasive ventilation	2.9	1.1	2.9	2.2–3.8
ECMO	0.7	0.3	2.4	1.5–4.0
Death	1.5	1.2	1.7	1.2–2.4

aRR, adjusted risk ration; *CI*, confidence interval; *ECMO*, extracorporeal membrane oxygenation; *ICU*, intensive care unit.

A characteristic pattern of placental pathology following placental COVID-19 infection is still being investigated (Algarroba et al., 2020; Baud et al., 2020; Hosier et al., 2020; Penfield et al., 2020; Shanes et al., 2020; Sisman et al., 2020; Vivanti et al., 2020). Histological examination may reveal histiocytic (CD68 positive) intervillositis and villitis associated with villous karyorrhexis and necrosis, focal basal chronic villitis, focal parabasal infarct, and features of meconium straining in the fetal membranes (Sisman et al., 2020). A possible explanation is a fetal under perfusion, secondary to fetal vascular thrombosis (Prabhu et al., 2020). Electron microscopy may reveal viral particles, which are 89–129 nm in diameter, within membrane-bound cisternal spaces in the syncytiotrophoblastic cells (Sisman et al., 2020).

Initially, vertical transmission was not detected in cohorts and case reports (Chen et al., 2020a; Zhu et al., 2020a; Penfield et al., 2020; Algarroba et al., 2020; Savirón-Cornudella et al., 2021; Rottenstreich et al., 2020). However, a metaanalysis, which analyzed 161 studies reporting 2059 cases of maternal infections with COVID-19, reported 61 neonatal cases with intrauterine COVID-19 infection (Rodrigues et al., 2020). Yet, only mild neonatal cases with a good outcome were reported until this date (Zeng et al., 2020).

A growing concern resides in the rising rate of preterm birth and still birth during the COVID-19 pandemic with subsequent lockdown. Two possible explanations exist. On one hand, the neonatal complications may be a direct consequence of maternal, placental, and/or neonatal infection, which may result into an iatrogenic preterm birth. On the other hand, patients tend to avoid or delay visiting hospitals with the onset of classical obstetrical complications, e.g., reduced fetal kicks, hypertensive diseases in pregnancy, and gestational diabetes; hence, many cases with possible adverse outcomes cannot be treated (Khalil et al., 2020). This may reduce the effectiveness of healthcare infrastructure in developed countries, rendering them less efficient with regard to neonatal outcome.

In summary, vertical transmission is rare but exists. It has a good outcome. The bigger concern remains the increased rate of preterm labor and still birth, which may be directly and indirectly related to COVID-19 infection. Yet, due to the low incidence of vertical infections and complicated maternal outcomes, it is still difficult to form a conclusive statement.

REFERENCES

Abbassi-Ghanavati, M., Greer, L.G., Cunningham, F.G., 2009. Pregnancy and laboratory studies: a reference table for clinicians. Obstet. Gynecol. 114, 1326–1331.

Abu-Raya, B., Michalski, C., Sadarangani, M., Lavoie, P.M., 2020. Maternal immunological adaptation during normal pregnancy. Front. Immunol. 11.

Aghaeepour, N., Ganio, E.A., Mcilwain, D., Tsai, A.S., Tingle, M., van Gassen, S., Gaudilliere, D.K., Baca, Q., Mcneil, L., Okada, R., Ghaemi, M.S., Furman, D., Wong, R.J., Winn, V.D., Druzin, M.L., EL-Sayed, Y.Y., Quaintance, C., Gibbs, R., Darmstadt, G.L., Shaw, G.M., Stevenson, D.K., Tibshirani, R., Nolan, G.P., Lewis, D.B., Angst, M.S., Gaudilliere, B., 2017. An immune clock of human pregnancy. Sci. Immunol. 2.

Algarroba, G.N., Rekawek, P., Vahanian, S.A., Khullar, P., Palaia, T., Peltier, M.R., Chavez, M.R., Vintzileos, A.M., 2020. Visualization of severe acute respiratory syndrome coronavirus 2 invading the human placenta using electron microscopy. Am. J. Obstet. Gynecol. 223, 275–278.

Alshukairi, A.N., Zheng, J., Zhao, J., Nehdi, A., Baharoon, S.A., Layqah, L., Bokhari, A., AL Johani, S.M., Samman, N., Boudjelal, M., Ten Eyck, P., AL-Mozaini, M.A., Zhao, J., Perlman, S., Alagaili, A.N., 2018. High prevalence of MERS-CoV infection in camel workers in Saudi Arabia. mBio 9.

Amino, N., Tanizawa, O., Miyai, K., Tanaka, F., Hayashi, C., Kawashima, M., Ichihara, K., 1978. Changes of serum immunoglobulins Igg, Iga, IgM, and IgE during pregnancy. Obstet. Gynecol. 52, 415–420.

Anand, S., Montez-Rath, M., Han, J., Bozeman, J., Kerschmann, R., Beyer, P., Parsonnet, J., Chertow, G.M., 2020. Prevalence of SARS-CoV-2 antibodies in a large nationwide sample of patients on dialysis in the USA: a cross-sectional study. Lancet 396, 1335–1344.

Baden, L.R., EL Sahly, H.M., Essink, B., Kotloff, K., Frey, S., Novak, R., Diemert, D., Spector, S.A., Rouphael, N., Creech, C.B., Mcgettigan, J., Kehtan, S., Segall, N., Solis, J., Brosz, A., Fierro, C., Schwartz, H., Neuzil, K., Corey, L., Gilbert, P., Janes, H., Follmann, D., Marovich, M., Mascola, J., Polakowski, L., Ledgerwood, J., Graham, B.S., Bennett, H., Pajon, R., Knightly, C., Leav, B., Deng, W., Zhou, H., Han, S., Ivarsson, M., Miller, J., Zaks, T., COVE Study Group, 2020. Efficacy and safety of the mRNA-1273 SARS-CoV-2 vaccine. N. Engl. J. Med. 384 (5), 403–416. https://doi.org/10.1056/NEJMoa2035389, 7787219.

Bailey, K., Herrod, H.G., Younger, R., Shaver, D., 1985. Functional aspects of T-lymphocyte subsets in pregnancy. Obstet. Gynecol. 66, 211–215.

Baud, D., Greub, G., Favre, G., Gengler, C., Jaton, K., Dubruc, E., Pomar, L., 2020. Second-trimester miscarriage in a pregnant woman with SARS-CoV-2 infection. JAMA 323, 2198–2200.

Belo, L., Santos-Silva, A., Rocha, S., Caslake, M., Cooney, J., Pereira-Leite, L., Quintanilha, A., Rebelo, I., 2005.

Fluctuations in C-reactive protein concentration and neutrophil activation during normal human pregnancy. Eur. J. Obstet. Gynecol. Reprod. Biol. 123, 46—51.

Berger, C.,N.-L.A., Bouvier Gallacchi, M., Brügger, D., Martinez de Tejada, B., Spaar Zographos, A., Surbek, D., 2018. Influenza- und Pertussis-Impfung in der Schwangerschaft (SGGG Experten Brief). Available: https://www.sggg.ch/fileadmin/user_upload/55_Impfen_in_der_Schwangerschaft.pdf.

Bhat, N.M., Mithal, A., Bieber, M.M., Herzenberg, L.A., Teng, N.N., 1995. Human CD5+ B lymphocytes (B-1 cells) decrease in peripheral blood during pregnancy. J. Reprod. Immunol. 28, 53—60.

Breslin, N., Baptiste, C., Gyamfi-Bannerman, C., Miller, R., Martinez, R., Bernstein, K., Ring, L., Landau, R., Purisch, S., Friedman, A.M., Fuchs, K., Sutton, D., Andrikopoulou, M., Rupley, D., Sheen, J.J., Aubey, J., Zork, N., Moroz, L., Mourad, M., Wapner, R., Simpson, L.L., D'alton, M.E., Goffman, D., 2020. Coronavirus disease 2019 infection among asymptomatic and symptomatic pregnant women: two weeks of confirmed presentations to an affiliated pair of New York city hospitals. Am. J. Obstet. Gynecol. MFM 2, 100118.

Budd, R.C., Fortner, K.A., 2013. 13 - T lymphocytes. In: Firestein, G.S., Budd, R.C., Gabriel, S.E., Mcinnes, I.B., O'dell, J.R. (Eds.), Kelley's Textbook of Rheumatology, ninth ed. W.B. Saunders, Philadelphia.

Buss, C., Davis, E.P., Shahbaba, B., Pruessner, J.C., Head, K., Sandman, C.A., 2012. Maternal cortisol over the course of pregnancy and subsequent child amygdala and hippocampus volumes and affective problems. Proc. Natl. Acad. Sci. U. S. A. 109, E1312.

Busse, M., Campe, K.J., Redlich, A., Oettel, A., Hartig, R., Costa, S.D., Zenclussen, A.C., 2020. Regulatory B cells are decreased and impaired in their function in peripheral maternal blood in pre-term birth. Front. Immunol. 11, 386.

CDC, 2019. Influenza (Flu) Vaccine and Pregnancy. Available: https://www.cdc.gov/vaccines/pregnancy/hcp-toolkit/flu-vaccine-pregnancy.html. (Accessed 28 December 2020).

Chang, R.-Q., Zhou, W.-J., Li, D.-J., Li, M.-Q., 2020. Innate lymphoid cells at the maternal-fetal interface in human pregnancy. Int. J. Biol. Sci. 16, 957—969.

Chen, H., Guo, J., Wang, C., Luo, F., Yu, X., Zhang, W., Li, J., Zhao, D., Xu, D., Gong, Q., Liao, J., Yang, H., Hou, W., Zhang, Y., 2020a. Clinical characteristics and intrauterine vertical transmission potential of COVID-19 infection in nine pregnant women: a retrospective review of medical records. Lancet 395, 809—815.

Chen, L., Li, Q., Zheng, D., Jiang, H., Wei, Y., Zou, L., Feng, L., Xiong, G., Sun, G., Wang, H., Zhao, Y., Qiao, J., 2020b. Clinical characteristics of pregnant women with covid-19 in Wuhan, China. N. Engl. J. Med. 382, e100.

Corman, V.M., Landt, O., Kaiser, M., Molenkamp, R., Meijer, A., Chu, D.K., Bleicker, T., Brünink, S., Schneider, J., Schmidt, M.L., Mulders, D.G., Haagmans, B.L., van der Veer, B., van den Brink, S., Wijsman, L., Goderski, G., Romette, J.L., Ellis, J., Zambon, M., Peiris, M., Goossens, H., Reusken, C., Koopmans, M.P., Drosten, C., 2020. Detection of 2019 novel coronavirus (2019-nCoV) by real-time RT-PCR. Euro Surveill. 25.

Coronaviridae Study Group of the International Committee on Taxonomy of Viruses, 2020. The species severe acute respiratory syndrome-related coronavirus: classifying 2019-nCoV and naming it SARS-CoV-2. Nat. Microbiol. 5, 536—544.

Coulam, C.B., Silverfield, J.C., Kazmar, R.E., Fathman, C.G., 1983. T-lymphocyte subsets during pregnancy and the menstrual cycle. Am. J. Reprod. Immunol. 4, 88—90.

Cyranoski, D., 2020. Why emergency COVID-vaccine approvals pose a dilemma for scientists. Nature 588, 18—19.

Deeks, J.J., Dinnes, J., Takwoingi, Y., Davenport, C., Spijker, R., Taylor-Phillips, S., Adriano, A., Beese, S., Dretzke, J., Ferrante di Ruffano, L., et al., 2020. Antibody tests for identification of current and past infection with SARS-CoV-2. Cochrane Database Syst. Rev. 6.

Efrati, P., Presentey, B., Margalith, M., Rozenszajn, L., 1964. Leukocytes of normal pregnant women. Obstet. Gynecol. 23, 429—432.

Faas, M.M., de vos, P., 2017. Maternal monocytes in pregnancy and preeclampsia in humans and in rats. J. Reprod. Immunol. 119, 91—97.

Fell, D.B., Savitz, D.A., Kramer, M.S., Gessner, B.D., Katz, M.A., Knight, M., Luteijn, J.M., Marshall, H., Bhat, N., Gravett, M.G., Skidmore, B., Ortiz, J.R., 2017. Maternal influenza and birth outcomes: systematic review of comparative studies. BJOG 124, 48—59.

Fell, D.B., Platt, R.W., Basso, O., Wilson, K., Kaufman, J.S., Buckeridge, D.L., Kwong, J.C., 2018. The relationship between 2009 pandemic H1N1 influenza during pregnancy and preterm birth: a population-based cohort study. Epidemiology 29, 107—116.

Ferrazzi, E.M., Frigerio, L., Cetin, I., Vergani, P., Spinillo, A., Prefumo, F., Pellegrini, E., Gargantini, G., 2020. COVID-19 Obstetrics Task Force, Lombardy, Italy: executive management summary and short report of outcome. Int. J. Gynaecol. Obstet. 149, 377—378.

González-Candelas, F., Astray, J., Alonso, J., Castro, A., Cantón, R., Galán, J.C., Garin, O., Sáez, M., Soldevila, N., Baricot, M., Castilla, J., Godoy, P., Delgado-Rodríguez, M., Martín, V., Mayoral, J.M., Pumarola, T., Quintana, J.M., Tamames, S., Domínguez, A., 2012. Sociodemographic factors and clinical conditions associated to hospitalization in influenza A (H1N1) 2009 virus infected patients in Spain, 2009—2010. PloS One 7, e33139.

Goodnight, W.H., Soper, D.E., 2005. Pneumonia in pregnancy. Crit. Care Med. 33.

Herrmann, B.L., 2020. The prevalence rate of anti-SARS-CoV-2-IgG is 1.2% - screening in asymptomatic outpatients in Germany (Northrhine-Westfalia). MMW - Fortschritte Med. 162, 44—46.

Hosier, H., Farhadian, S.F., Morotti, R.A., Deshmukh, U., LU-Culligan, A., Campbell, K.H., Yasumoto, Y., Vogels, C.B., Casanovas-Massana, A., Vijayakumar, P., Geng, B., Odio, C.D., Fournier, J., Brito, A.F., Fauver, J.R., Liu, F., Alpert, T., Tal, R., Szigeti-Buck, K., Perincheri, S.,

Larsen, C., Gariepy, A.M., Aguilar, G., Fardelmann, K.L., Harigopal, M., Taylor, H.S., Pettker, C.M., Wyllie, A.L., Cruz, C.D., Ring, A.M., Grubaugh, N.D., Ko, A.I., Horvath, T.L., Iwasaki, A., Reddy, U.M., Lipkind, H.S., 2020. SARS-CoV-2 infection of the placenta. J. Clin. Invest. 130, 4947–4953.

Huang, C., Wang, Y., Li, X., Ren, L., Zhao, J., Hu, Y., Zhang, L., Fan, G., Xu, J., Gu, X., Cheng, Z., Yu, T., Xia, J., Wei, Y., Wu, W., Xie, X., Yin, W., Li, H., Liu, M., Xiao, Y., Gao, H., Guo, L., Xie, J., Wang, G., Jiang, R., Gao, Z., Jin, Q., Wang, J., Cao, B., 2020. Clinical features of patients infected with 2019 novel coronavirus in Wuhan, China. Lancet 395, 497–506.

Indenbaum, V., Koren, R., Katz-Likvornik, S., Yitzchaki, M., Halpern, O., Regev-Yochay, G., Cohen, C., Biber, A., Feferman, T., Cohen Saban, N., Dhan, R., Levin, T., Gozlan, Y., Weil, M., Mor, O., Mandelboim, M., Sofer, D., Mendelson, E., Lustig, Y., 2020. Testing IgG antibodies against the RBD of SARS-CoV-2 is sufficient and necessary for COVID-19 diagnosis. PloS One 15, e0241164.

Iuliano, A.D., Roguski, K.M., Chang, H.H., Muscatello, D.J., Palekar, R., Tempia, S., Cohen, C., Gran, J.M., Schanzer, D., Cowling, B.J., Wu, P., Kyncl, J., Ang, L.W., Park, M., Redlberger-Fritz, M., Yu, H., Espenhain, L., Krishnan, A., Emukule, G., van Asten, L., Pereira da Silva, S., Aungkulanon, S., Buchholz, U., Widdowson, M.A., Bresee, J.S., 2018. Estimates of global seasonal influenza-associated respiratory mortality: a modelling study. Lancet 391, 1285–1300.

Khalil, A., von Dadelszen, P., Draycott, T., Ugwumadu, A., O'brien, P., Magee, L., 2020. Change in the incidence of stillbirth and preterm delivery during the COVID-19 pandemic. JAMA 324, 705–706.

Ko, J.H., Müller, M.A., Seok, H., Park, G.E., Lee, J.Y., Cho, S.Y., Ha, Y.E., Baek, J.Y., Kim, S.H., Kang, J.M., Kim, Y.J., Jo, I.J., Chung, C.R., Hahn, M.J., Drosten, C., Kang, C.I., Chung, D.R., Song, J.H., Kang, E.S., Peck, K.R., 2017. Serologic responses of 42 MERS-coronavirus-infected patients according to the disease severity. Diagn. Microbiol. Infect. Dis. 89, 106–111.

Kourtis, A.P., Read, J.S., Jamieson, D.J., 2014. Pregnancy and infection. N. Engl. J. Med. 370, 2211–2218.

Kühnert, M., Strohmeier, R., Stegmüller, M., Halberstadt, E., 1998. Changes in lymphocyte subsets during normal pregnancy. Eur. J. Obstet. Gynecol. Reprod. Biol. 76, 147–151.

Lampé, R., Kövér, Á., Szűcs, S., Pál, L., Árnyas, E., Ádány, R., Póka, R., 2015. Phagocytic index of neutrophil granulocytes and monocytes in healthy and preeclamptic pregnancy. J. Reprod. Immunol. 107, 26–30.

Lima, J., Martins, C., Leandro, M.J., Nunes, G., Sousa, M.J., Branco, J.C., Borrego, L.M., 2016. Characterization of B cells in healthy pregnant women from late pregnancy to postpartum: a prospective observational study. BMC Pregnancy Childbirth 16, 139.

Lissauer, D., Goodyear, O., Khanum, R., Moss, P.A., Kilby, M.D., 2014. Profile of maternal CD4 T-cell effector function during normal pregnancy and in women with a history of recurrent miscarriage. Clin. Sci. 126, 347–354.

Lomauro, A., Aliverti, A., 2015. Breathe 11, 297–301.

Luppi, P., Haluszczak, C., Betters, D., Richard, C.A., Trucco, M., Deloia, J.A., 2002. Monocytes are progressively activated in the circulation of pregnant women. J. Leukoc. Biol. 72, 874–884.

Luteijn, J.M., Brown, M.J., Dolk, H., 2014. Influenza and congenital anomalies: a systematic review and meta-analysis. Hum. Reprod. 29, 809–823.

Luteijn, J.M., Addor, M.C., Arriola, L., Bianchi, F., Garne, E., Khoshnood, B., Nelen, V., Neville, A., Queisser-Luft, A., Rankin, J., Rounding, C., Verellen-Dumoulin, C., de Walle, H., Wellesley, D., Wreyford, B., Yevtushok, L., de Jong-van den Berg, L., Morris, J., Dolk, H., 2015. The association of H1N1 pandemic influenza with congenital anomaly prevalence in Europe: an ecological time series study. Epidemiology 26, 853–861.

Ma, Y., Kong, L.R., Ge, Q., Lu, Y.Y., Hong, M.N., Zhang, Y., Ruan, C.C., Gao, P.J., 2018. Complement 5a-mediated trophoblasts dysfunction is involved in the development of pre-eclampsia. J. Cell Mol. Med. 22, 1034–1046.

Mahmoud, F., Abul, H., Omu, A., AL-Rayes, S., Haines, D., Whaley, K., 2001. Pregnancy-associated changes in peripheral blood lymphocyte subpopulations in normal Kuwaiti women. Gynecol. Obstet. Invest. 52, 232–236.

Marzi, M., Vigano, A., Trabattoni, D., Villa, M.L., Salvaggio, A., Clerici, E., Clerici, M., 1996. Characterization of type 1 and type 2 cytokine production profile in physiologic and pathologic human pregnancy. Clin. Exp. Immunol. 106, 127–133.

Matsumoto, K., Ogasawara, T., Kato, A., Homma, T., Iida, M., Akasawa, A., Wakiguchi, H., Saito, H., 2003. Eosinophil degranulation during pregnancy and after delivery by cesarean section. Int. Arch. Allergy Immunol. 131 (Suppl. 1), 34–39.

Mattern, J., Vauloup-Fellous, C., Zakaria, H., Benachi, A., Carrara, J., Letourneau, A., Bourgeois-Nicolaos, N., DE Luca, D., Doucet-Populaire, F., Vivanti, A.J., 2020. Post lockdown COVID-19 seroprevalence and circulation at the time of delivery, France. PloS One 15, e0240782.

Matthiesen, L., Ekerfelt, C., Berg, G., Ernerudh, J., 1998. Increased numbers of circulating interferon-gamma- and interleukin-4-secreting cells during normal pregnancy. Am. J. Reprod. Immunol. 39, 362–367.

Melgert, B.N., Spaans, F., Borghuis, T., Klok, P.A., Groen, B., Bolt, A., de Vos, P., van Pampus, M.G., Wong, T.Y., VAN Goor, H., Bakker, W.W., Faas, M.M., 2012. Pregnancy and preeclampsia affect monocyte subsets in humans and rats. PloS One 7, e45229.

Mendes, J., Areia, A.L., Rodrigues-Santos, P., Santos-Rosa, M., Mota-Pinto, A., 2020. Innate lymphoid cells in human pregnancy. Front. Immunol. 11.

Merle, N.S., Church, S.E., Fremeaux-Bacchi, V., Roumenina, L.T., 2015. Complement system part I - molecular mechanisms of activation and regulation. Front. Immunol. 6, 262.

Mertz, D., Geraci, J., Winkup, J., Gessner, B.D., Ortiz, J.R., Loeb, M., 2017. Pregnancy as a risk factor for severe outcomes from influenza virus infection: a systematic review and meta-analysis of observational studies. Vaccine 35, 521–528.

Miller, E.C., Abel, W., 1984. Changes in the immunoglobulins Igg, IgA and IgM in pregnancy and the puerperium. Zentralblatt für Gynäkol. 106, 1084–1091.

Moghadas, S.M., Vilches, T.N., Zhang, K., Wells, C.R., Shoukat, A., Singer, B.H., Meyers, L.A., Neuzil, K.M., Langley, J.M., Fitzpatrick, M.C., Galvani, A.P., 2020. The impact of vaccination on COVID-19 outbreaks in the United States. medRxiv.

Mor, G., Cardenas, I., 2010. The immune system in pregnancy: a unique complexity. Am. J. Reprod. Immunol. 63, 425–433.

Mudd, P.A., Remy, K.E., 2021. Prolonged adaptive immune activation in COVID-19: implications for maintenance of long-term immunity? J. Clin. Invest. 131.

Mulligan, M.J., Lyke, K.E., Kitchin, N., Absalon, J., Gurtman, A., Lockhart, S., Neuzil, K., Raabe, V., Bailey, R., Swanson, K.A., Li, P., Koury, K., Kalina, W., Cooper, D., Fontes-Garfias, C., Shi, P.-Y., Türeci, Ö., Tompkins, K.R., Walsh, E.E., Frenck, R., Falsey, A.R., Dormitzer, P.R., Gruber, W.C., Şahin, U., Jansen, K.U., 2020. Phase I/II study of COVID-19 RNA vaccine BNT162b1 in adults. Nature 586, 589–593.

Panchaud, A., Favre, G., Pomar, L., Vouga, M., Aebi-Popp, K., Baud, D., 2020. An international registry for emergent pathogens and pregnancy. Lancet 395, 1483–1484.

Pecks, U., Kuschel, B., Mense, L., Oppelt, P., Rüdiger, M., CRONOS-Network, 2020. Pregnancy and SARS CoV-2 infection in Germany—the CRONOS registry. Dtsch. Arztebl. Int. 121, 841–842.

Penfield, C.A., Brubaker, S.G., Limaye, M.A., Lighter, J., Ratner, A.J., Thomas, K.M., Meyer, J.A., Roman, A.S., 2020. Detection of severe acute respiratory syndrome coronavirus 2 in placental and fetal membrane samples. Am. J. Obstet. Gynecol. MFM 2, 100133.

Piroth, L., Cottenet, J., Mariet, A.S., Bonniaud, P., Blot, M., Tubert-Bitter, P., Quantin, C., 2020. Comparison of the characteristics, morbidity, and mortality of COVID-19 and seasonal influenza: a nationwide, population-based retrospective cohort study. Lancet Respir. Med. 9.

Polack, F.P., Thomas, S.J., Kitchin, N., Absalon, J., Gurtman, A., Lockhart, S., Perez, J.L., Pérez Marc, G., Moreira, E.D., Zerbini, C., Bailey, R., Swanson, K.A., Roychoudhury, S., Koury, K., Li, P., Kalina, W.V., Cooper, D., Frenck, R.W., Hammitt, L.L., Türeci, Ö., Nell, H., Schaefer, A., Ünal, S., Tresnan, D.B., Mather, S., Dormitzer, P.R., Şahin, U., Jansen, K.U., Gruber, W.C., 2020. Safety and efficacy of the BNT162b2 mRNA covid-19 vaccine. N. Engl. J. Med. 383, 2603–2615.

Prabhu, M., Cagino, K., Matthews, K.C., Friedlander, R.L., Glynn, S.M., Kubiak, J.M., Yang, Y.J., Zhao, Z., Baergen, R.N., Dipace, J.I., Razavi, A.S., Skupski, D.W., Snyder, J.R., Singh, H.K., Kalish, R.B., Oxford, C.M., Riley, L.E., 2020. Pregnancy and postpartum outcomes in a universally tested population for SARS-CoV-2 in New York city: a prospective cohort study. BJOG 127, 1548–1556.

Premkumar, L., Segovia-Chumbez, B., Jadi, R., Martinez, D.R., Raut, R., Markmann, A., Cornaby, C., Bartelt, L., Weiss, S., Park, Y., Edwards, C.E., Weimer, E., Scherer, E.M., Rouphael, N., Edupuganti, S., Weiskopf, D., Tse, L.V., Hou, Y.J., Margolis, D., Sette, A., Collins, M.H., Schmitz, J., Baric, R.S., DE Silva, A.M., 2020. The receptor binding domain of the viral spike protein is an immunodominant and highly specific target of antibodies in SARS-CoV-2 patients. Sci. Immunol. 5.

Raghupathy, R., Makhseed, M., Azizieh, F., Omu, A., Gupta, M., Farhat, R., 2000. Cytokine production by maternal lymphocytes during normal human pregnancy and in unexplained recurrent spontaneous abortion. Hum. Reprod. 15, 713–718.

Rasmussen, I.S., Mortensen, L.H., Krause, T.G., Nybo Andersen, A.M., 2018. The association between seasonal influenza-like illness cases and foetal death: a time series analysis. Epidemiol. Infect. 147, 1–7.

Ray, J.G., Burows, R.F., Ginsberg, J.S., Burrows, E.A., 2000. Paroxysmal nocturnal hemoglobinuria and the risk of venous thrombosis: review and recommendations for management of the pregnant and nonpregnant patient. Haemostasis 30, 103–117.

Regal, J.F., Gilbert, J.S., Burwick, R.M., 2015. The complement system and adverse pregnancy outcomes. Mol. Immunol. 67, 56–70.

Rodrigues, C., Baía, I., Domingues, R., Barros, H., 2020. Pregnancy and breastfeeding during COVID-19 pandemic: a systematic review of published pregnancy cases. Front. Public Health 8, 558144.

Rottenstreich, A., Tsur, A., Braverman, N., Kabiri, D., Porat, S., Benenson, S., Oster, Y., Kam, H.A., Walfisch, A., Bart, Y., Meyer, R., Lifshitz, S.J., Amikam, U., Biron-Shental, T., Cohen, G., Sciaky-Tamir, Y., Shachar, I.B., Yinon, Y., Reubinoff, B., 2020. Vaginal delivery in SARS-CoV-2-infected pregnant women in Israel: a multicenter prospective analysis. Arch. Gynecol. Obstet. 1–5.

Saito, S., Sakai, M., Sasaki, Y., Tanebe, K., Tsuda, H., Michimata, T., 1999. Quantitative analysis of peripheral blood Th0, Th1, Th2 and the Th1:Th2 cell ratio during normal human pregnancy and preeclampsia. Clin. Exp. Immunol. 117, 550–555.

Savirón-Cornudella, R., Villalba, A., Zapardiel, J., Andeyro-Garcia, M., Esteban, L.M., Pérez-López, F.R., 2021. Severe acute respiratory syndrome coronavirus 2 (SARS-CoV-2) universal screening in gravids during labor and delivery. Eur. J. Obstet. Gynecol. Reprod. Biol. 256, 400–404.

Sendag, F., Itil, I.M., Terek, M.C., Yilmaz, H., 2002. The changes of circulating lymphocyte sub-populations in women with preterm labour: a case-controlled study. Aust. N. Z. J. Obstet. Gynaecol. 42, 358–361.

Shanes, E.D., Mithal, L.B., Otero, S., Azad, H.A., Miller, E.S., Goldstein, J.A., 2020. Placental pathology in COVID-19. Am. J. Clin. Pathol. 154, 23–32.

Siegel, I., Gleicher, N., 1981. Peripheral white blood cell alterations in early labor. Diagn. Gynecol. Obstet. 3, 123–126.

Sisman, J., Jaleel, M.A., Moreno, W., Rajaram, V., Collins, R.R.J., Savani, R.C., Rakheja, D., Evans, A.S., 2020. Intrauterine transmission of SARS-COV-2 infection in a preterm infant. Pediatr. Infect. Dis. J. 39, e265–e267.

Speake, H.A., Pereira, G., Regan, A.K., 2020. Risk of adverse maternal and foetal outcomes associated with inactivated influenza vaccination in first trimester of pregnancy. Paediatr. Perinat. Epidemiol. 35. (2), 196–205.

Swieboda, D., Littauer, E.Q., Beaver, J.T., Mills, L.K., Bricker, K.M., Esser, E.S., Antao, O.Q., Williams, D.T., Skountzou, I., 2020. Pregnancy downregulates plasmablast metabolic gene expression following influenza without altering long-term antibody function. Front. Immunol. 11, 1785.

Szekeres-Bartho, J., Wegmann, T.G., 1996. A progesterone-dependent immunomodulatory protein alters the Th1/Th2 balance. J. Reprod. Immunol. 31, 81–95.

Tanacan, A., Erol, S.A., Turgay, B., Anuk, A.T., Secen, E.I., Yegin, G.F., Ozyer, S., Kirca, F., Dinc, B., Unlu, S., YAPAR Eyi, E.G., Keskin, H.L., Sahin, D., Surel, A.A., Tekin, O.M., 2020. The rate of SARS-CoV-2 positivity in asymptomatic pregnant women admitted to hospital for delivery: experience of a pandemic center in Turkey. Eur. J. Obstet. Gynecol. Reprod. Biol. 253, 31–34.

Thellin, O., Coumans, B., Zorzi, W., Igout, A., Heinen, E., 2000. Tolerance to the foeto-placental 'graft': ten ways to support a child for nine months. Curr. Opin. Immunol. 12, 731–737.

Trahan, M.J., Mitric, C., Malhamé, I., Abenhaim, H.A., 2021. Screening and testing pregnant patients for SARS-CoV-2: first-wave experience of a designated COVID-19 hospitalization centre in Montreal. J. Obstet. Gynaecol. Can. 43 (5), 571–575.

Vivanti, A.J., Vauloup-Fellous, C., Prevot, S., Zupan, V., Suffee, C., DO Cao, J., Benachi, A., DE Luca, D., 2020. Transplacental transmission of SARS-CoV-2 infection. Nat. Commun. 11, 3572.

Vivier, E., Artis, D., Colonna, M., Diefenbach, A., DI Santo, J.P., Eberl, G., Koyasu, S., Locksley, R.M., Mckenzie, A.N.J., Mebius, R.E., Powrie, F., Spits, H., 2018. Innate lymphoid cells: 10 years on. Cell 174, 1054–1066.

Walsh, E.E., Frenck, R.W., Falsey, A.R., Kitchin, N., Absalon, J., Gurtman, A., Lockhart, S., Neuzil, K., Mulligan, M.J., Bailey, R., Swanson, K.A., Li, P., Koury, K., Kalina, W., Cooper, D., Fontes-Garfias, C., Shi, P.-Y., Türeci, Ö., Tompkins, K.R., Lyke, K.E., Raabe, V., Dormitzer, P.R., Jansen, K.U., Şahin, U., Gruber, W.C., 2020. Safety and immunogenicity of two RNA-based covid-19 vaccine candidates. N. Engl. J. Med. 383, 2439–2450.

Watanabe, M., Iwatani, Y., Kaneda, T., Hidaka, Y., Mitsuda, N., Morimoto, Y., Amino, N., 1997. Changes in T, B, and NK lymphocyte subsets during and after normal pregnancy. Am. J. Reprod. Immunol. 37, 368–377.

Wegmann, T.G., 1984. Foetal protection against abortion: is it immunosuppression or immunostimulation? Ann. Immunol. 135d, 309–312.

Wegmann, T.G., Lin, H., Guilbert, L., Mosmann, T.R., 1993. Bidirectional cytokine interactions in the maternal-fetal relationship: is successful pregnancy a TH2 phenomenon? Immunol. Today 14, 353–356.

Wölfel, R., Corman, V.M., Guggemos, W., Seilmaier, M., Zange, S., Müller, M.A., Niemeyer, D., Jones, T.C., Vollmar, P., Rothe, C., Hoelscher, M., Bleicker, T., Brünink, S., Schneider, J., Ehmann, R., Zwirglmaier, K., Drosten, C., Wendtner, C., 2020. Virological assessment of hospitalized patients with COVID-2019. Nature 581, 465–469.

Wu, L.P., Wang, N.C., Chang, Y.H., Tian, X.Y., Na, D.Y., Zhang, L.Y., Zheng, L., Lan, T., Wang, L.F., Liang, G.D., 2007. Duration of antibody responses after severe acute respiratory syndrome. Emerg. Infect. Dis. 13, 1562–1564.

Xia, Y.Q., Zhao, K.N., Zhao, A.D., Zhu, J.Z., Hong, H.F., Wang, Y.L., Li, S.H., 2019. Associations of maternal upper respiratory tract infection/influenza during early pregnancy with congenital heart disease in offspring: evidence from a case-control study and meta-analysis. BMC Cardiovasc. Disord. 19, 277.

Yan, J., Guo, J., Fan, C., Juan, J., Yu, X., Li, J., Feng, L., Li, C., Chen, H., Qiao, Y., Lei, D., Wang, C., Xiong, G., Xiao, F., He, W., Pang, Q., Hu, X., Wang, S., Chen, D., Zhang, Y., Poon, L.C., Yang, H., 2020. Coronavirus disease 2019 in pregnant women: a report based on 116 cases. Am. J. Obstet. Gynecol. 223, 111.e1–111.e14.

Yasuhara, M., Tamaki, H., Iyama, S., Yamaguchi, Y., Tachi, J., Amino, N., 1992. Reciprocal changes in serum levels of immunoglobulins (Igg, Iga, IgM) and complements (C3, C4) in normal pregnancy and after delivery. J. Clin. Lab. Immunol. 38, 137–141.

Zambrano, L.D., Ellington, S., Strid, P., Galang, R.,R.,, Oduyebo, T., Tong, V.T., Woodworth, K.R., Nahabedian 3rd, J.F., Azziz-Baumgartner, E., Gilboa, S.M., Meaney-Delman, D., 2020. CDC COVID-19 Response Pregnancy and Infant Linked Outcomes Team. Update: Characteristics of Symptomatic Women of Reproductive Age with Laboratory-Confirmed SARS-CoV-2 Infection by Pregnancy Status — United States, January 22–October 3, 2020. Available from: https://www.cdc.gov/mmwr/volumes/69/wr/mm6944e3.htm#suggestedcitation.

Zeng, L., Xia, S., Yuan, W., Yan, K., Xiao, F., Shao, J., Zhou, W., 2020. Neonatal early-onset infection with SARS-CoV-2 in 33 neonates born to mothers with COVID-19 in Wuhan, China. JAMA Pediatr. 174, 722–725.

Zhang, Y., Kremsdorf, R.A., Sperati, C.J., Henriksen, K.J., Mori, M., Goodfellow, R.X., Pitcher, G.R., Benson, C.L., Borsa, N.G., Taylor, R.P., Nester, C.M., Smith, R.J.H., 2020. Mutation of complement factor B causing massive fluid-phase dysregulation of the alternative complement pathway can result in atypical hemolytic uremic syndrome. Kidney Int. 98, 1265–1274.

Zhao, J., Zhao, J., Perlman, S., 2010. T cell responses are required for protection from clinical disease and for virus clearance in severe acute respiratory syndrome coronavirus-infected mice. J. Virol. 84, 9318–9325.

Zhu, H., Wang, L., Fang, C., Peng, S., Zhang, L., Chang, G., Xia, S., Zhou, W., 2020a. Clinical analysis of 10 neonates born to mothers with 2019-nCoV pneumonia. Transl. Pediatr. 9, 51–60.

Zhu, N., Zhang, D., Wang, W., Li, X., Yang, B., Song, J., Zhao, X., Huang, B., Shi, W., Lu, R., Niu, P., Zhan, F., Ma, X., Wang, D., Xu, W., Wu, G., Gao, G.F., Tan, W., 2020b. A novel coronavirus from patients with pneumonia in China, 2019. N. Engl. J. Med. 382, 727−733.

Zimmer, J.P., Garza, C., Butte, N.F., Goldman, A.S., 1998. Maternal blood B-cell (CD19+) percentages and serum immunoglobulin concentrations correlate with breast-feeding behavior and serum prolactin concentration. Am. J. Reprod. Immunol. 40, 57−62.

Zöllkau, J., Baier, M., Scherag, A., Schleussne, E., Groten, T., 2020. Period prevalence of SARS-CoV-2 in an unselected sample of pregnant women in Jena, Thuringia. Z. Geburtshilfe Neonatol. 224, 194−198.

CHAPTER 3

Diagnosis of COVID-19 Infection in Pregnancy

AHMED M. MAGED EL-GOLY, MD • AHMED A. METWALLY, MD

(Dr. Prof. M Fadel Shaltout. Prof. of Obstetrics and Gynecology, Cairo University, Faculty of Medicine)

Covid-19 Infection and Pregnancy. https://doi.org/10.1016/B978-0-323-90595-4.00001-7

INTRODUCTION

Pregnancy is a well-known risk factor for development of respiratory infections. Pregnant women are at high risk of development of complications and severe forms of infections caused by coronaviruses including severe acute respiratory syndrome (SARS), Middle East respiratory syndrome (MERS), and severe acute respiratory syndrome coronavirus-2 (SARS-CoV-2). Pregnant women were identified as a vulnerable group and were advised to take additional precautions during the COVID-19 pandemic. To reduce transmission risks for both pregnant women and healthcare workers, the International Federation of Gynecology and Obstetrics recommended the suspension of much routine antenatal care and replacement with video or telephone consultations whenever possible.

PHYSIOLOGICAL CHANGES WITH PREGNANCY

Many physiological changes that occur during pregnancy increase the susceptibility of the women to respiratory viral infections including COVID-19 infection. These changes include immunological, respiratory, coagulation, and endothelial cell responses.

Immunological Response

The immune system adapts during pregnancy to allow for the growth of a semiallogenic fetus, resulting in an altered immune response to infections during pregnancy (Cascella et al., 2020). The modulations of the maternal immune system in pregnancy may affect the response to infections, and specifically to viruses. The altered inflammatory response to viruses during pregnancy is thought to be mediated, at least in part, by the following:

1) A shift in CD4+ T cell population toward the Th2 phenotype over Th1 during pregnancy (a response that promotes humoral responses over cellular immune responses). For the immune response to viral infections, a decrease in Th1 reactivity can result in an altered clearance of infected cells (Zhou et al., 2020).

2) A decrease in circulating natural killer (NK) cells during pregnancy that may alter the ability to clear viruses (Bergsbaken et al., 2009).

3) A decrease in circulating plasmacytoid dendritic cells (pDCs). These cells are key for type 1 interferon production against viruses. Moreover, pDCs from pregnant women have also been shown to have an attenuated inflammatory response to the H1N1 virus. This is thought to be one of the reasons why pregnant women were more severely affected by the H1N1 pandemic in 2009 (Vanders et al., 2013).

4) An increase in circulating progesterone levels that has immunomodulatory properties. Progesterone also can enhance lung repair of damage induced by influenza virus, making high levels during pregnancy potentially beneficial for the recovery after viral lung infections. However, in a mouse model of influenza A infection, treatment with progesterone or the progestin, levonorgestrel, also resulted in a decrease in virus-specific antibody levels, as well as a decrease in virus-specific CD8+ T cells in mice. When these mice were rechallenged with influenza A, this resulted in more severe disease (Hall et al., 2017).

5) Alterations in the innate immune system, including the pattern recognition receptors Toll-like receptors (TLRs) during pregnancy. COVID-19 infection causes pyroptosis (inflammation mediated) of host cells and release of damage-associated molecular patterns, which can be TLR ligands and further enhance inflammation (Young et al., 2014).

Respiratory Response

Physiological alterations to the chest shape and elevation of the diaphragm due to diaphragmatic splinting by the gravid uterus cause changes to the respiratory function. Although there is a 30%–40% increase in tidal volume, the reduction in chest volume leads to a decrease in functional residual capacity, end-expiratory volumes, and residual volumes from early in pregnancy. The reduction in total lung capacity and inability to clear secretions can make pregnant women more susceptible to severe respiratory infections (Goodnight and Soper, 2005).

Coagulation Response

In the general population, COVID-19 is associated with high rates of thromboembolic complications. One study in 184 critically ill COVID-19 patients (24% of them were female) reported thrombotic events in 31% of them (Ji et al., 2020). This is due to activation of coagulation pathways and potential progression to disseminated vascular coagulopathy (DIC) and fibrinolysis with resultant dynamic hypercoagulation occurring alongside thrombocytopenia. Pregnancy is a hypercoagulable state with increased thrombin production and an increase in intravascular inflammation. During pregnancy, there are higher levels of circulating coagulation and fibrinolytic factors, such as plasmin, and these may be implicated in the pathogenesis of SARS-CoV-2 infection (Creanga et al., 2017). Pregnant women are at increased risk of thromboembolic events with associated mortality. Therefore, pregnant women with COVID-19 may have additive or synergistic risk factors for thrombosis. Current guidelines recommend that all pregnant women with confirmed COVID-19 should have thromboprophylaxis until 10 days postnatal and that their clinicians have a low threshold for investigation of possible thromboembolism (Royal College of Obstetricians and Gynecologists, 2020).

Endothelial Cell Function

Mortality in COVID-19 is predominantly due to acute respiratory distress syndrome (ARDS). Emerging evidence suggests that pulmonary endothelial cell dysfunction has an important role in the onset and progression of ARDS. In health, endothelial cells are surrounded by mural cells (pericytes) and limit inflammation by restricting immune cell entry and prevent coagulation via expression of anticoagulant factors. In ARDS, this endothelial barrier is damaged, leading to tissue edema, excessive inflammation, and hypercoagulability. Risk factors for COVID-19 (increasing age, obesity, diabetes mellitus, and cardiovascular disease) are all associated with endothelial cell dysfunction (Li et al., 2019). Maternal vascular adaptation to pregnancy is critical for optimal pregnancy outcomes. At implantation, the specialized uterine spiral arterioles are remodeled to form sinuses that become placental villi. Systemic vascular physiology also undergoes significant adaptations to pregnancy. Given the potential importance of endothelial cell function in the development and progression of COVID19, these women may be at particular risk, if infected, and an early systematic review found higher rates of preeclampsia in pregnant women hospitalized with COVID-19 (Wu and McGoogan, 2020).

CLINICAL FINDINGS

Diagnosis of SARS-COV-2 infection in pregnant women is more difficult than the general population as some of its symptoms are similar to pregnancy-induced manifestations (Dashraath et al., 2020). These manifestations include nasal congestion, which is present in 5% of COVID-19-infected patients and 20% of normal pregnant women in late pregnancy (estrogen induced nasopharyngeal hyperemia) and shortness of breath which is presented in 18% of COVID-19, while most pregnant women have this symptom as a result of increased oxygen needs from higher metabolism, hyperdynamic circulation, physiological anemia, and fetal consumption (Nelson-Piercy, 2015).

Incubation Period

The incubation period for COVID-19 infection is about 14 days after viral exposure. However, in most cases, it is approximately 4−5 days after exposure (Guan et al., 2020a). One study in 1099 reported a median incubation period of 4 days (Guan et al., 2020b). Another study that included 181 confirmed cases in China with clearly identified exposure reported a median incubation period of 5.1 days (2.5% and 97.5% of cases symptomatize within 2.2 and 11.5 days, respectively) (Lauer et al., 2020). A third one conducted on 1084 patients who had traveled or resided in Wuhan reported a median incubation period of 7.8 days (5−10% of cases developed symptoms after 14 or more days after exposure) (Qin et al., 2020b).

Signs and Symptoms

There are no specific clinical characteristics that can reliably distinguish between SARS-CoV-2 and other viral respiratory infections (Struyf et al., 2020). However, some features raise suspicion of COVID-19 infection as development of dyspnea several days after the onset of initial symptoms (Cohen et al., 2020).

Anosmia, myalgia, general malaise, headache, extreme tiredness, and fever were the most linked nonrespiratory manifestations to positive testing for SARS-CoV-2 in one study of healthcare workers (mainly 20- to 40-year-old women) (Tostmann et al., 2020). Other, unusual findings, as new-onset pernio-like lesions, also raise suspicion for COVID-19 infection. However, none of these findings definitively establish the diagnosis without microbiologic testing (Caliendo Angela, 2020).

All pregnant persons should be monitored for development of symptoms and signs of COVID-19 (which are similar to those in nonpregnant individuals), particularly if they have had close contact with a confirmed case or persons under investigation.

Asymptomatic infections are well documented in many studies (World Health Organization, 2020d). However, the proportion of asymptomatic infected patients vary among them. The difference was related to the absence of the longitudinal follow-up to assess for symptom development and the use of different definitions of "asymptomatic" depending on which specific symptoms were assessed in most studies. In one study based on data from three large cohorts that identified cases through population-based testing infections, the prevalence of asymptomatic cases was 30%−40% (Lavezzo et al., 2020; Oran and Topol, 2020). A study reported that 58% of the 712 confirmed COVID-19 cases were asymptomatic at the time of diagnosis

(Sakurai et al., 2020). In another study, among the 1271 cases confirmed to have COVID-19 infection, 88% were asymptomatic at the time of testing and 43% remained asymptomatic throughout the observation period (Kasper et al., 2020). High rates of asymptomatic infection have also been reported among pregnant women presenting for delivery (Campbell et al., 2020; Sutton et al., 2020).

Absence of symptoms does not mean absence of the disease or its objective clinical abnormalities (Hu et al., 2020; Wang et al., 2020c). In one study that included 24 asymptomatic COVID-19 patients, 50% of them had typical ground-glass opacities or patchy shadowing detected in chest computed tomography (CT) examination and another 20% had atypical imaging abnormalities (Hu et al., 2020). In another study of 55 asymptomatic COVID-19 patients, 37 of them (67%) had CT evidence of pneumonia on admission; 2 developed hypoxia; and all were recovered (Wang et al., 2020c).

Many classifications were suggested for COVID-19 infection severity. The National Institutes of Health categorized nonpregnant COVID-19 patients into asymptomatic, mild illness, moderate, severe, and critical illness. Asymptomatic or presymptomatic disease include those having no symptoms but tested positive for SARS-CoV-2. Mild illness includes patients with any signs and symptoms (as fever, cough, sore throat, malaise, headache, muscle pain) without shortness of breath, dyspnea, or abnormal chest imaging. Moderate illness includes patients with evidence of lower respiratory disease by clinical assessment or imaging and ≥94% oxygen saturation of on-room air at sea level. Severe illness includes patients with respiratory rate >30 breaths per minute, oxygen saturation <94% on room air at sea level, the ratio of arterial partial pressure of oxygen to fraction of inspired oxygen (PaO/FiO) < 300, or lung infiltrates >50%. Critical illness includes patients with respiratory failure, septic shock, and/or multiple organ dysfunction (National Institutes of Health, 2020).

Another classification was suggested by Wu categorized COVID-19 patients into mild, severe, and critical illness. Mild illness includes patients with no or mild symptoms (fever, fatigue, cough, and/or less common features of COVID-19). Severe illness includes patients with tachypnea (respiratory rate >30 breaths per minute), hypoxia (oxygen saturation ≤93% on room air or PaO/FiO <300 mmHg), or >50% lung involvement on imaging. Critical illness includes patients with respiratory failure, shock, or multiorgan dysfunction (Wu and McGoogan, 2020).

The Centers for Disease Control and Prevention (CDC) evaluated the prevalence of different symptoms in more than 386,000 nonpregnant and over 23,000 pregnant females. It reported the following.

Cough was present in 50.3% and 51.3% of pregnant and nonpregnant women, respectively (Zambrano et al., 2020). Cough was present in 28% of 55 pregnant women in another study. The latter compared the presence of cough in COVID-19-, SARS-, and MERS-infected pregnant women and found a prevalence of 28%, 76%, and 67%, respectively (Dashraath et al., 2020).

Fever was one of the presenting symptoms in 32% and 39.3% of pregnant and nonpregnant women, respectively, in the CDC report. Dashraath and colleagues reported fever in 84% of COVID-19 patients compared with 100% and 58% of patients infected with SARS and MERS, respectively (Dashraath et al., 2020). Although fever is a common finding in COVID-19 infection, it is not a universal finding at presentation even in hospitalized patients. In one study that included 1099 patients, fever was present in 44% of patients on admission and increased to 89% during hospitalization (Guan et al., 2020a). Another study in 5000 hospitalized COVID-19 patients reported an incidence of 31 for fever (Richardson et al., 2020). Even though the study reported fever in all patients, 20% of them had a low grade fever <100.4°F/38°C (Huang et al., 2020).

Shortness of breath was present in 25.9% and 24.8% of pregnant and nonpregnant women, respectively, in CDC report (Zambrano et al., 2020). Dashraath and colleagues reported dyspnea in 18% of COVID-19 patients compared with 35% and 58% of patients infected with SARS and MERS, respectively (Dashraath et al., 2020).

Smell and taste disorders as anosmia and dysgeusia are common manifestations of COVID-19 infection. New loss of taste or smell were reported by CDC in 21.5% and 24.8% of pregnant and nonpregnant women, respectively (Zambrano et al., 2020). In one metaanalysis, abnormalities in smell were reported in 52%, while taste abnormalities were reported in 44% of patients (Tong et al., 2020). In 202 patients with mild COVID-19 infection, changes in taste and/or smell were reported in 64%, severe alterations were documented in 24%, it was the only symptom in 3%, and it preceded other symptoms in 12% of patients (Spinato et al., 2020). However, the rate of objective smell or taste abnormalities may be lower than the self-reported rates. On objective testing, 38% of patients who reported complete anosmia had a normal smell function (Lechien et al., 2020). These changes in taste and smell are temporary. In one study, 89% of patients with smell or taste alterations reported complete recovery or improvement within 4 weeks (Boscolo-Rizzo et al., 2020).

Other symptoms in CDC report included the following:

- Headache in 42.7% and 54.9% of pregnant and nonpregnant women, respectively
- Muscle aches in 36.7% and 45.2% of pregnant and nonpregnant women, respectively
- Sore throat in 28.4% and 34.6% of pregnant and nonpregnant women, respectively
- Other symptoms that occurred in >10% of each group included nausea or vomiting, fatigue, diarrhea, and rhinorrhea

Gastrointestinal (GI) symptoms including nausea, vomiting, diarrhea, and abdominal pains are not uncommon in COVID-19 patients. The CDC reported diarrhea in 19% and nausea and/or vomiting in 12% nonpregnant women (Zambrano et al., 2020). In a systematic review, the prevalence of overall GI symptoms was 18%, and the prevalence of diarrhea, nausea/vomiting, and abdominal pain were 13%, 10%, and 9%, respectively (Cheung et al., 2020).

Other manifestations include conjunctivitis (Ma et al., 2020), falls, general health decline, delirium, (especially in elderly and patients with neurocognitive impairments) (Annweiler et al., 2020), and cutaneous manifestations such as maculopapular, urticarial, and vesicular eruptions and transient livedo reticularis (Galván Casas et al., 2020). Sometimes, reddish-purple nodules may appear on the distal digits similar in appearance to pernio (chilblains) (mainly in children and young adults) named by some as COVID toes (de Masson et al., 2020; Galván Casas et al., 2020).

Patients with nonsevere illness may continue with the same grade of the disease or progress to more severe illness. Progression may occur over 7 days (Cohen et al., 2020). One study conducted on 138 hospitalized COVID-19 patients in Wuhan reported development of pneumonia and hospital admission after a median of 5 and 7 days since the onset of symptoms, respectively (Wang et al., 2020a). Another study reported a median time of 8 days between onset of symptoms and development of dyspnea (Huang et al., 2020).

Although pregnancy does not appear to increase the risk for acquiring COVID-19 infection, it appears to worsen the clinical course of the disease when compared with nonpregnant patients of the same sex and age (Allotey et al., 2020; Badr et al., 2020). Clinical deterioration may be rapid. Intensive care unit (ICU) admission was reported in 1%–3% pregnant patients with COVID-19 infection (Centers for Disease Control and Prevention, 2021). Although most (>90%) infected pregnant persons recover spontaneously without hospitalization or delivery, they are at increased risk of death when compared with symptomatic nonpregnant females of reproductive age (Vincenzo Berghella and Hughes, 2020).

COVID-19 Versus Flu

Influenza is a very common disease; it should be differentiated from COVID-19 infection. Both diseases have many similarities and differences. Both COVID-19 and influenza have airborne spread and share common manifestations such as fever, cough, shortness of breath, fatigue, sore throat, rhinorrhea, muscle pains, and headache. Many patients infected with influenza or COVID-19 have mild symptoms and recover spontaneously without hospitalization after rest and fluids or even no treatment at all. However, both infections may have serious sequelae as pneumonia, ARDS, multiple organ failure, stroke, and even death (both morbidity and mortality are higher in COVID-19 compared with influenza) (MayoClinic, 2020).

COVID-19 and influenza have several differences being caused by different viruses (SARS-CoV-2 and influenza A and B viruses). COVID-19 infection is more contagious with longer incubation period (symptoms appear 2–14 days and 1–4 days after exposure to SARS-CoV-2 and influenza virus, respectively). Some symptoms may differ as loss of taste and smell, which is common in COVID-19 not in influenza. COVID-19 may have different complications from flu as hypercoagulability and multisystem inflammatory syndrome in children (MayoClinic, 2020).

Laboratory testing and imaging of COVID-19

A real-time reverse-transcription polymerase chain reaction (rRT-PCR) assay is the current gold standard for diagnosis of SARS-CoV-2 infection. Other laboratory investigations can be used for diagnosis, determination of the possible treatment, evaluation of treatment efficacy, and prognosis of the infection. Chest imaging is of special importance for diagnosis and prognosis of the disease.

Whom to Test

Both symptomatic and asymptomatic individuals need testing for SARS-CoV-2 infection according to the following guidelines.

Asymptomatic individuals

Indications of testing included the following (Center for Disease Control and Prevention, 2020; Infectious Diseases Society of America, 2020b):

- Close contact with a known case (including neonates born to infected mothers). Testing 5–7 days after exposure is suggestive based on the average incubation period (Infectious Diseases Society of America, 2020a).
- Individuals at risk for severe disease living in congregate facilities that (e.g., long-term care facilities, correctional and detention facilities, homeless shelters). They should be screened in response to identified COVID-19 cases within the facility beside routine intermittent screening of employees and residents.
- Hospitalized patients at communities with high prevalence (+ve PCR in ≥10% of the community).
- Before aerosol-generating procedures and time-sensitive surgical procedures.
- Before immunosuppressive therapy.

The CDC recommended against retesting asymptomatic individuals who were previously diagnosed with SARS-CoV-2 within the prior 3 months because of the low likelihood that a repeat positive test during this interval represents an active reinfection.

Symptomatic individuals

Cases with clinical suspicion infections should be tested. Clinical suspicion is high in patients who developed new-onset fever and/or respiratory tract symptoms (e.g., cough, dyspnea), patients with severe lower respiratory tract (LRT) illness without any clear cause, those who developed dyspnea several days after the onset of initial symptoms (Cohen et al., 2020), and patients who present with extrapulmonary complications that may be linked to SARS-CoV-2 infection as cardiac injury, ischemic stroke, and other thromboembolic events (Caliendo Angela, 2020).

Clinicians should have a low threshold for suspicion of COVID-19, and the likelihood is further increased if the patient lives in or is traveled to allocation with high prevalence of COVID-19 infection within the prior 14 days or is in close contact with a suspected or confirmed COVID-19 case in the prior 14 days (Caliendo Angela, 2020).

Many recommendations were described for COVID-19 testing priorities. The Infectious Diseases Society of America (IDSA) categorized priorities for RT-PCR testing (or antigen testing, if available) when testing capacity is limited into four levels (Infectious Diseases Society of America, 2020a).

First (high) priority was suggested for critically ill patients receiving ICU-level care with unexplained viral pneumonia or respiratory failure; individuals having fever or features of an LRT illness and close contact with patients with laboratory-confirmed COVID-19 within 14 days of symptom onset (including all residents of long-term care facilities with a confirmed case); individuals with fever or features of an LRT illness who are also immunosuppressed (including patients with HIV), are older, or have underlying chronic health conditions; and individuals with fever or features of an LRT illness who are critical to the pandemic response, including healthcare workers, public health officials, and other essential leaders.

Second priority was suggested for non-ICU hospitalized patients and long-term care residents with unexplained fever and features of an LRT illness with consideration of the number of confirmed cases in the community.

Third priority for outpatients who meet criteria for influenza testing (e.g., symptoms such as fever, cough, and other suggestive respiratory symptoms plus comorbid conditions, such as diabetes mellitus, chronic obstructive pulmonary disease, congestive heart failure, age >50 years, immunocompromising conditions) and testing of outpatient pregnant women and symptomatic children with similar risk factors is also included in this priority level with consideration of the number of confirmed cases in the community.

Fourth priority for community surveillance is directed by public health and/or infectious diseases authorities.

The WHO suggested that in the presence of limited resources in areas with community transmission, prioritization for testing should be given to people at risk of developing severe disease and vulnerable populations, who will require hospitalization and advanced care for COVID-19 health workers (including emergency services and nonclinical staff) regardless of whether they are a contact of a confirmed case (to protect health workers and reduce the risk of nosocomial transmission), the first symptomatic individuals in a closed setting (e.g., schools, long-term living facilities, prisons, hospitals) to quickly identify outbreaks and ensure containment measures. All other individuals with symptoms related to the close settings may be considered probable cases and isolated without additional testing if testing capacity is limited (World Health Organization, 2020f).

Diagnostic testing of SARS-CoV-2 is done through detection of the virus itself (viral RNA or antigen) or detection of the human immune response to the virus (antibodies or other biomarkers). Standard infection confirmation is based on the detection of unique viral sequences by nucleic acid amplification tests (NAATs), such as rRT-PCR. The assays' targets include regions on the E, RdRP, N, and S genes (World Health Organization, 2020e).

The virus may be detectable in the upper respiratory tract (URT) 1–3 days before the onset of symptoms. The concentration of SARS-CoV-2 in the URT is highest around the time of symptom onset and then gradually decreases (Zou et al., 2020). There is a controversy about the association between virus loads and severity of infection (Lavezzo et al., 2020). The presence of viral RNA in the LRT, and for a subset of individuals in the feces, increases during the second week of illness (Weiss et al., 2020). The viral RNA detection is not consistent among individuals. The period of detection may be several days, weeks, or months (Li et al., 2020b). Prolonged presence of viral RNA does not necessarily signify prolonged infectiousness.

The optimum site for obtaining the specimen is dependent on the clinical presentation and interval since symptom onset. Specimens include respiratory, fecal, serum, semen, ocular fluid, and others.

Respiratory Specimen

These include upper and lower respiratory specimens. The variability of respiratory secretion composition and the adequacy of sampling efforts may result in false-negative PCR results (Tang et al., 2020).

Upper respiratory specimens are adequate for testing the early stage of infections, especially in asymptomatic or mild cases. Combined nasopharyngeal and oropharyngeal swabs increase the sensitivity for detection of respiratory viruses and improve the reliability of the result (Sutjipto et al., 2020). Two individual swabs can be combined in one collection tube, or a combined nasopharyngeal and oropharyngeal swab can be taken (Lieberman et al., 2010). A few studies have found that individual nasopharyngeal swabs yield a more reliable result than oropharyngeal swabs (Sutjipto et al., 2020).

Lower respiratory specimens are recommended during the delayed course of the disease and in patients with a negative upper respiratory sampling with strong clinical suspicion of COVID-19 (Liu et al., 2020). LRT specimens can consist of spontaneously produced sputum (induced sputum is not recommended due to high risk of aerosol transmission (World Health Organization, 2020b)) and/or endotracheal aspirate or bronchoalveolar lavage in patients with more severe respiratory disease. Caution should be exercised due to the high risk of aerosolization; therefore, strict adherence to IPC procedures during sample collection is required. The indication for an invasive procedure should be evaluated by a physician.

Fecal Specimens

Feces or rectal swabs have been shown to be positive for SARS-CoV-2 RNA from the second week after symptom onset and onward. It can be considered if both upper and lower respiratory specimens were negative along with high suspicion of infection (Ng et al., 2020). When testing faces, ensure that the intended extraction method and NAAT has been validated for this type of sample. Some suggested that fecal positivity for SARS-CoV-2 is prolonged compared with that of respiratory tract specimens (Wong et al., 2020).

Serum Specimens

If negative NAAT results are obtained from a patient in whom SARS-CoV-2 infection is strongly suspected, a paired serum specimen could be collected. One specimen taken in the acute phase and one in the convalescent phase 2–4 weeks later can be used to look for seroconversion or a rise in antibody titers. These two samples can be used retrospectively to determine whether the individual has had COVID-19, especially when the infection could not be detected using NAAT. Some studies linked the detection of virus in blood to the severity of the disease (Corman et al., 2020a,b).

Other specimens include oral fluid specimens (with variable detection rate compared with upper respiratory) (Williams et al., 2020), gargling/mouth washes (have limited data) (Guo et al., 2020b), ocular fluids (in patients with and without conjunctivitis) (Zhang et al., 2020b), urine (only in a limited number of patients) (Nomoto et al., 2020), semen (Li et al., 2020a), and cerebrospinal fluid (Moriguchi et al., 2020). Thus, SARS-CoV-2 can be detected in a wide range of other body fluids and compartments, but it is most frequently detected in respiratory material, and therefore, respiratory samples remain the sample type of choice for diagnosis of SARS-CoV-2. Postmortem specimens can be obtained through postmortem swab, needle biopsy, or tissue specimens from the autopsy, including lung tissue for further pathological and microbiological testing (Tian et al., 2020).

Specimens for virus detection should reach the laboratory as soon as possible after collection. Correct handling of specimens during transportation and in the laboratory is essential.

SARS-CoV-2 infections should be tested with NAAT. A rRT-PCR assay is the current gold standard for

SARS-CoV-2 detection. The test uses specific primers and probes that target the RNA-dependent RNA polymerase (RdRp), envelope, and nucleocapsid genes of SARS-CoV-2, among which the RdRp assay has the highest analytical sensitivity (3.8 RNA copies/reaction at 95% detection probability) (Corman et al., 2020a,b).

Optimal diagnostics consist of a NAAT assay with at least two independent targets on the SARS-CoV-2 genome. When using a one-target assay, it is recommended to have a strategy in place to monitor for mutations that might affect performance (World Health Organization, 2020e).

RT-PCR is a quantitative method in which the amplification of DNA is detected in real time. It can detect the viral load. However, this usually requires laboratories to develop in-house test kits and to validate them with internal controls (Pan et al., 2020).

The practical limitations of RT-PCR testing include the need for a biosafety level-2 (BSL-2) facility, a requirement for kits with specific reagents and primers, the need to maintain a cold chain (as the specimens require storage at $2-8°C$), and the use of strict, validated protocols for testing; consequently, countries with resource limitations or acute spikes in the number of suspected cases may not be able to meet these demands. However, there are no good alternatives: antigen_antibody detection tests are not validated, and viral culture is impractical, as it takes at least 3 days for SARS-CoV-2 to cause cytopathic effects in selected cell lines (VeroE6 and Huh7 cells) (Zhou et al., 2020c). In addition, viral culture will require a BSL-3 facility, which is usually found only in tertiary medical or university research centers (Dashraath et al., 2020).

Pooling of Specimens for Nucleic Acid Amplification Test

Pooling of samples from multiple individuals can increase the diagnostic capacity for detecting SARS-CoV-2 when the rate of testing does not meet the demand in some settings (Abdalhamid et al., 2020). One strategy for pooling stated that if the pooled result is negative, all individual specimens in the pool are regarded as negative. If the pool test is positive, the follow-up steps depend on the strategy, but, in general, each specimen needs to undergo individual testing (pool deconvolution) to identify the positive specimen(s). Another strategy is matrix pooling in which the pools are made per row and per column, and tested by PCR; the position in the matrix identifies the positive specimen without additional testing if prevalence is sufficiently low. Depending on how robust the matrix testing method is in the specific context, it might still be advisable to retest the identified positive samples for confirmation. Pooling of specimens could be considered in population groups with a low/very low expected prevalence of SARS-CoV-2 infection, but not for cases or cohorts that are more likely to be infected with SARS-CoV-2. Routine use of the pooling of specimens from multiple individuals in clinical care and for contact tracing purposes is not recommended (World Health Organization, 2020e).

To perform reliable pooling, adequate automation is key (e.g., robotic systems, software supporting the algorithms to identify positive samples, laboratory information systems, and middleware that can work with sample pooling). Based on currently available data, intraindividual pooling (multiple specimens from one individual that are pooled and tested as a single sample) from URT samples can be used. Intraindividual pooling of sputum and feces with URT samples is not recommended because the former may contain compounds that inhibit rRT-PCR (World Health Organization, 2020e).

Sensitivity of the test depends on the type and quality of the specimen obtained, the time of testing in relation to the course of the disease, and the specific assay:

Test performance by specimen type

LRT specimens are more likely to yield positive tests than URT specimens as they may have higher viral loads. In one study of 205 COVID-19 patients, the highest rates of positive viral RNA tests were reported from bronchoalveolar lavage (95%, 14 of 15 specimens) and sputum (72%, 72 of 104 specimens), compared with oropharyngeal swab (32%, 126 of 398 specimens) (Wang et al., 2020b). Some studies have suggested that viral RNA levels are higher and more frequently detected in nasal compared with oropharyngeal specimens (Kujawski et al., 2020; Wang et al., 2020b).

Self-collected specimens can reduce the need for personal protective equipment with relatively good accuracy with certain self-collected specimens (i.e., nasal swabs and saliva specimens). The sensitivity of NAAT with self-collected nasal or nasal midturbinate specimens may be similar to that with nasopharyngeal specimens collected by a healthcare provider (Tu et al., 2020).

Studies suggest that the relative sensitivity of self-collected saliva specimens compared with nasopharyngeal specimens is 85% or higher (Wyllie et al., 2020). In one study of 70 patients hospitalized with COVID-19 who underwent repeat testing every few days, RT-PCR testing of first-morning saliva (enough to fill one-third of a sterile cup) yielded on average one-half log higher viral RNA levels than that of nasopharyngeal specimens and was more frequently positive in the first 10 days (Wyllie et al., 2020).

Test performance by illness duration: The detection of SARS-CoV-2 RNA is variable among the course of the disease (Zhou et al., 2020c; Guo et al., 2020a,b). Analysis of seven studies reported that estimated rates of false-negative results were 100%, 38% , 20%, and 66% on the day of exposure, day 5 (the first day of symptoms), day 8, and day 21, respectively (Kucirka et al., 2020). Heterogeneity across studies and assumptions made in the analysis (e.g., about incubation period and time of exposure) reduce confidence in these results. Other studies have also suggested that viral RNA levels are high prior to the development of symptoms (Furukawa et al., 2020).

Test performance by assay type: There are also differences in the limit of detection among the major commercial NAAT assays, and retesting samples on different platforms may yield conflicting results (Nalla et al., 2020). Additionally, point-of-care NAAT assays may not be as sensitive as laboratory-based tests (Dinnes et al., 2020). In an August 2020 systematic review of 11 studies evaluating four rapid, point-of-care molecular tests using confirmed SARS-CoV-2 samples, sensitivity ranged from 68% to 100%, with an average of 95.2% (Dinnes et al., 2020). However, most studies were judged to be at high risk of bias.

Cycle threshold: The cycle threshold refers to the number of cycles in an RT-PCR assay needed to amplify viral RNA to reach a detectable level. The cycle threshold value indicates the relative viral RNA level in a specimen (with lower Ct values reflective of higher viral levels). Resulting laboratories generally do not provide the cycle threshold value with the qualitative NAAT result, although it can be obtained upon request for some testing platforms. However, the clinical application of the cycle threshold is uncertain. Ct values are not standardized across RT-PCR platforms, so results cannot be compared across different tests (Caliendo Angela, 2020).

Test interpretation and additional testing (Caliendo Angela, 2020).

Positive NAAT result confirms the diagnosis of SARS-CoV-2 infections without any need for additional diagnostic testing (additional testing may be required for management in hospitalized patients). Some patients may have positive NAATs following documented viral RNA clearance. It is unknown whether this finding indicated relapse or recurrent infection.

Negative NAAT result: In most cases, a single negative NAAT excludes the diagnosis of SARS-CoV-2 infection.

In case of high suspicion of SARS-CoV-2 infection and confirming the presence of infection is important for management or infection control and the initial testing was negative, it is suggested to repeat the test 24−48 h after the initial test. Repeat testing within 24 h is not recommended.

The IDSA and WHO recommend to reserve LRT specimen NAAT for hospitalized patients who have an initial negative test on an URT specimen but for whom suspicion for LRT SARS-CoV-2 infection remains.

For patients who present 3−4 weeks into the course of illness and have negative NAAT, checking a serologic test may be informative.

In many cases, because of the limited availability of testing and concern for false-negative results, the diagnosis of COVID-19 is made presumptively based on a compatible clinical presentation in the setting of an exposure risk (residence in or travel to an area with widespread community transmission or known contact) in the absence of other identifiable causes.

False-negative tests can occur in the following conditions:

- Poor quality of the specimen with insufficient material
- Late collection of the specimen.
- The specimen was obtained from a body compartment that did not contain the virus at that given time
- Inappropriate handling and/or shipping of the specimen
- Technical reasons inherent in the test, e.g., PCR inhibition or virus mutation

Indeterminate NAAT result: The interpretation of an inconclusive or indeterminate result depends on the specific NAAT performed; the clinician should confer with the performing laboratory about additional testing.

In some cases, an inconclusive or indeterminate result indicates that only one of the two or more genes that the NAAT test targets was identified. These results can be considered presumptive positive results, given the high specificity of NAAT assays. If the patient is early in the disease course, repeat testing can be helpful to confirm.

Rapid Diagnostic Tests Based on Antigen Detection

Rapid diagnostic tests (RDTs) that detect the presence of SARS-CoV-2 viral proteins (antigens) in respiratory tract specimens are being developed and commercialized. Most of these are lateral flow immunoassays (LFIs), which are typically completed within 30 min. In contrast to NAATs, there is no amplification of the target that is detected, making antigen tests less sensitive. Additionally, false-positive results may occur if the antibodies on the test strip also recognize antigens of viruses other than SARS-CoV-2, such as other human coronaviruses. The sensitivity of different RDTs compared with rRT-PCR in specimens from URT (nasopharyngeal swabs) appears to be highly variable (Dinnes et al., 2020; Porte et al., 2020), but specificity is consistently reported to be high. Currently, data on antigen performance in the clinical setting are still limited: paired NAAT and antigen validations in clinical studies are encouraged to identify which of the antigen detection tests that are either under development or have already been commercialized demonstrate acceptable performance in representative field studies. When performance is acceptable, antigen RDTs could be implemented in a diagnostic algorithm to reduce the number of molecular tests that need to be performed and to support rapid identification and management of COVID-19 cases. How antigen detection would be incorporated into the testing algorithm depends on the sensitivity and specificity of the antigen test and on the prevalence of SARS-CoV-2 infection in the intended testing population. Higher viral loads are associated with improved antigen test performance; therefore, test performance is expected to be best around symptom onset and in the initial phase of a SARS-CoV-2 infection (World Health Organization, 2020e).

WHO recommended against the use of SARS-CoV-2 Ag-RDTs in (1) asymptomatic individuals unless he had contact with a confirmed case as pretest probability is low; (2) communities with sporadic cases as positive test results would likely be false positives; molecular testing is preferred; (3) if patient management is not dependent on test results; (4) airport or border screening at points of entry as disease prevalence is highly variable among travelers and both positive and negative tests would require confirmatory testing to increase positive predictive value and negative predictive value for decision-making; and (5) screening before blood donation as positive result would not necessarily correlate with presence of viremia. Asymptomatic blood donors do not meet the definition of a suspect case (World Health Organization, 2020a).

Antibody Testing

Serological assays that detect antibodies produced by the human body in response to infection with the SARS-CoV-2 can be useful in various settings. For example, serosurveillance studies can be used to support the investigation of an ongoing outbreak and to support the retrospective assessment of the attack rate or the size of an outbreak (World Health Organization, 2020c). Commercial and noncommercial tests measuring binding antibodies (total immunoglobulins [Ig], IgG, IgM, and/or IgA in different combinations) utilizing various techniques including LFI, enzyme-linked immunosorbent assay (ELISA), and chemiluminescence immunoassay have become available. As SARS-CoV-2 is a novel pathogen, our understanding of the antibody responses it engenders is still emerging, and therefore, antibody detection tests should be used with caution and not used to determine acute infections. Nonquantitative assays (e.g., lateral flow assays) cannot detect an increase in antibody titers, in contrast to (semi)quantitative or quantitative assays. Lateral flow antibody detection assays (or other nonquantitative assays) are currently not recommended for acute diagnosis and clinical management, and their role in epidemiologic surveys is being studied. Serology should not be used as a stand-alone diagnostic to identify acute cases in clinical care or for contact tracing purposes. Interpretations should be made by an expert and are dependent on several factors including the timing of the disease, clinical morbidity, the epidemiology and prevalence within the setting, the type of test used, the validation method, and the reliability of the results. Seroconversion (development of measurable antibody response after infection) has been observed to be more robust and faster in patients with severe disease compared with those with milder disease or asymptomatic infections. Antibodies have been detected as early as in the end of the first week of illness in a fraction of patients but can also take weeks to develop in patients with subclinical/mild infection (Deeks et al., 2020; Okba et al., 2020; Zhao et al., 2020). A reliable diagnosis of COVID-19 infection based on patients' antibody response will often only be possible in the recovery phase, when opportunities for clinical intervention or interruption of disease transmission have passed. Therefore, serology is not a suitable replacement for virological assays to inform contact tracing or clinical management. The duration of the persistence of antibodies generated in response to SARS-CoV-2 is still under study (Seow et al., 2020). Furthermore, the presence of antibodies that bind to SARS-CoV-2 does not guarantee that they are neutralizing antibodies or that they offer protective immunity.

The predictive ability of serologic assays varies widely according to the severity of the disease, timing of the test, the target viral protein, and characteristics of the studied individual (e.g., young or old). Antibody detection tests for coronavirus may also cross-react with other pathogens, including other human coronaviruses (Okba et al., 2020; Gorse et al., 2020), or with preexisting conditions (e.g., pregnancy, autoimmune diseases), and thus yield false-positive results. Virus neutralization assays are the gold standard test for detecting the presence of functional antibodies. These tests require highly skilled staff and BSL-3 culture facilities and, therefore, are unsuitable for use in routine diagnostic testing. Diagnostic testing for SARS-CoV-2: Implementation and interpretation of antibody testing in the clinical laboratory. When implementing serological assays in the clinical laboratory, an in-house validation or verification of the specific assays is advisable. Even if commercial tests have been authorized for use in emergencies, an in-house verification (or if required by local authorities a validation) is still required. Protocols and examples with suggestions as to how to do this are now available (Theel et al., 2020). Each serological test is different. With regard to commercial tests, follow the manufacturer's instructions for use. Studies show that several commercial assays measuring total Ig or IgG have performed well. Most of these studies show no advantage of IgM over IgG, as IgM does not appear much earlier than IgG (Deeks et al., 2020). The additional role of IgA testing in routine diagnostics has not been established. For confirmation of a recent infection, acute and convalescent sera must be tested using a validated (semi)quantitative or quantitative assay. The first sample should be collected during the acute phase of illness, and the second sample should be collected at least 14 days after the initial sera were collected. Maximum antibody levels are expected to occur in the third/fourth week after symptom onset. Seroconversion or a rise in antibody titers in paired sera will help to confirm whether the infection is recent and/or acute. If the initial sample test is positive, this result could be due to a past infection that is not related to the current illness. The first known case of reinfection with SARS-CoV-2 has been documented [182]. Only limited information is available on the interpretation of SARS-CoV-2 antibody tests after a previous infection with SARS-CoV-2 and on the dynamics of SARS-CoV-2 serology if a subsequent infection with another coronavirus occurs. In these two sets of circumstances, interpretation of serology may be extremely challenging (World Health Organization, 2020e).

Viral Isolation

Virus isolation is not recommended as a routine diagnostic procedure. All procedures involving viral isolation in cell culture require trained staff and BSL-3 facilities.

A thorough risk assessment should be carried out when culturing specimens from potential SARSCoV-2 patients for other respiratory viruses because SARS-CoV-2 has been shown to grow on a variety of cell lines (Chu et al., 2020).

Genomic Sequencing for SARS-CoV-2

Genomic sequencing for SARS-CoV-2 can be used to investigate the dynamics of the outbreak, including changes in the size of an epidemic over time, its spatiotemporal spread, and testing hypotheses about transmission routes. In addition, genomic sequences can be used to decide which diagnostic assays, drugs, and vaccines may be suitable candidates for further exploration. Analysis of SARS-CoV-2 virus genomes can, therefore, complement, augment, and support strategies to reduce the disease burden of COVID19. However, the potentially high cost and volume of the work required for genomic sequencing means that laboratories should have clarity about the expected returns from such investment and what is required to maximize the utility of such genomic sequence data. WHO guidance on SARS-CoV-2 genomic sequencing is currently being developed (World Health Organization, 2020e).

Pathogen Detection in Secondary Infection

Patients with COVID-19 infection are at high risk for catching secondary bacterial and/or fungal infections especially in those having the severe form of the disease. Microbiological examination should be considered in these cases. The site of the obtained specimen for microbiological examination differs according to clinical suspicion. Sites include respiratory secretions, blood or serum, GIT secretions, cerebrospinal fluid, and others.

The elevation of C-reactive protein (CRP) has poor specificity for the diagnosis of secondary infection. Elevated procalcitonin levels are of great significance for clinical diagnosis of sepsis. In a cohort study that included 731 hospitalized COVID-19 patients, secondary bacterial infection was diagnosed in 9.3%, 7.9% had one or more secondary bloodstream infections (BSIs), and 3% had at least one possible lower respiratory tract infections (pLRTIs). The overall 28-day cumulative incidence was 16.4% (95% CI: 12.4%−21.0%). Most of the BSIs were due to gram-positive pathogens (76/106 isolates, 71.7%), specifically coagulase-negative staphylococci (53/76%, 69.7%), while among gram-negatives (23/106%, 21.7%) *Acinetobacter baumanii* (7/23%, 30.4%) and *Escherichia coli* (5/23%, 21.7%) predominated. pLRTIs were caused mainly by gram-negative pathogens (14/26%, 53.8%). Eleven patients were diagnosed with putative invasive aspergillosis. At multivariable analysis, factors associated with secondary infections were low baseline lymphocyte count (≤ 0.7 vs. >0.7 per 10^9/L, subdistribution hazard ratios (sdHRs) 1.93, 95% CI: 1.11−3.35), baseline PaO_2/FiO_2 (per 100 points lower: sdHRs 1.56, 95% CI: 1.21−2.04), and ICU admission in the first 48 h (sdHR 2.51, 95% CI: 1.04−6.05). They concluded that patients hospitalized with COVID-19 had a high incidence of secondary infections and identified the early need for ICU, respiratory failure, and severe lymphopenia as risk factors for secondary infections (Ripa et al., 2021).

In case of the suspected fungal infection, in addition to fungal culture, G test, and GM test, Cryptococcus antigen detection can also be performed.

Laboratory Investigations
Total leukocytic count
At the onset of the disease, the total peripheral blood leukocyte is normal or decreased, and the lymphocyte count is decreased. Patients with a lower absolute lymphocyte value generally have a poor prognosis, and peripheral blood lymphocytes in critical patients show a progressive decrease. Elevated neutrophil to lymphocyte ratio (NLR) is an independent risk factor affecting the occurrence of severe illness (Liu et al., 2020b).

A retrospective study found several differences in WCC between severe and nonsevere COVID-19 patients (Qin et al., 2020a). Both groups experienced an increase in leukocytes with the severe group having a significantly greater rise (5.6 vs. 4.9 × 109/L; $P < .001$). Neutrophils were predominantly driving this increase as the severe set (4.3 vs. 3.2 × 109/L; $P < .001$). Interestingly, the levels of lymphocytes, monocytes, basophils, and eosinophils were less, resulting in a greater neutrophil-to-lymphocyte ratio (NLR; 5.5 vs. 3.2; $P < .001$). Another study conducted in China concludes similar findings of high neutrophil and low LC count in severely affected patients, suggesting NLR could be a potential biomarker for early detection of severe COVID-19 (Wang et al., 2020a). However, other factors may disrupt the accuracy of the WCC results observed. These include glucocorticoid therapy and other underlying viral/bacterial infections (Yip et al., 2005). A descriptive study in China reported depleted lymphocytes levels in the majority of COVID-19 patients (Chen et al., 2020). Another study has found low blood lymphocyte percentage (LYM%) in critically ill patients, suggesting low lymphocytes count indicates poor prognosis. However, since the virus can target lymphoid tissue and mechanisms of IL-6, other causes of low lymphocytes count must be investigated (Tan et al., 2020b).

Similar to NLR, the clinical benefits of LC count as a biomarker for COVID-19 remain uncertain (Kermali et al., 2020).

Platelet Count
A metaanalysis of 1799 patients reveals those with severe COVID-19 infection had significantly lower platelet counts (WMD −31 × 109/L; 95% CI: −35 to −29 × 109/L) (Lippi et al., 2020). When using mortality as an endpoint, nonsurvivors evidently had a significantly lower platelet count (WMD, −48 × 109/L; 95% CI: −57 to −39 × 109/L). Using thrombocytopenia as an endpoint also revealed a fivefold greater risk of COVID-19 (OR, 5.13; 95% CI: 1.81−14.58). A retrospective study that used Cox proportional hazard regression analysis found that platelet count is an independent risk factor for mortality among COVID-19 patients, where a 50 × 109/L increase is associated with 40% deceased mortality (HR 0.60, 95% CI: 0.43, 0.84) (Liu et al., 2020d). Here, thrombocytopenia at admission was more likely to occur in nonsurvivors than in survivors. Although many risk factors were accounted for in this study, the possibility for unmeasured confounder cannot be excluded. Another study corroborates the previously documented work. The nadir platelet count was significantly associated with mortality—and the lower the nadir, the stronger the association (Yang et al., 2020). Again, thrombocytopenia was more likely to occur in nonsurvivors than survivors. This study is from adequate sample sizes providing statistical power; however, similar to the previous studies, they are all retrospective making the correlation seen difficult to extrapolate from. Testing the platelet count is a routine part of laboratory tests, and the literature suggests it has inherent value in providing more detail on the patient's condition (Kermali et al., 2020).

Elevated C-Reactive Protein
Elevation of CRP (plasma protein produced by the liver) occurs in various inflammatory conditions. A rise in CRP level is associated with an increase in disease severity. In a retrospective study, most patients with severe COVID-19 infection showed significantly higher levels compared with the nonsevere cohort (57.9 mg/L vs. 33.2 mg/L, $P < .001$) (Qin et al., 2020a). A second retrospective cohort study found the likelihood of progressing to severe COVID-19 disease increased in patients with CRP levels >41.8 mg/L (Liu et al., 2020a). CRP is suggested as one of the first biomarkers within blood plasma that changes to reflect physiological complications. Compared with erythrocyte sedimentation rate (ESR), CRP levels are significantly greater during

early periods of severe cases and proved to be a more sensitive biomarker in reflecting disease development. Compared with CT scans alone, CRP values are more reliable for earlier identification of case severity (Tan et al., 2020a).

Interleukin-6

Cytokine release syndrome (CRS) is an overexaggerated immune response involving an overwhelming release of proinflammatory mediators. This mechanism underlies several pathological processes including ARDS (Mahajan et al., 2019). CRS was linked to disease severity in SARS and MERS infections (Mahallawi et al., 2018). Understanding their role in COVID-19 disease may help facilitate the design of novel immunotherapies. Studies have revealed that levels of IL-6, the most common type of cytokine released by activated macrophages, rise sharply in severe manifestations of COVID-19 (Chen et al., 2020). One metaanalysis reviewing six studies shows mean IL-6 concentrations were 2.9-fold higher in patients with complicated COVID-19 compared with those with noncomplicated disease (n = 1302; 95% CI: 1.17−7.19) (Coomes and Haghbayan, 2020). In its analysis, the outcomes of the studies include ICU admission, onset of ARDS, and mortality. Since the proportionate rise of IL-6 is correlated with disease severity, this study can prove groundbreaking. Although clinicians can use this to identify severity earlier and commence oxygen therapy sooner, the varying outcomes make it somewhat difficult to ascertain what level of IL-6 corresponds to what negative outcome. Furthermore, many studies recruited participants from the same center, giving rise to the potential of selection bias (Kermali et al., 2020).

Indicators reflecting body's inflammation and immune status, such as CRP, procalcitonin, ferritin, erythrocyte sedimentation, total lymphocytes and subpopulations, IL-6, and blood lactic acid, can facilitate the clinical stages judgment, which can be used as a clinical warning indicator for severe and critical cases, and provide a basis for the formulation of treatment strategies (National Health Commission & State Administration of Traditional Chinese Medicine, 2020). Normal procalcitonin and significantly elevated ESR and CRP are seen in most patients. There are significant elevation of IL-6 and IL-10 expressions in patients with severe disease and significant reduction of the numbers of CD8+ T lymphocytes. Serial measures of IL-6, IL-10, and CD8+ T lymphocyte levels can be used in the assessment of prognosis of the severity of the disease (Liu et al., 2020b).

Lactate Dehydrogenase

Lactate dehydrogenase (LDH) secretion is triggered by necrosis of the cell membrane, signaling viral infection or lung damage, such as the pneumonia induced by SARS-CoV-2 (Han et al., 2020). There is convincing evidence linking LDH levels to the development of COVID-19 disease (Ferrari et al., 2020). A multicenter study involving 1099 patients reported supporting evidence correlating extent of tissue damage and inflammation with increasing levels of LDH (Guan et al., 2020b). Furthermore, when LDH levels were correlated with CT scans, significantly higher levels reflected the severity of pneumonia (Xiong et al., 2020). There is increasing confidence in using LDH as a biomarker to measure severity of COVID-19 infection. Another study found that there was a significant rise in LDH levels among refractory COVID-19 patients (Kermali et al., 2020).

D-Dimer

D-dimer is released in the blood as a result of lysis of cross-linked fibrin, and its increasing levels indicate activation of coagulation and fibrinolysis (Zhang et al., 2018). A retrospective study was conducted on 191 COVID-19 patients and found that D-dimer levels >1.0 µg/mL ($P = .0033$) were associated with higher mortality, and levels ≥2.0 µg/mL on admission were the optimum cutoff to predict in-hospital mortality for COVID-19 (Zhang et al., 2020a). Studies have reported that nearly 90% of inpatients with pneumonia had increased coagulation activity marked rising D-dimer levels (Milbrandt et al., 2009). Furthermore, Huang et al. found that levels of D-dimer on admission could be used to triage patients into critical care (Huang et al., 2020). The researchers found that median D-dimer levels were higher in ICU patients compared with non-ICU patients (2.4 mg/L vs. 0.5 mg/L; $P = .0042$). D-dimer levels can be used as a prognostic marker and help clinicians monitor those who are likely to deteriorate earlier. However, this study confirmed the diagnosis of COVID-19 using LRT specimens and did not use paired nasopharyngeal swabs to investigate the viral RNA detection rate between the URT and LRT specimens. Secondly, with a cohort size of 41 patients, it is difficult to assess predictors of disease severity and mortality with multivariable-adjusted methods (Kermali et al., 2020).

Cardiac Troponin

There is growing evidence of higher mortality rates among those with underlying cardiovascular disease due to COVID-19 infection [22,52,53]. Some have investigated the use of high-sensitivity cardiac troponin I (hs-TnI) as a marker of disease progression and mortality. A retrospective study performed in China of patients with confirmed COVID-19 based on SARS-CoV-2 RNA detection revealed a univariable odds ratio for death at 80.1 (95% CI: 10.3–620.4, $P < .0001$) for hs-TnI (Zhou et al., 2020a). This risk was higher compared with other biomarkers such as D-dimer and lymphocyte count. Another study of 416 hospitalized patients with COVID-19 reported that hs-TnI was elevated in one in five patients on presentation (Shi et al., 2020). These patients were more likely to require invasive (22% vs. 4%, $P < .001$) or noninvasive (46% vs. 4%, $P < .001$) ventilation and develop ARDS (59% vs. 15%, $P < .001$) or acute kidney injury (9% vs. 0%, $P < .001$). Early recognition of myocardial injury indicated by elevated hs-TnI aids in appropriate triage to a critical care area and informs the use of inotropes and vasopressors. However, elevated levels are common in hospitalized patients and are likely to be due to nonischemic causes of myocardial injury. This may lead to inappropriate use of cardiology consultation and downstream testing and increased risk to cardiac physiology staff (Kermali et al., 2020).

Renal Markers

There is also evidence that chronic kidney disease is associated with severe forms of COVID-19 infection (Henry and Lippi, 2020). Studies have demonstrated significantly higher levels of renal biomarkers such as serum urea, creatinine, and markers of glomerular filtration rate in severe cases (Xiang et al., 2020). A study of 701 patients revealed that elevated serum creatinine levels on admission correlated with severity due to significant abnormalities in the coagulation pathway (Xiang et al., 2020). They also found that these patients were more likely to require mechanical ventilation or be placed in ICU. Univariate Cox regression analysis found elevated creatinine levels were also associated with in-hospital mortality (HR 2.99, 95% CI: 2.00, 4.47). Proteinuria, hematuria, and elevated urea levels had similar, if not larger, hazard ratios. Interestingly, another study showed a potential role for urinalysis over serum markers of kidney function (Zhou et al., 2020b). Here, abnormalities in the routine urine test on admission correlated strongly with disease severity. They go on to suggest that urinalysis may reveal kidney impairment more readily than evaluation of serum renal biomarkers. However, these tests were only carried out on admission and so patients in earlier stages of the infection had changes in serum levels obscured by compensatory kidney function. Hence, renal abnormalities on admission may indicate higher risks of deterioration, ensuring appropriate triaging (Kermali et al., 2020).

Elevated alanine aminotransferase, aspartate aminotransferase, LDH, phosphocreatine kinase (CK), and myoglobin (Mb) in COVID-19 patients are suggestive of multiple organ dysfunctions.

The WHO reported that the laboratory features associated with severe COVID-19 infection include the increased levels of D-dimer >1000 ng/mL, CRP >100 mg/L, LDH >245 units/L, troponin >2× the upper limit of normal, ferritin >500 mcg/L, CPK >2× the upper limit of normal, and decreased absolute lymphocyte count <800/mL (Guan et al., 2020a; World Health Organization, 2020f).

A systematic review of pregnant women with suspected or confirmed COVID-19 revealed the occurrence of lymphopenia in 35%, leukocytosis in 27%, elevated procalcitonin level in 21%, and abnormal liver chemistries in 11% (Allotey et al., 2020).

Screening for Disseminated Intravascular Coagulation

DIC is a clinical syndrome in which pathogenic factors damage the microvascular system, leading to activation of coagulation cascade, systemic formation of microvascular thrombosis, massive consumption coagulopathy, and secondary hyperfibrinolysis.

The diagnosis of DIC depends on clinical characteristics and laboratory findings. Clinically, the patient has bleeding tendency from multiple sites. However, it rarely happens to the critical ill COVID-19 patient in whom microcirculatory disturbances and multiple organ failure are commonly observed. The specific manifestations included progressive and rapid deterioration of pulmonary functions, deterioration of the liver and kidney functions, disturbed conscious level, myocardial damage, and shock, which cannot be explained by other causes. No single indicator can diagnose DIC. The value of laboratory parameters for the early warning on COVID-19 with DIC varies. The sensitivities of various tests arranged in decreasing pattern include increased D-dimer, thrombocytopenia, prolonged prothrombin time, decreased fibrinogen, and prolonged activated partial thromboplastin time. Their values regarding specificities are arranged as follows: progressive decreases in fibrinogen > progressive decreases in platelets > prolonged APTT > increased D-dimer (Mei and Hu, 2020).

Blood Gas Analysis

Blood gas analysis includes the measurement of the pH, partial pressure of oxygen (PO_2), partial pressure of carbon dioxide (PCO_2), and electrolyte concentration in the blood. It provides references for a quick judgment on the presence of respiratory dysfunction and acid–base imbalance in COVID-19 patients. It is used for assessing the severity of the disease and helps to guide the prognosis and treatment of patients with COVID-19 (Dukić et al., 2016). The radial artery is the preferred site for sample collection, followed by the brachial artery, dorsalis pedis artery, and femoral artery. The blood should be sampled when the patient is rested and quiet to prevent over breathing or breathe holding; if the body temperature is not easy to control, it is necessary to input the body temperature value for correction during the test. If the patient's oxygen administration mode changes, a stable oxygen state shall be ensured for at least 20–30 min before blood collection, and the oxygen inhalation parameters shall be input during the test to ensure the accuracy of the test results (Dukić et al., 2016).

Chest Imaging

Chest imaging has been considered as part of the diagnostic workup of patients with suspected COVID-19 (Manna et al., 2020). Imaging has been also considered to complement clinical evaluation and laboratory parameters in the management of patients already diagnosed with COVID-19 (World Health Organization, 2020b). or the use of chest imaging in acute care of adult patients with suspected, probable, or confirmed COVID-19, including chest radiography, CT, and lung ultrasound.

The WHO provided the following recommendations for the use of chest imaging in COVID-19 patients (World Health Organization, 2020g):

1. Chest imaging is not recommended for the diagnosis of COVID-19 in asymptomatic contacts of patients with COVID-19. RT-PCR should be done to confirm diagnosis.
2. Chest imaging is not recommended for the diagnosis of COVID-19 in symptomatic patients with suspected COVID-19 infection when RT-PCR testing is available with timely results. RT-PCR should be done to confirm diagnosis.
3. Chest imaging for diagnosis of COVID-19 in symptomatic patients with suspected COVID-19 infection is used when RT-PCR testing is not available, but results are delayed or initial RT-PCR testing is negative, but with high clinical of suspicion of COVID-19.

4. For nonhospitalized patients with suspected or confirmed COVID-19 with mild symptoms, the WHO suggests using chest imaging in addition to clinical and laboratory assessment to decide on hospital admission versus home discharge.
5. For nonhospitalized patients with suspected or confirmed COVID-19 with moderate-to-severe symptoms, the WHO suggests using chest imaging in addition to clinical and laboratory assessment to decide on regular ward admission versus ICU admission.
6. For hospitalized patients with suspected or confirmed COVID-19 with moderate-to-severe symptoms, the WHO suggests using chest imaging in addition to clinical and laboratory assessment to inform the therapeutic management.
7. For hospitalized patients with COVID-19 whose symptoms are resolved, the WHO suggests not using chest imaging in addition to clinical and/or laboratory assessment to inform the decision regarding discharge.

In the general population, CT of chest has a sensitivity of 97%, specificity of 25%, positive predictive value of 65%, and negative predictive value of 83% (Ai et al., 2020). CT can be done during pregnancy as it exposes the fetus to 0.03 mGy dose and doses up to 50 mGY are not associated with fetal teratogenicity (American College of Obstetricians and Gynecologists, 2017).

Chest Computed Tomography Scan
Preparation prior to admission

1. Reserve a CT scanner for suspected or confirmed cases, if available. Preference is given to movable CT scanner (if available) or the CT scanner that can lift the examination bed through the console, a separate control room (operating room) is required; if not, when disinfecting after examination, air disinfection of other computer rooms connected to the control room (operating room) is also required.
2. If a central air-conditioning fresh air system is used in the examination room, adjust the air supply and exhaust to the maximum; if an ordinary central air-conditioning is used, turn off the central air-conditioning in the examination room and operation room, and turn on the standby separate air-conditioning; if no spare separate air conditioner is available, turn on the central air conditioner after examination and disinfecting.
3. To reduce the viral transmission, a disposable medical middle sheet is needed during the examination to isolate the equipment from patients.

4. Two technicians are required, with one operating the CT scanner, and the other one enter to be in the examination room for positioning. (According to the requirements of the National Center for Disease Control and Prevention, both technicians for the operation and positioning require secondary or higher protection.)

Preparation for patient

The patient must wear a mask and lie down in a supine position. The technician trains the patient to hold his or her breath at the end of inspiration during the examination.

Scope and direction of scanning

Scan from apex pulmonis to costophrenic angle. For severe and critical patients (who are difficult to hold their breath), the scanning can be from costophrenic angle to apex pulmonis to reduce respiratory motion artifacts caused by difficulty in holding breath in the lower lung field, so as to ensure image quality.

Scanning parameters

The technician uses a low-dose chest CT protocol to scan the patient. The automatic tube voltage selected 100−120 kV of tube voltage, smart mAs of 20−50 mAs, collimator with 0.5−1.5 mm width, layer thickness, and layer spacing of 1−5 mm. For severe and critical patients, a larger pitch (1.0−1.5 pitches) can be used to reduce scanning time and respiratory motion artifacts.

Keys for CT Diagnosis (Xing et al., 2020).:

Early-stage manifestations. The presence of Multiple lesions affecting both lungs is common and unilateral affection is rare. Peripheral affection of the lungs or under the pleura is more common in the lower fields. Lesions are fan-shaped and irregular, but flaky or nearly round lesions can be noticed. Lesions generally do not affect the whole lung segments (patchy distribution). The density is uneven, often limited to small patches or large ground glass opacities, in which thickened blood vessels and thick-walled bronchi are seen, with or without localized grid-like interlobular septal thickening. The consolidation range is small and limited, with air bronchial signs visible.

Advanced stage manifestations. There is increase in the distributions of the lesions, and fused lesions can be seen, involving larger parts of lung lobes. There is increased lesion density due consolidation patches that are irregular, wedge-shaped, or fan-shaped, and the boundary is unclear. Bronchial vascular bundle thickening or multifocal lung consolidation can be seen under the pleura. The lesion progresses and changes rapidly, and the morphological changes are evident in the short-term review, which can be combined with the necrosis of lung tissue to form a small cavity. There are air bronchograms, usually with no pleural effusion and mediastinal and hilar lymph node enlargement.

Severe-stage manifestations. Sign of "white lungs" with elevation of diaphragm can be seen when most of the lungs are affected. Other features such as air bronchial signs and bronchiectasis are seen with uneven distribution of the lesions. The areas without consolidation can be patchy with ground glass opacities, bilateral pleural thickening, and thickened interlobular pleura, with pleural effusion.

Absorption-stage manifestations. The clinical improvement usually precedes the characteristic imaging changes. The lesions become narrower, decrease in size, and become less in density; the consolidations gradually vanish, with complete absorption of the ground glass opacities; and the pleural effusion is absorbed or organized by the body (Fig. 3.1).

3-2 Bedside X-ray

Bedside chest plain X-ray has become the main imaging method for patients with severe and critical COVID-19.

The findings are like those in CT in patients with severe pneumonia, including the shadows of consolidation, which may be patchy, reticular, stripe, hilar and mediastinal changes, pneumothorax, pleural effusions, thickening of the pleura, etc.

COVID-19 follow-up using imaging techniques

Examination status. Primary examination or reexamination is included.

Areas of the lesions. The method of 18-segment segmentation is used in lungs to record the number of involved lung segment regardless the lesion of the size. However, the term of significantly involved lung segment is defined when the lesion is at least 1/2 of the lung segment in size. Lesion area of 7/15 means there are 15 involved lung segments, of which there are seven significantly involved lung segments.

Evolution of pulmonary lesions. Using a serial chest CT scan presented different patterns with six manifestations: progression, stability, stalemate, improvement, sequelae, and complete radiological resolution.

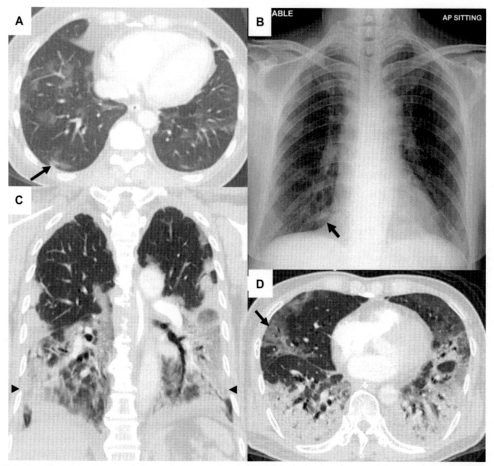

FIG. 3.1 Imaging of two patients with COVID-19. **(A)**, Contrast-enhanced CT of one patient in the axial plane across the lower lobes of the lungs shows patchy GGO in a lobular distribution. Early changes of consolidation are present in the posterior segment of the right lower lobe (arrow). **(B)** Corresponding chest radiograph does not reveal significant abnormality other than a small focus of consolidation in the medial right lower zone (arrow), which would have been easily missed owing to projection adjacent to the right cardiophrenic angle and overlapping rib shadow. **(C and D)** CT pulmonary angiogram of a different patient with severe pneumonia in the axial and coronal planes showing extensive multilobar GGO (arrows) with areas of confluent consolidation (arrowheads) mostly distributed in the posterior and basal regions of the lower lobes. No pulmonary embolism was detected. These findings are not specific to COVID-19 and may be seen in other viral and atypical pneumonias. COVID-19, coronavirus disease 2019; CT, computed tomography; GGO, ground-glass opacity (Ashokka et al., 2020).

Progression: Increased lung lesions with increased consolidation.

Stability: No evident changes from previous chest CT scans.

Stalemate: New lesions could be seen with partial absorption of old lesions.

Improvement: Decreased lung involvement with decreased density of the old lesions.

Sequelae: The patient improved clinically but with the typical CT changes, such as bronchiectasis and pleural effusion.

Complete radiological resolution: All lesions disappeared completely on chest CT scan (Fig. 3.2). Criteria for discharge

1. Pulmonary lesions are significantly decreased in size or completely absorbed.

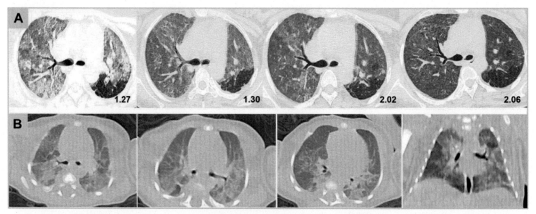

FIG. 3.2 Chest computed tomography (CT) screening from the mother (patient 6) and her twin neonate. **(A)** CT findings from the mother. The first axial image showed extensive ground-glass opacities (GGOs) and nodules. On the following days, the intensity decreased, indicating the lesions were gradually absorbed after effective treatment. **(B)** CT findings from the twin neonate. The axial and coronal images at the 19th day of birth presented with extensive GGO along the bronchovascular bundle or in the peripheral area, and the localized consolidation in dorsal segment of right lower lobe (Huang et al., 2020).

2. Only a few shadows of fibrotic stripes seen in the follow-up images.
3. No de novo lesion is seen.

The follow-up is recommended after discharge after 14 days or according to the clinical needs.

Ultrasonography
COVID-19 disease shows rapid transmission, progression, and high rate of critical cases (Xu et al., 2020). Ultrasound plays an important role in the diagnosis, efficacy assessment, and follow-up of COVID-19 patients.

Ultrasonography. Ultrasound for COVID-19 patients is highly recommended to be performed as bedside service of isolation ward area and fever clinics to decrease transmitted infections caused by machine transport. The ultrasound assessment for COVID-19 patients is focused on lungs and heart and for the fetal assessment in pregnant ladies with COVID-19. Moreover, critically ill patients may suffer from multiple organ dysfunction that needs imaging of these organs such as liver and kidneys, and this is of help for assessment of the condition of the patients.

To reduce the exposure time of the technician, relevant image data could be collected fast and fully in an isolation ward area and analyzed after the technician leaves the infectious ward area. To reduce the time of exposure of the sonograoher, ultrasound evaluation should be targeted through rapid assessment not comprehensive routine evaluation.

Thoracic examination
Diagnosis and localization of pleural effusion. It also helps in identification and localization for the patients who are candidates for catheter drainage, which can be ultrasound-guided.

Additional diagnosis of pneumonia. The consolidations caused by COVID-19 will enhance the ultrasound penetration of the lung tissue.

Abnormal signs of pneumonia by ultrasound include disappearance of line A which is a horizontal artifact indicating normal lung surface and lung sliding sign, disappearance of line B which is a kind of comet-tail artifact indicating subpleural interstitial edema, lung pneumonic patches, air bronchogram, and pleural effusion (Rouby et al., 2018).

Abnormal signs of pneumothorax by ultrasound. Abnormal signs of pneumothorax by ultrasound include disappearance of lung sliding sign of the pleura and appearance of "lung point," which is highly specific ultrasound sign of pneumothorax. It involves visualizing the point where the visceral pleura begins to separate from the parietal pleura at the margin of a pneumothorax (Lichtenstein and Mezière, 2008).

Echo heart. Cardiac insult occurs in approximately 31% of critically ill patients. Rapid assessment of the functions of the right and left sides of the heart includes the following:

(1) Visual measurement of left heart function is recommended in patients with normal ventricular wall motion.

(2) M-mode assessment of left heart function is recommended for diffuse attenuation of ventricular wall motion.

(3) Abnormal regional wall motion can be assessed using the uniplanar or biplanar Simpson method.

(4) The maximum systolic excursion of the tricuspid annular plane is measured by visual inspection of right ventricular wall motion or M-mode method, if necessary, and the right ventricular fractional area change rate is estimated by the two-dimensional method.

(5) Rapid identification of pericardial effusion and localization: Observation of subxiphoid and parasternal sections is recommended.

(6) Exclusion of other cardiac structural abnormalities: valvular heart disease, cardiomyopathy, myocardial infarction, infective endocarditis, aortic dissection, and other diseases. Comprehensive routine measurement is not necessary.

Assessment of pulmonary artery pressure. Pulmonary artery pressure changes can be dynamically observed via ultrasound so as to adjust diagnostic and therapeutic strategies in a timely manner. Pulmonary artery systolic pressure is estimated using tricuspid regurgitation method or pulmonary venous reflux in the absence of right ventricular outflow tract stenosis.

Ultrasound monitoring supported by intensive care unit and extracorporeal membrane oxygenation. In ICU patients, left atrial pressure and vein width are dynamically monitored to determine whether fluid therapy shall be terminated. During the extracorporeal membrane oxygenation support, echocardiography can detect the size of the cardiac chamber, monitor whether the blood flow is emptied, and evaluate cardiac function and lung changes; ultrasound is used to determine the presence or absence of lung recruitment before weaning.

Examination for peripheral vascular thrombosis. Ultrasound shall diagnose deep venous thrombosis and arterial embolism in the early phase and determine the distribution range of thrombosis, etc., so as to decrease the possibility of systemic and pulmonary emboli, keeping in mind the increased risk of deep venous thrombosis in critically ill patients with COVID-19. Severe patients with COVID-19, especially the elderly and those with underlying diseases, may have multiple other risk factors that cause more increased risk of embolism.

Fetal assessment. The ultrasound assessment of fetal well-being and fetal growth remains the corner stone of assessment of fetus, being the second susceptible patient in the pregnant lady with COVID-19 disease.

REFERENCES

Abdalhamid, B., et al., 2020. Assessment of specimen pooling to conserve SARS CoV-2 testing resources. Am. J. Clin. Pathol. 153 (6), 715–718. https://doi.org/10.1093/ajcp/aqaa064.

Ai, T., et al., 2020. Correlation of chest CT and RT-PCR testing for coronavirus disease 2019 (COVID-19) in China: a report of 1014 cases. Radiology 296 (2), E32–E40. https://doi.org/10.1148/radiol.2020200642.

Allotey, J., et al., 2020. Clinical manifestations, risk factors, and maternal and perinatal outcomes of coronavirus disease 2019 in pregnancy: living systematic review and meta-analysis. Br. Med. J. 370 https://doi.org/10.1136/bmj.m3320.

American College of Obstetricians and Gynecologists, 2017. Guidelines for Diagnostic Imaging During Pregnancy and Lactation. ACOG Committee Opinion, Number 723, October 2017. Available at: https://www.acog.org/Clinical-Guidance-and-Publications/Committee-Opinions/Committee-on-Obstetric-Practice/Guidelines-for-Diagnostic-Imaging-During-Pregnancy-and-Lactation?IsMobileSet=false. (Accessed 14 January 2020).

Annweiler, C., et al., 2020. National French survey of coronavirus disease (COVID-19) symptoms in people aged 70 and over. Clin. Infect. Dis. https://doi.org/10.1093/cid/ciaa792.

Ashokka, B., Loh, M.H., Tan, C.H., Su, L.L., Young, B.E., Lye, D.C., Biswas, A., Illanes, S.E., Choolani, M., 2020. Care of the pregnant woman with coronavirus disease 2019 in labor and delivery: anesthesia, emergency cesarean delivery, differential diagnosis in the acutely ill parturient, care of the newborn, and protection of the healthcare personnel. Am. J. Obstet. Gynecol. 223 (1), 66–74.e3. https://doi.org/10.1016/j.ajog.2020.04.005.

Badr, D.A., et al., 2020. 'Are clinical outcomes worse for pregnant women at ≥20 weeks' gestation infected with coronavirus disease 2019? A multicenter case-control study with propensity score matching'. Am. J. Obstet. Gynecol. 223 (5), 764–768. https://doi.org/10.1016/j.ajog.2020.07.045.

Bergsbaken, T., Fink, S.L., Cookson, B.T., 2009. Pyroptosis: host cell death and inflammation. Nat. Rev. Microbiol. 7 (2), 99–109. https://doi.org/10.1038/nrmicro2070.

Boscolo-Rizzo, P., et al., 2020. Evolution of altered sense of smell or taste in patients with mildly symptomatic COVID-19. JAMA Otolaryngol. Head Neck Surg. 146 (8), 729–732. https://doi.org/10.1001/jamaoto.2020.1379.

Caliendo Angela, M.,H.K.E., 2020. Coronavirus Disease 2019 (COVID-19): Diagnosis, UpToDate. Available at: https://www.uptodate.com/contents/coronavirus-disease-2019-covid-19-diagnosis?search=covid. (Accessed 12 January 2020).

Campbell, K.H., et al., 2020. Prevalence of SARS-CoV-2 among patients admitted for childbirth in Southern Connecticut. J. Am. Med. Assoc. 323 (24), 2520–2522. https://doi.org/10.1001/jama.2020.8904.

Cascella, M., et al., 2020. Features, evaluation, and treatment of coronavirus. Treasure Island (FL).

Center for Disease Control and Prevention, 2020. Overview of Testing for SARS-CoV-2. Available at: https://www.cdc.gov/coronavirus/2019-ncov/hcp/testing-overview.html. (Accessed 26 September 2020).

Center for Disease Control and Prevention, 2021. Data on COVID-19 during Pregnancy: Severity of Maternal Illness. Available at: https://www.cdc.gov/coronavirus/2019-ncov/cases-updates/special-populations/pregnancy-data-on-covid-19.html. (Accessed 11 January 2020).

Chen, N., et al., 2020. Epidemiological and clinical characteristics of 99 cases of 2019 novel coronavirus pneumonia in Wuhan, China: a descriptive study. Lancet 395 (10223), 507–513. https://doi.org/10.1016/S0140-6736(20)30211-7.

Cheung, K.S., et al., 2020. Gastrointestinal manifestations of SARS-CoV-2 infection and virus load in fecal samples from a Hong Kong cohort: systematic review and meta-analysis. Gastroenterology 159 (1), 81–95. https://doi.org/10.1053/j.gastro.2020.03.065.

Chu, H., et al., 2020. Comparative tropism, replication kinetics, and cell damage profiling of SARS-CoV-2 and SARS-CoV with implications for clinical manifestations, transmissibility, and laboratory studies of COVID-19: an observational study. Lancet Microbe 1 (1), e14–e23. https://doi.org/10.1016/S2666-5247(20)30004-5.

Cohen, P.A., et al., 2020. The early natural history of SARS-CoV-2 infection: clinical observations from an Urban, ambulatory COVID-19 clinic. Mayo Clin. Proc. 95 (6), 1124–1126. https://doi.org/10.1016/j.mayocp.2020.04.010.

Coomes, E.A., Haghbayan, H., 2020. Interleukin-6 in Covid-19: a systematic review and meta-analysis. Rev. Med. Virol. 30 (6), 1–9. https://doi.org/10.1002/rmv.2141.

Corman, V.M., et al., 2020a. Detection of 2019 novel coronavirus (2019-nCoV) by real-time RT-PCR. Euro Surveill. 25 (3), 1–8. https://doi.org/10.2807/1560-7917.ES.2020.25.3.2000045.

Corman, V.M., et al., 2020b. SARS-CoV-2 asymptomatic and symptomatic patients and risk for transfusion transmission. Transfusion 60 (6), 1119–1122. https://doi.org/10.1111/trf.15841.

Creanga, A.A., et al., 2017. Pregnancy-related mortality in the United States, 2011–2013. Obstet. Gynecol. 130 (2), 366–373. https://doi.org/10.1097/AOG.0000000000002114.

Dashraath, P., et al., 2020. Coronavirus disease 2019 (COVID-19) pandemic and pregnancy. Am. J. Obstet. Gynecol. 222 (6), 521–531. https://doi.org/10.1016/j.ajog.2020.03.021.

Deeks, J.J., et al., 2020. Antibody tests for identification of current and past infection with SARS-CoV-2. Cochrane Database Syst. Rev. 6 (6), CD013652. https://doi.org/10.1002/14651858.CD013652.

Dinnes, J., et al., 2020. Rapid, point-of-care antigen and molecular-based tests for diagnosis of SARS-CoV-2 infection. Cochrane Database Syst. Rev. 8 https://doi.org/10.1002/14651858.CD013705.

Dukić, L., et al., 2016. Blood gas testing and related measurements: national recommendations on behalf of the Croatian Society of Medical Biochemistry and Laboratory Medicine. Biochem. Med. 26 (3), 318–336. https://doi.org/10.11613/BM.2016.036.

Ferrari, D., et al., 2020. Routine blood tests as a potential diagnostic tool for COVID-19. Clin. Chem. Lab. Med. 58 (7), 1095–1099. https://doi.org/10.1515/cclm-2020-0398.

Furukawa, N.W., Brooks, J.T., Sobel, J., 2020. Evidence supporting transmission of severe acute respiratory syndrome coronavirus 2 while presymptomatic or asymptomatic. Emerg. Infect. Dis. 26 (7) https://doi.org/10.3201/eid2607.201595.

Galván Casas, C., et al., 2020. Classification of the cutaneous manifestations of COVID-19: a rapid prospective nationwide consensus study in Spain with 375 cases. Br. J. Dermatol. 183 (1), 71–77. https://doi.org/10.1111/bjd.19163.

Goodnight, W.H., Soper, D.E., 2005. Pneumonia in pregnancy. Crit. Care Med. 33 (10 Suppl. l), S390–S397. https://doi.org/10.1097/01.ccm.0000182483.24836.66.

Gorse, G.J., Donovan, M.M., Patel, G.B., 2020. Antibodies to coronaviruses are higher in older compared with younger adults and binding antibodies are more sensitive than neutralizing antibodies in identifying coronavirus-associated illnesses. J. Med. Virol. 92 (5), 512–517. https://doi.org/10.1002/jmv.25715.

Guan, W., Ni, Z., Hu, Y., Liang, W., Ou, C., He, J., Liu, L., Shan, H., Lei, C., Hui, D.S.C., et al., 2020a. Clinical characteristics of 2019 novel coronavirus infection in China. medRxiv. https://doi.org/10.1101/2020.02.06.20020974.

Guan, W., Ni, Z., Hu, Y., Liang, W., Ou, C., He, J., Liu, L., Shan, H., Lei, C., Hui, D.S.C., et al., 2020b. Clinical characteristics of coronavirus disease 2019 in China. N. Engl. J. Med. 382 (18), 1708–1720. https://doi.org/10.1056/nejmoa2002032.

Guo, L., et al., 2020a. Profiling early humoral response to diagnose novel coronavirus disease (COVID-19). Clin. Infect. Dis. 71 (15), 778–785. https://doi.org/10.1093/cid/ciaa310.

Guo, W.-L., et al., 2020b. Effect of throat washings on detection of 2019 novel coronavirus. Clin. Infect. Dis. 71 (8), 1980–1981. https://doi.org/10.1093/cid/ciaa416.

Hall, O.J., et al., 2017. Progesterone-based contraceptives reduce adaptive immune responses and protection against sequential influenza A virus infections. J. Virol. 91 (8) https://doi.org/10.1128/JVI.02160-16.

Han, Y., et al., 2020. Lactate dehydrogenase, an independent risk factor of severe COVID-19 patients: a retrospective and observational study. Aging 12 (12), 11245–11258. https://doi.org/10.18632/aging.103372.

Henry, B.M., Lippi, G., 2020. Chronic kidney disease is associated with severe coronavirus disease 2019 (COVID-19) infection. Int. Urol. Nephrol. 1193–1194. https://doi.org/10.1007/s11255-020-02451-9.

Hu, Z., et al., 2020. Clinical characteristics of 24 asymptomatic infections with COVID-19 screened among close contacts in Nanjing, China. Sci. China Life Sci. 63 (5), 706–711. https://doi.org/10.1007/s11427-020-1661-4.

Huang, C., et al., 2020. Clinical features of patients infected with 2019 novel coronavirus in Wuhan, China. Lancet 395 (10223), 497–506. https://doi.org/10.1016/S0140-6736(20)30183-5.

Infectious Diseases Society of America, 2020a. COVID-19 Prioritization of Diagnostic Testing. Available at: https://www.idsociety.org/globalassets/idsa/public-health/covid-19-prioritization-of-dx-testing.pdf. (Accessed 14 November 2020).

Infectious Diseases Society of America, 2020b. Guidelines on the Diagnosis of COVID-19. Available at: https://www.idsociety.org/practice-guideline/covid-19-guideline-diagnostics/. (Accessed 11 November 2020).

Ji, H.-L., et al., 2020. Elevated plasmin(ogen) as a common risk factor for COVID-19 susceptibility. Physiol. Rev. 100 (3), 1065–1075. https://doi.org/10.1152/physrev.00013.2020.

Kasper, M.R., et al., 2020. An outbreak of covid-19 on an aircraft carrier. N. Engl. J. Med. 383 (25), 2417–2426. https://doi.org/10.1056/NEJMoa2019375.

Kermali, M., et al., 2020. The role of biomarkers in diagnosis of COVID-19 – a systematic review. Life Sci. 254 (117788).

Kucirka, L.M., et al., 2020. Variation in false-negative rate of reverse transcriptase polymerase chain reaction-based SARS-CoV-2 tests by time since exposure. Ann. Intern. Med. 173 (4), 262–267. https://doi.org/10.7326/M20-1495.

Kujawski, S.A., et al., 2020. Clinical and virologic characteristics of the first 12 patients with coronavirus disease 2019 (COVID-19) in the United States. Nat. Med. 26 (6), 861–868. https://doi.org/10.1038/s41591-020-0877-5.

Lauer, S.A., et al., 2020. The incubation period of coronavirus disease 2019 (COVID-19) from publicly reported confirmed cases: estimation and application. Ann. Intern. Med. 172 (9), 577–582. https://doi.org/10.7326/M20-0504.

Lavezzo, E., et al., 2020. Suppression of a SARS-CoV-2 outbreak in the Italian municipality of Vo'. Nature 584 (7821), 425–429. https://doi.org/10.1038/s41586-020-2488-1.

Lechien, J.R., et al., 2020. Loss of smell and taste in 2013 European patients with mild to moderate COVID-19. Ann. Intern. Med. 173 (8), 672–675. https://doi.org/10.7326/M20-2428.

Li, D., et al., 2020a. Clinical characteristics and results of semen tests among men with coronavirus disease 2019. JAMA Netw Open 3 (5), e208292. https://doi.org/10.1001/jamanetworkopen.2020.8292.

Li, N., Wang, X., Lv, T., 2020b. Prolonged SARS-CoV-2 RNA shedding: not a rare phenomenon. J. Med. Virol. 92 (11), 2286–2287. https://doi.org/10.1002/jmv.25952.

Li, X., Sun, X., Carmeliet, P., 2019. Hallmarks of endothelial cell metabolism in health and disease. Cell Metabol. 30 (3), 414–433. https://doi.org/10.1016/j.cmet.2019.08.011.

Lichtenstein, D.A., Mezière, G.A., 2008. Relevance of lung ultrasound in the diagnosis of acute respiratory failure: the BLUE protocol. Chest 134 (1), 117–125. https://doi.org/10.1378/chest.07-2800.

Lieberman, D., et al., 2010. Pooled nasopharyngeal and oropharyngeal samples for the identification of respiratory viruses in adults. Eur. J. Clin. Microbiol. Infect. Dis. 29 (6), 733–735. https://doi.org/10.1007/s10096-010-0903-5.

Lippi, G., Plebani, M., Henry, B.M., 2020. Thrombocytopenia is associated with severe coronavirus disease 2019 (COVID-19) infections: a meta-analysis. Clin. Chim. Acta 506, 145–148. https://doi.org/10.1016/j.cca.2020.03.022.

Liu, F., et al., 2020a. Prognostic value of interleukin-6, C-reactive protein, and procalcitonin in patients with COVID-19. J. Clin. Virol. 127, 104370. https://doi.org/10.1016/j.jcv.2020.104370.

Liu, J., et al., 2020b. Longitudinal characteristics of lymphocyte responses and cytokine profiles in the peripheral blood of SARS-CoV-2 infected patients. EBioMedicine 55. https://doi.org/10.1016/j.ebiom.2020.102763.

Liu, R., et al., 2020c. Positive rate of RT-PCR detection of SARS-CoV-2 infection in 4880 cases from one hospital in Wuhan, China, from Jan to Feb 2020. Clin. Chim. Acta 505, 172–175. https://doi.org/10.1016/j.cca.2020.03.009.

Liu, Y., et al., 2020d. Association between platelet parameters and mortality in coronavirus disease 2019: retrospective cohort study. Platelets 31 (4), 490–496. https://doi.org/10.1080/09537104.2020.1754383.

Ma, N., et al., 2020. Ocular manifestations and clinical characteristics of children with laboratory-confirmed COVID-19 in Wuhan, China. JAMA Ophthalmol. 138 (10), 1079–1086. https://doi.org/10.1001/jamaophthalmol.2020.3690.

Mahajan, S., et al., 2019. Plcγ2/Tmem178 dependent pathway in myeloid cells modulates the pathogenesis of cytokine storm syndrome. J. Autoimmun. 100, 62–74. https://doi.org/10.1016/j.jaut.2019.02.005.

Mahallawi, W.H., et al., 2018. MERS-CoV infection in humans is associated with a pro-inflammatory Th1 and Th17 cytokine profile. Cytokine 104, 8–13. https://doi.org/10.1016/j.cyto.2018.01.025.

Manna, S., et al., 2020. COVID-19: a multimodality review of radiologic techniques, clinical utility, and imaging features. Radiology 2 (3), e200210. https://doi.org/10.1148/ryct.2020200210.

de Masson, A., et al., 2020. Chilblains is a common cutaneous finding during the COVID-19 pandemic: a retrospective nationwide study from France. J. Am. Acad. Dermatol. 667–670. https://doi.org/10.1016/j.jaad.2020.04.161.

MayoClinic, 2020. Coronavirus vs. Flu: Similarities and Differences. Available at: https://www.mayoclinic.org/diseases-conditions/coronavirus/in-depth/coronavirus-vs-flu/art-20490339. (Accessed 10 December 2020).

Mei, H., Hu, Y., 2020. Characteristics, causes, diagnosis and treatment of coagulation dysfunction in patients with COVID-19. Zhonghua Xue Ye Xue Za Zhi 41 (3), 185–191. https://doi.org/10.3760/cma.j.issn.0253-2727.2020.0002.

Milbrandt, E.B., et al., 2009. Prevalence and significance of coagulation abnormalities in community-acquired pneumonia. Mol. Med. 15 (11), 438–445. https://doi.org/10.2119/molmed.2009.00091.

Moriguchi, T., et al., 2020. A first case of meningitis/encephalitis associated with SARS-Coronavirus-2. Int. J. Infect. Dis. 55–58. https://doi.org/10.1016/j.ijid.2020.03.062.

Nalla, A.K., et al., 2020. Comparative performance of SARS-CoV-2 detection assays using seven different primer-probe sets and one assay kit. J. Clin. Microbiol. 58 (6) https://doi.org/10.1128/JCM.00557-20.

National Health Commission & State Administration of Traditional Chinese Medicine, 2020. Diagnosis and Treatment Protocol for Novel Coronavirus Pneumonia. National Health Commission of the People's Republic of China, pp. 1–17. Available at: https://www.chinadaily.com.cn/pdf/2020/1.Clinical.Protocols.for.the.Diagnosis.and.Treatment.of.COVID-19.V7.pdf%0A.

National Institutes of Health, 2020. Clinical Spectrum of SARS-CoV-2 Infection. COVID-19 Treatment Guidelines. Available at: https://www.covid19treatmentguidelines.nih.gov/overview/clinical-spectrum/. (Accessed 17 December 2020).

Nelson-Piercy, C., 2015. Respiratory disease. In: Handbook of Obstetric Medicine. CRC Press, Boca Raton FL.

Ng, S.C., Chan, F.K.L., Chan, P.K.S., 2020. Screening FMT donors during the COVID-19 pandemic: a protocol for stool SARS-CoV-2 viral quantification. Lancet Gastroenterol. Hepatol. 642–643. https://doi.org/10.1016/S2468-1253(20)30124-2.

Nomoto, H., et al., 2020. Cautious handling of urine from moderate to severe COVID-19 patients. Am. J. Infect. Contr. 969–971. https://doi.org/10.1016/j.ajic.2020.05.034.

Okba, N.M.A., et al., 2020. Severe acute respiratory syndrome coronavirus 2-specific antibody responses in coronavirus disease patients. Emerg. Infect. Dis. 26 (7), 1478–1488. https://doi.org/10.3201/eid2607.200841.

Oran, D.P., Topol, E.J., 2020. Prevalence of asymptomatic SARS-CoV-2 infection. Ann. Intern. Med. 173 (5), 362–367. https://doi.org/10.7326/M20-3012.

Pan, Y., et al., 2020. Viral load of SARS-CoV-2 in clinical samples. Lancet Infect. Dis. 20 (4), 411–412. https://doi.org/10.1016/S1473-3099(20)30113-4.

Porte, L., et al., 2020. Evaluation of a novel antigen-based rapid detection test for the diagnosis of SARS-CoV-2 in respiratory samples. Int. J. Infect. Dis. 99, 328–333. https://doi.org/10.1016/j.ijid.2020.05.098.

Qin, C., et al., 2020a. Dysregulation of immune response in patients with coronavirus 2019 (COVID-19) in Wuhan, China. Clin. Infect. Dis. 71 (15), 762–768. https://doi.org/10.1093/cid/ciaa248.

Qin, J., et al., 2020b. Estimation of incubation period distribution of COVID-19 using disease onset forward time: a novel cross-sectional and forward follow-up study. medRxiv. https://doi.org/10.1101/2020.03.06.20032417.

Richardson, S., et al., 2020. Presenting characteristics, comorbidities, and outcomes among 5700 patients hospitalized with COVID-19 in the New York city area. J. Am. Med. Assoc. 323 (20), 2052–2059. https://doi.org/10.1001/jama.2020.6775.

Ripa, M., et al., 2021. Secondary infections in patients hospitalized with COVID-19: incidence and predictive factors. Clin. Microbiol. Infect. https://doi.org/10.1016/j.cmi.2020.10.021.

Rouby, J.-J., et al., 2018. Training for lung ultrasound score measurement in critically ill patients. Am. J. Respir. Crit.

Care Med. 398–401. https://doi.org/10.1164/rccm.201802-0227LE.

Royal College of Obstetricians and Gynaecologists, 2020. Coronavirus (COVID-19) Infection and Abortion Care. Available at: https://www.rcog.org.uk/globalassets/documents/guidelines/2020-03-21-abortion-guidance.pdf.

Sakurai, A., et al., 2020. Natural history of asymptomatic SARS-CoV-2 infection. N. Engl. J. Med. 383 (9), 885–886. https://doi.org/10.1056/NEJMc2013020.

Seow, J., et al., 2020. Longitudinal evaluation and decline of antibody responses in SARS-CoV-2 infection. medRxiv. https://doi.org/10.1101/2020.07.09.20148429.

Shi, S., et al., 2020. Association of cardiac injury with mortality in hospitalized patients with COVID-19 in Wuhan, China. JAMA Cardiol. 5 (7), 802–810. https://doi.org/10.1001/jamacardio.2020.0950.

Spinato, G., et al., 2020. Alterations in smell or taste in mildly symptomatic outpatients with SARS-CoV-2 infection. J. Am. Med. Assoc. 323 (20), 2089–2090. https://doi.org/10.1001/jama.2020.6771.

Struyf, T., et al., 2020. Signs and symptoms to determine if a patient presenting in primary care or hospital outpatient settings has COVID-19 disease. Cochrane Database Syst. Rev. 7 (7), CD013665. https://doi.org/10.1002/14651858.CD013665.

Sutjipto, S., et al., 2020. The effect of sample site, illness duration, and the presence of pneumonia on the detection of SARS-CoV-2 by real-time reverse transcription PCR. Open Forum Infect. Dis. 7 (9) https://doi.org/10.1093/ofid/ofaa335.

Sutton, D., et al., 2020. Universal screening for SARS-CoV-2 in women admitted for delivery. N. Engl. J. Med. 382 (22), 2163–2164. https://doi.org/10.1056/NEJMc2009316.

Tan, C., et al., 2020a. C-reactive protein correlates with computed tomographic findings and predicts severe COVID-19 early. J. Med. Virol. 92 (7), 856–862. https://doi.org/10.1002/jmv.25871.

Tan, L., et al., 2020b. Lymphopenia predicts disease severity of COVID-19: a descriptive and predictive study. Signal Transduct. Target. Ther. 5 (1), 16–18. https://doi.org/10.1038/s41392-020-0148-4.

Tang, X., et al., 2020. Positive RT-PCR tests among discharged COVID-19 patients in Shenzhen, China. Infect. Contr. Hosp. Epidemiol. 41 (9), 1110–1112. https://doi.org/10.1017/ice.2020.134.

Theel, E., Filkins, L., Palavecino, E., Mitchell, S., Campbell, S., Pentella, M., Butler-Wu, S., Jerke, K., Dharmarha, V., Couturier, M., Schuetz, A., 2020. Verification procedure for commercial serologic tests with emergency use authorization for detection of antibodies to SARS-CoV-2. Am. Soc. Microbiol. https://asm.org/Protocols/VerifyEmergency-Use-Authorization-EUA-SARS-CoV-2.

Tian, S., et al., 2020. Pathological study of the 2019 novel coronavirus disease (COVID-19) through postmortem core biopsies. Mod. Pathol. 33 (6), 1007–1014. https://doi.org/10.1038/s41379-020-0536-x.

Tong, J.Y., et al., 2020. The prevalence of olfactory and gustatory dysfunction in COVID-19 patients: a systematic review and meta-analysis. Otolaryngol. Head Neck Surg 163 (1), 3–11. https://doi.org/10.1177/0194599820926473.

Tostmann, A., et al., 2020. Strong associations and moderate predictive value of early symptoms for SARS-CoV-2 test positivity among healthcare workers, The Netherlands, March 2020. Euro Surveill. 25 (16), 2000508. https://doi.org/10.2807/1560-7917.ES.2020.25.16.2000508.

Tu, Y.-P., et al., 2020. Swabs collected by patients or health care workers for SARS-CoV-2 testing. N. Engl. J. Med. 383 (5), 494–496. https://doi.org/10.1056/NEJMc2016321.

Vanders, R.L., et al., 2013. Plasmacytoid dendritic cells and CD8 T cells from pregnant women show altered phenotype and function following H1N1/09 infection. J. Infect. Dis. 208 (7), 1062–1070. https://doi.org/10.1093/infdis/jit296.

Vincenzo Berghella, Hughes, B., 2020. Coronavirus Disease 2019 (COVID-19): Pregnancy Issues and Antenatal Care - UpToDate. Available at: https://www.uptodate.com/contents/coronavirus-disease-2019-covid-19-pregnancy-issues-and-antenatal-care. (Accessed 14 November 2020).

Wang, D., et al., 2020a. Clinical characteristics of 138 hospitalized patients with 2019 novel coronavirus-infected pneumonia in Wuhan, China. J. Am. Med. Assoc. 323 (11), 1061–1069. https://doi.org/10.1001/jama.2020.1585.

Wang, W., et al., 2020b. Detection of SARS-CoV-2 in different types of clinical specimens. J. Am. Med. Assoc. 323 (18), 1843–1844. https://doi.org/10.1001/jama.2020.3786.

Wang, Y., et al., 2020c. Clinical outcomes in 55 patients with severe acute respiratory syndrome coronavirus 2 who were asymptomatic at hospital admission in Shenzhen, China. J. Infect. Dis. 221 (11), 1770–1774. https://doi.org/10.1093/infdis/jiaa119.

Weiss, A., Jellingsø, M., Sommer, M.O.A., 2020. Spatial and temporal dynamics of SARS-CoV-2 in COVID-19 patients: a systematic review and meta-analysis. EBioMedicine 58. https://doi.org/10.1016/j.ebiom.2020.102916.

Williams, E., et al., 2020. Saliva as a noninvasive specimen for detection of SARS-CoV-2. J. Clin. Microbiol. https://doi.org/10.1128/JCM.00776-20.

Wong, M.C., et al., 2020. Detection of SARS-CoV-2 RNA in fecal specimens of patients with confirmed COVID-19: a meta-analysis. J. Infect. 81 (2), e31–e38. https://doi.org/10.1016/j.jinf.2020.06.012.

World Health Organization, 2020a. Antigen-Detection in the Diagnosis of SARS-CoV-2 Infection Using Rapid Immunoassays. Available at: file:///C:/Users/prof. Ahmed Maged/Downloads/WHO-2019-nCoV-Antigen_Detection-2020.1-eng.pdf. (Accessed 13 January 2020).

World Health Organization, 2020b. Clinical Management of COVID-19 (Interim Guidance).

World Health Organization, 2020c. Coronavirus Disease (COVID-19) Technical Guidance: The Unity Studies: WHO Sero-Epidemiological Investigations Protocols. Available at: https://www.who.int/emergencies/diseases/novel-coronavirus-2019/technical-guidance/early-investigations. (Accessed 20 December 2020).

World Health Organization, 2020d. Coronavirus Disease 2019 (COVID-19) Situation Report − 28. Available at: https://www.who.int/docs/default-source/coronaviruse/situation-reports/20200217-sitrep-28-covid-19.pdf?sfvrsn=a19cf2ad_2. (Accessed 18 February 2020).

World Health Organization, 2020e. Diagnostic Testing for SARS-CoV-2. Interim Guidance. Available at: file:///C:/Users/prof. Ahmed Maged/Downloads/WHO-2019-nCoV-laboratory-2020.6-eng.pdf. (Accessed 12 January 2020).

World Health Organization, 2020f. Laboratory Testing Strategy Recommendations for COVID-19. Interim Guidance. Available at: https://www.who.int/publications/i/item/laboratory-biosafety-guidance-related-to-coronavirus-disease-(covid-19. (Accessed 13 January 2020).

World Health Organization, 2020g. Use of Chest Imaging in Covid-19. A Rapid Advice Guide. Available at: file:///C:/Users/prof. Ahmed Maged/Downloads/WHO-2019-nCoV-Clinical-Radiology_imaging-2020.1-eng (1).pdf. (Accessed 14 January 2020).

Wu, Z., McGoogan, J.M., 2020. Characteristics of and important lessons from the coronavirus disease 2019 (COVID-19) outbreak in China: summary of a report of 72,314 cases from the Chinese center for disease control and prevention. J. Am. Med. Assoc. 323 (13), 1239–1242. https://doi.org/10.1001/jama.2020.2648.

Wyllie, A.L., et al., 2020. Saliva or nasopharyngeal swab specimens for detection of SARS-CoV-2. N. Engl. J. Med. 383 (13), 1283–1286. https://doi.org/10.1056/NEJMc2016359.

Xiang, J., et al., 2020. Potential biochemical markers to identify severe cases among COVID-19 patients. medRxiv. https://doi.org/10.1101/2020.03.19.20034447.

Xiong, Y., et al., 2020. Clinical and high-resolution CT features of the COVID-19 infection: comparison of the initial and follow-up changes. Invest. Radiol. 55 (6). Available at: https://journals.lww.com/investigativeradiology/Fulltext/2020/06000/Clinical_and_High_Resolution_CT_Features_of_the.2.aspx.

Xing, H., Yang, L., Mingxing Xie, J.W., Wang, Y., Cheng, F., Hu, Y., 2020. COVID-19 diagnosis. In: Cheng, F., Zhang, Y. (Eds.), The Clinical Diagnosis and Treatment for New Coronavirus Pneumonia, first ed. Springer Nature, Singapore, pp. 35–60.

Xu, X., Wu, X., Jiang, X., Xu, K., Ying, L., Ma, C., 2020. Clinical findings in a group of patients infected with the 2019 novel coronavirus (SARS-Cov-2) outside of Wuhan, China: retrospective case series. BMJ 368, m606. https://doi.org/10.1136/bmj.m606.

Yang, X., et al., 2020. Thrombocytopenia and its association with mortality in patients with COVID-19. J. Thromb. Haemostasis 18 (6), 1469–1472. https://doi.org/10.1111/jth.14848.

Yip, T.T.C., et al., 2005. Protein chip array profiling analysis in patients with severe acute respiratory syndrome identified serum amyloid a protein as a biomarker potentially useful in monitoring the extent of pneumonia. Clin. Chem. 51 (1), 47–55. https://doi.org/10.1373/clinchem.2004.031229.

Young, B.C., et al., 2014. Longitudinal expression of toll-like receptors on dendritic cells in uncomplicated pregnancy and postpartum. Am. J. Obstet. Gynecol. 210 (5) https://doi.org/10.1016/j.ajog.2013.11.037, 445.e1–6.

Zambrano, L.D., et al., 2020. Update: characteristics of symptomatic women of reproductive age with laboratory-confirmed SARS-CoV-2 infection by pregnancy status — United States, January 22–October 3, 2020. Morb. Mortal. Wkly. Rep. 69 (44), 1641–1647. https://doi.org/10.15585/mmwr.mm6944e3.

Zhang, L., et al., 2018. Use of D-dimer in oral anticoagulation therapy. Int. J. Lit. Humanit. 40 (5), 503–507. https://doi.org/10.1111/ijlh.12864.

Zhang, L., et al., 2020a. D-dimer levels on admission to predict in-hospital mortality in patients with Covid-19. J. Thromb. Haemostasis 18 (6), 1324–1329. https://doi.org/10.1111/jth.14859.

Zhang, X., et al., 2020b. The evidence of SARS-CoV-2 infection on ocular surface. Ocul. Surf. 360–362. https://doi.org/10.1016/j.jtos.2020.03.010.

Zhao, J., et al., 2020. Antibody responses to SARS-CoV-2 in patients with novel coronavirus disease 2019. Clin. Infect. Dis. 71 (16), 2027–2034. https://doi.org/10.1093/cid/ciaa344.

Zhou, F., et al., 2020a. Clinical course and risk factors for mortality of adult inpatients with COVID-19 in Wuhan, China: a retrospective cohort study. Lancet 395 (10229), 1054–1062. https://doi.org/10.1016/S0140-6736(20)30566-3.

Zhou, H., et al., 2020b. Urinalysis, but not blood biochemistry, detects the early renal-impairment in patients with COVID-19. medRxiv. https://doi.org/10.1101/2020.04.03.2005 1722.

Zhou, P., et al., 2020c. A pneumonia outbreak associated with a new coronavirus of probable bat origin. Nature 579 (7798), 270–273. https://doi.org/10.1038/s41586-020-2012-7.

Zhou, Y., et al., 2020d. Pathogenic T cells and inflammatory monocytes incite inflammatory storm in severe COVID-19 patients. Natl. Sci. Rev. nwaa041. https://doi.org/10.1093/nsr/nwaa041.

Zou, L., et al., 2020. SARS-CoV-2 viral load in upper respiratory specimens of infected patients. N. Engl. J. Med. 382 (12), 1177–1179. https://doi.org/10.1056/NEJMc2001737.

CHAPTER 4

Management of COVID-19 Infection During Pregnancy, Labor, and Puerperium

AHMED M. MAGED EL-GOLY, MD

(Dr. Prof. M Fadel Shaltout. Prof. of Obstetrics and Gynecology, Cairo University, Faculty of Medicine)

Covid-19 Infection and Pregnancy. https://doi.org/10.1016/B978-0-323-90595-4.00004-2

INTRODUCTION

There is no specific treatment for COVID-19 infection in the general population till the present time. Treatment of such infection is more challenging during pregnancy as successful drugs that can be used in nonpregnant women may have a hazardous effect on the growing fetus. Most clinical trials for a particular therapy do not include pregnant women for safety reasons. This theory of "protection by exclusion"—although providing maternal and fetal safety—deprives the pregnant women from using a potentially effective drug at early stages of its use. Also, exclusion of pregnant women infected with COVID-19 from interventions with possible benefits makes the assessment of drug safety for the mother, the fetus, and the neonates, and their efficacy during pregnancy is not possible. For these reasons, the current largest clinical trial assessing the different treatment modalities for COVID-19 infection named randomized evaluation of COVID-19 therapy "RECOVERY" is recruiting pregnant women.

PREVENTION OF INFECTION

As there is no definitive treatment for COVID-19 infection and pregnant women are more vulnerable to viral infections than the general population, prevention of catching the infection is the most important step to avoid its hazardous effects.

The Centers for Disease Control and Prevention (CDC) recommended preventive measures that everyone should follow:

1. Pregnant women should know the main ways of spread of infection through direct contact and through respiratory droplets and the less common ways as contact with contaminated surfaces and airborne transmission in poorly ventilated areas. They must realize that asymptomatic infected personnel could also spread the infection as the symptomatic ones.
2. Avoid contact with those who are infected or exposed to infections even if they were within women household.
3. When interacting with others,
   a. wear a mask;
   b. ask your contacts to wear a mask;
   c. avoid close contact such as shaking hands, hugs, and kissing; and
   d. try to keep 6 feet or more from others, if possible.
4. Wash your hands for 20 s or more with water and soap. Use a hand sensitizer with 60% alcohol if water and soap are not reachable.
5. Avoid unnecessary activities and poorly ventilated areas.
6. Ask for at least 30 days' supply of your needed medications to avoid unnecessary visits to the pharmacy.

Pregnant women must contact their healthcare provider immediately in case of having any medical concerns including doubts in catching the infection, felling depressed, or having any query about your health.

There is no need to skip your medical appointments. These visits are important to provide medical and psychological support especially in hard times. Discuss the safest place for your delivery. Be aware of the safety measures to protect you and your kid during and after delivery and precautions you have to follow during delivery and those who will be in contact with you during your stay at hospital.

Worried women should ask their healthcare provider about safety measures they follow to separate healthy from suspected or confirmed cases. They can make a telemedicine visits through phone or video calls if the healthcare providers have no concerns. These behaviors can be modified according to the situation of the community and risk level.

Pregnant women are instructed to receive their vaccines as scheduled and not to delay it during the epidemic. The recommended vaccines related to COVID-19 infection are the influenza and whooping cough vaccines. Both diseases can produce similar symptoms as COVID-19 and that is confusing for proper management of either infections.

Currently, there are no specific vaccines for COVID-19 infection neither for pregnant women nor for the general population.

The CDC suggested some criteria for effective mask. It must have at least two layers of washable breathable fabric that cover the mouth and nose completely and fit the sides of the face with no gaps. The proper way to wear the mask is to wash the hands before putting it, to place it over the nose and mouth, to fit it below the chin, to fit it against face sides, and then to ensure easy breathing through it. Take off the mask through untying of the strings behind the head, handle it through the ear loops only, fold the outer edges together, and then wash it. The woman should take care not to touch her face during mask removal and be instructed to wash her hands immediately after removal of the mask. Masks with exhalation valve or vent are not recommended.

If the mask will not be washed immediately, it must be stored in a sealed elastic bag if wet and in a paper bag if dry.

Women are advised, if they have to take off the mask to eat or drink, to keep it in a clean safe place as a paper bag and, after finishing, to put the mask with the same side facing out. They are also instructed to wash or sanitize their hands after removal and putting back the mask.

Only cloth masks can be washed and reused, whereas disposable masks should not be reused. Cloth masks can be washed using the washing machine with the regular laundry using the settings appropriate for the fabric label or the tap water and soap by hands. The mask must be dried completely before use in a hot dryer or air.

Hand sanitizer must be rubbed over all parts of the hands and fingers till it dries. It must be kept away from the eyes and of course not to be swallowed as suggested by some politicians.

Gloves are to be used during cleaning and disinfecting homes and when caring for a diseased person, and hands must be washed after taking off the gloves. Gloves are not needed when using outdoor machines such as ATM as it is not protective, which may help spreading of the germs. Handwashing with water and soap or sanitizers provides the needed protection in these cases (Center for Disease Control and Prevention, 2020d).

There are more than 120 developing vaccines. Human trials are involved in 10 of them (https://www.who.int/emergencies/diseases/novel-coronavirus-2019/covid-19-vaccines). Vaccines included RNA-based types such as BNT162b1/2 and mRNA-1273, DNA types, inactivated types such as BBIBP-CorV, and adenovirus-vectored ones such as ChAdOx1 nCoV-19 and nonreplicating adenovirus type-5/26 (Yuan et al., 2020).

To fasten vaccine production, the World Health Organization (WHO) follows the ACT Accelerator. Classically, development of vaccines passes through consecutive steps that may take many years. During the current pandemic, the WHO allowed to proceed in the various steps of development in parallel with precautions to maintain safety standards. Once the safety and efficacy of a vaccine is proven, COVAX (guided by the WHO, GAVI, and CEPI) will help its production and distribution all over the world giving priority to high-risk personnel.

The main aim of the developing vaccine is to produce antibodies against the S protein in amounts that can stop viral replication inside the vaccinated person. That can be achieved through assembling the S protein of the virus or its subunits with the adjuvants directly leading to production of neutralizing antibodies and T cell−mediated immune response (Ren et al., 2020; Folegatti et al., 2020; Li et al., 2020) or through using nucleoside-modified RNA (modRNA) as BNT162b1(Mulligan et al., 2020; Sahin et al., 2020) that encodes the virus receptor-binding domain and BNT162b2 (Walsh et al., 2020) encoding the complete S protein with two proline mutations aiming to keep the prefusion conformation. Targeting the S protein is also tried through DNA-based vaccine (Smith et al., 2020) and attenuated virus with NS1-deleted RBD domain (Kaur and Gupta, 2020).

Pfizer company and its partner BioNTech evaluated the efficacy of their vaccine on 43,000 participants and reported 95% and 94% efficacy in general and elder people (>65 years old), respectively. They reported 162 versus 8 symptomatic cases with confirmed infection and 9 versus 1 severe case in the control versus study group, respectively. Only 3.7% of the vaccinated personnel reported fatigue with no serious side effects. Similar efficacy and safety were reported by Moderna in evaluation of their developing vaccine conducted on 30,000 participants. The press release described a consistently high safety and efficacy within all the studied subgroups. Both vaccines used a novel technique using messenger RNA that codes for the virus surface protein. The main challenge facing these two vaccines is their durability. Degradation of the RNA and lipid particles occurs in warm temperature, so these vaccines require a cold chain to transport them from manufacturing plants to distribution areas.

Both mRNA vaccines need to improve their statistical certainty that can be achieved when recruiting 150 to 165 cases of infected participants.

Comparing these two vaccines, Moderna vaccine has a larger dose of mRNA (100 vs. 30 µg), which can lead to serious side effects. Moderna vaccine can be stored at −20°C compared with −70°C needed to store Pfizer/BioNTech vaccine (Cohen, (n.d.)a).

A Russian vaccine "Sputnik V" uses two different shots using nonharmful adenoviruses as gene delivery "vectors" that hold the gene for SARS-CoV-2 surface protein. The first shot used adenovirus 26 (Ad26), called spike, while the second used adenovirus 5 (Ad5). Testing was done 21 days after the first dose as they attend to receive the second dose. However, the trial covered only 20 confirmed cases, which is too few to prove its efficacy (Cohen, (n.d.)b).

Available Safety Information Related to the Use of COVID-19 Vaccines in Pregnancy

Despite American College of Obstetricians and Gynecologists' (ACOG's) persistent advocacy for the inclusion of pregnant individuals in COVID-19 vaccine trials, none of the COVID-19 vaccines approved under EUA have been tested in pregnant individuals. However, studies in pregnant women are planned.

A combined developmental and perinatal/postnatal reproductive toxicity (DART) study of Moderna's mRNA-1273 in rats was submitted to the FDA on December 4, 2020. FDA review of this study concluded that mRNA1273 given prior to mating and during gestation periods at dose of 100 μg did not have any adverse effects on female reproduction, fetal/embryonal development, or postnatal developmental except for skeletal variations, which are common and typically resolve postnatally without intervention. These DART studies provide the first safety data to help inform the use of the vaccine in pregnancy until there are more data in this population (American Colleague of Obstetrics and Gynecology, 2021).

The ACOG recommends that COVID-19 vaccines should not be withheld from pregnant individuals who meet criteria for vaccination based on Advisory Committee on Immunization Practices (ACIP)—recommended priority groups. While safety data on the use of COVID-19 vaccines in pregnancy are not currently available, there are also no data to indicate that the vaccines should be contraindicated, and no safety signals were generated from DART studies for the Pfizer-BioNtech and Moderna COVID-19 vaccines. Therefore, in the interest of allowing pregnant individuals who would otherwise be considered a priority population for vaccines approved for use under EUA to make their own decisions regarding their health, the ACOG recommends that pregnant individuals should be free to make their own decision in conjunction with their clinical care team.

Discussions between the patient and their healthcare provider may assist with decisions regarding the use of vaccines approved under EUA for the prevention of COVID-19 by pregnant patients. Important considerations include the level of the pandemic activity in the local community, the potential efficacy of the vaccine, and the potential risk and severity of maternal disease, including fetal and neonatal risks of both the disease and the vaccine. Considerations also included the woman current health status and risk of exposure (at work or at home and the possibility for exposing high-risk household members) (American Colleague of Obstetrics and Gynecology, 2021).

Vaccination considerations (American Colleague of Obstetrics and Gynecology, 2021):
- Pregnant women who experience fever following vaccination should be counseled to take acetaminophen. Acetaminophen has been proven to be safe for use in pregnancy and does not appear to impact antibody response to COVID-19 vaccines.
- There is currently no preference for the use of one COVID-19 vaccine over another except for 16—17 year olds who are only eligible for the Pfizer-BioNtech vaccine.
- Individuals should complete their two-dose series with the same vaccine product.
- COVID-19 vaccines should not be administered within 14 days of receipt of another vaccine. For pregnant individuals, vaccines including Tdap and influenza should be deferred for 14 days after the administration of COVID-19 vaccines.
- Anti-D immunoglobulin (i.e., Rhogam) should not be withheld from an individual who is planning or has recently received a COVID-19 vaccine as it will not interfere with the immune response to the vaccine.

Pregnant patients who decline vaccination should be supported in their decision.

Lactating Individuals

The ACOG recommends COVID-19 vaccines be offered to lactating individuals similar to nonlactating individuals when they meet criteria for receipt of the vaccine based on prioritization groups outlined by the ACIP. While lactating individuals were not included in most clinical trials, COVID-19 vaccines should not be withheld from lactating individuals who otherwise meet criteria for vaccination. Theoretical concerns regarding the safety of vaccinating lactating individuals do not outweigh the potential benefits of receiving the vaccine. There is no need to avoid initiation or discontinue breastfeeding in patients who receive a COVID-19 vaccine (Academy of Breastfeeding Medicine, 2021).

INDIVIDUALS CONTEMPLATING PREGNANCY

Vaccination is strongly encouraged for nonpregnant individuals within the ACIP prioritization group(s). Furthermore, the ACOG recommends vaccination of individuals who are actively trying to become pregnant or are contemplating pregnancy and meet the criteria for vaccination based on ACIP prioritization recommendations. Additionally, it is not necessary to delay pregnancy after completing both doses of the COVID-19 vaccine.

Given the mechanism of action and the safety profile of the vaccine in nonpregnant individuals, COVID-19 mRNA vaccines are not thought to cause an increased risk of infertility.

If an individual becomes pregnant after the first dose of the COVID-19 vaccine series, the second dose should be administered as indicated. If an individual receives a COVID-19 vaccine and becomes pregnant within 30 days of receipt of the vaccine, participation in CDC's V-SAFE program should be encouraged.

Importantly, routine pregnancy testing is not recommended prior to receiving any EUA-approved COVID-19 vaccine (American Colleague of Obstetrics and Gynecology, 2021).

The ACIP has made the following recommendations for prioritization of COVID-19 vaccine allocation:

Phase 1a: Healthcare workers and long-term care facility residents (McClung et al., 2020). Phase 1b: Persons aged ≥ 75 years and frontline essential workers (Dooling, 2020). Phase 1c: Persons aged 65–75 years, persons aged 16–64 years with high-risk medical conditions (including pregnancy), and other essential workers (Dooling, 2020).

*High-risk medical conditions outlined by the CDC include the following:

- Pregnancy
- Cancer
- Chronic kidney disease
- COPD (chronic obstructive pulmonary disease)
- Heart conditions, such as heart failure, coronary artery disease, or cardiomyopathies
- Immunocompromised state (weakened immune system) from solid organ transplant
- Obesity (body mass index [BMI] of 30 kg/m^2 or higher but < 40 kg/m^2)
- Severe Obesity (BMI $\geq$ 40 kg/m2)
- Sickle cell disease
- Smoking (current or history)
- Type 2 diabetes mellitus

(ACIP, 2020; Center for Disease Control and Prevention, 2020e)

TREATMENT PLAN

All treatments suggested for COVID-19 are suggestive with no evidence. Analysis of treatment received for pregnant women infected with SARS-CoV-2 revealed marked diversity among different institutions. A global interim guidance from International Federation of Gynaecology and Obstetrics (FIGO) and allied partners suggested the use of symptomatic treatment in suspected cases as antipyretics for fever along with general treatment for maintenance of fluid and electrolyte balance with proper maternal and fetal monitoring. Mild confirmed cases should be managed as the suspected cases with proper monitoring for superimposed bacterial infections. Antiviral treatment may be considered after proper evaluation of its risks and benefits. In severe and critical cases, immediate aggressive treatment should be started to reduce maternal and perinatal morbidities and mortalities (Rasmussen et al., 2020). ICU admission in a negative pressure isolation room with proper monitoring of vital signs, fluid and electrolyte balance, and antibacterial and antiviral treatment along with oxygen therapy through noninvasive or invasive ventilation to keep a minimum of 95% O_2 saturation should be done (Poon et al., 2020; Bhatia et al., 2016). A multidisciplinary team (MDT) including expert obstetricians, anesthesiologist, neonatologist, respiratory internist, and others should be available (Rasmussen et al., 2020).

The ACOG and the Society of Maternal–Fetal Medicine suggested an Outpatient Assessment and Management for Pregnant Women with Suspected or Confirmed Novel Coronavirus (COVID-19) as follows:

1. Assessment of patients regarding both symptoms and exposure. Symptoms include fever of 38°C or more, chills, repeated shaking with chills, cough, shortness of breath, sore throat, nasal congestion or rhinorrhea, fatigue, muscle or body aches, headache, new change or loss of taste or smell, nausea or vomiting, and diarrhea. Exposure to known COVID-positive individual especially if unprotected is reported.
2. In the absence of symptoms and exposure history, follow routine prenatal care.
3. In the presence of symptoms and exposure history, test for SARS-CoV-2 infection according to facility and/or local guidance, community spread, and availability of testing.
4. Conduct illness severity assessment by asking about difficulty breathing or shortness of breath, difficulty completing a sentence without gasping for air, or needing to stop to catch breath frequently when walking across the room, cough more than 1 teaspoon of blood, new pain or pressure in the chest other than pain with coughing, inability to keep liquids down, signs of dehydration such as dizziness when standing, delayed response than usual, or confusion during talking.
5. If severity assessment revealed no positive findings, assess clinical and social risks as comorbidities (hypertension, diabetes, asthma, HIV, chronic heart disease, chronic liver disease, chronic lung disease, chronic kidney disease, blood dyscrasia,

and people on immunosuppressive medications), Obstetric issues (e.g., preterm labor) and inability to care for self or arrange follow-up if necessary.

6. If no comorbidities, normal obstetric issues, and able to care for women with herself and arrange for follow-up, consider as low risk. So:
   a. Refer patient for symptomatic care at home including hydration and rest.
   b. Monitor for development of any symptoms above and restart algorithm if new symptoms present.
   c. Routine obstetric precautions.

7. Women with positive illness severity assessment are considered as elevated risk. So, recommend her to immediately seek care in an emergency department or equivalent unit that treats pregnant women. When possible, send the patient to a setting where she can be isolated. Notifying the facility that you are referring a person under investigation is recommended to minimize the chance of spreading infection to other patients and/or healthcare workers at the facility Adhere to local infection control practices including personal protective equipment (PPE).

8. Women with positive clinical and social risks are considered as moderate risk. So, see the patient as soon as possible in an ambulatory setting with resources to determine severity of illness. When possible, send the patient to a setting where she can be isolated. Clinical assessment for respiratory compromise includes physical examination and tests such as pulse oximetry, chest X-ray, or ABG as clinically indicated. Pregnant women (with abdominal shielding) should not be excluded from chest computed tomography (CT) if clinically recommended.

9. Moderate-risk patients with no respiratory compromise or complications and able to follow up with care are managed as low risk as in number 5.

10. Admit the moderate-risk patients with respiratory compromise or complications for further evaluation and treatment. Review hospital or health system guidance on infection control measures to minimize the patient and provider exposure.

The Royal College of Obstetricians and Gynecologists recommended the following points of caring of pregnant women with suspected or confirmed COVID-19 infection:

- Any pregnant woman with suspected COVID-19 infection should be managed as confirmed case till confirmation or exclusion of diagnosis occurs.
- Obstetricians should be oriented with and follow the local protocols for dealing with possible COVID-19

cases. Stabilization of the general condition is the first line of treatment taking into consideration that young healthy women can resist deterioration in respiratory functions with maintenance of adequate oxygen saturation for a certain time and that can be followed by sudden decompensation.

- Obstetrician should consider both the absolute values and the degree of changes in the needed observations and investigations as the absolute oxygen saturation and its changes.
- Warning signs for decompensations include oxygen requirement increase above 40%, respiratory rate higher than 30 breaths/min on oxygen therapy, decreased urine output, or occurrence of drowsiness even with normal oxygen saturation.
- Oxygen saturation should be kept above 94% all the times.
- Intravenous fluid management needs monitoring fluid balancer charts especially in nonmild cases and to maintain neutral fluid balance during labor 250–500 mL of bolus fluids and check for volume overload before administration of further fluids.
- Antibiotics can be started early at first presentation of the disease to control any coexisting or secondary bacterial infections.
- An MDT includes expert obstetrician, anesthesiologist, neonatologist, intensivist familiar with obstetric care, midwife in charge, and neonatal nurse in charge working together with an infection control team.
- All pregnant women should be assessed for risk of thromboembolism and receive prophylactic dose of anticoagulants or therapeutic dose if there is high suspicion of venous thromboembolism (VTE).
- Women with thrombocytopenia (in some cases of severe COVID-19 infection) should stop any thromboprophylaxis and antiplatelet drugs as aspirin and consider using mechanical aids (e.g., intermittent calf compressors) and referred to an expert hematologist 96.
- The need for emergency termination of pregnancy by cesarean delivery (CD) or induction of labor to support maternal resuscitation or to save deteriorating fetal condition is individualized, and the decision needs discussion with all members of the MDT.
- Maternal stabilization before any intervention must be achieved.
- Fetal indications for termination of pregnancy follow the same guidelines of routine obstetrics if the maternal condition is stable (World Health Organization, 2020b; Royal Colleague of Obstetrics and Gynecology, 2020).

The WHO advised management of COVID-19 cases according to severity.

Mild cases are advised to have self-isolation at home or at health or community facility. The decision is individualized according to the local COVID-19 care pathway, the presenting manifestations, the need for supportive care, the degree of risk for development of severe form, and the available conditions at home. Symptomatic treatments such as antipyretics for fever, analgesics for pain (nonsteroidal antiinflammatory can be used), proper nutrition, and hydration are recommended (World Health Organization, 2020c).

No need for widespread use of antibiotics in mild cases. Widespread use of antibiotics increases bacterial resistance with subsequent increase in morbidity and mortality during and after the pandemic (Llor and Bjerrum, 2014; Goossens et al., 2005).

Precautions taken during home isolation include the following:
1. Mild cases should be warned about manifestations of complications and advised to seek immediate medical advice if they encounter any of them.
2. Cases with high risk should be monitored very closely.

Counsel patients with mild COVID-19 about signs and symptoms of complications that should prompt urgent care.

Remark

Patients with risk factors for severe illness should be monitored closely, given the possible risk of deterioration. If they develop any worsening symptoms (such as light headedness, difficulty breathing, chest pain, dehydration, and so on), they should seek urgent care through the established COVID-19 care pathway. Caregivers of children with mild COVID-19 should monitor for signs and symptoms of clinical deterioration requiring urgent reevaluation. These include difficulty breathing/fast or shallow breathing (for infants: grunting, inability to breastfeed), blue lips or face, chest pain or pressure, new confusion, inability to awaken/not interacting when awake, and inability to drink or keep down any liquids. Consider alternative delivery platforms such as home-based, phone, telemedicine, or community outreach teams to assist with monitoring (Greenhalgh et al., 2020).

Moderate cases with pneumonia should be isolated according to local care pathway at either home, community, or health facility and depending on the clinical presentation, presence of risk factors for development of severe disease, the need for supportive care, and the conditions at home. They should receive the same treatment as mild cases. Antibiotics are given only when bacterial infection is suspected (only 8% of hospitalized cases with COVID-19 have concomitant bacterial or fungal infections) (Rawson et al., 2020). Antibiotics not broad-spectrum ones can be used in elderly cases and children under 5 years old (World Health Organization, 2019).

Home-isolated cases should be monitored for symptoms and signs of severe disease, and they are instructed to seek urgent care through the established care pathway. They should have continuous follow-up through phone calls, telemedicine, or community outreach teams. Pulse oximeters follow-up at home is not recommended. Hospitalized moderate cases are monitored for vital signs, oxygen saturation, and early warning scores as NEWS2 and PEWS (Duncan et al., 2006).

Severe cases should be cared in areas of health facilities equipped with oxygen delivery systems with disposable oxygen delivery interfaces (such as nasal cannula, masks with reservoir bags and venturi mask, and oxygen saturation monitoring through pulse oximeters). Any patient with emergency manifestations (such as respiratory manifestations including dyspnea, apnea, or severe respiratory distress; vascular manifestations including central cyanosis or shock; neurological manifestations including coma and/or convulsions) or with oxygen saturation below 90% should receive oxygen therapy immediately till reaching 94% or more during resuscitation and 92%–95% or more in pregnant women after stabilization. That can be achieved through adjustment of oxygen flow rates using the appropriate delivery devices (nasal cannula, venturi mask, and face mask with reservoir bag are used to deliver rates up to 5, 6–10, and 10–15 L/min, respectively) (World Health Organization, 2016, 2018).

Positioning as high supported sittings may facilitate breathing, decrease energy expenditure, and optimize oxygenation. Prone position in awake individuals with spontaneous breathing is under investigation to prove its effect in improving oxygenation and ventilation/perfusion ratio. However, that could be difficult to be applied in pregnant women especially those in late trimester (Thomas et al., 2020).

Severe cases with excess respiratory secretions, retained secretions, and/or inadequate cough should have their airway cleared. That can be achieved through gravity-assisted drainage and active breathing cycles with avoidance of devices as inspiratory positive pressure breathing and mechanical insufflation–exsufflation as much as possible(Thomas et al., 2020).

Cases with severe COVID-19 infection should be closely monitored for vital signs, oxygen saturation, development of complications such as major organ failure (including respiratory, liver, kidney, and heart failures), and venous and arterial thromboembolism.

Cautions should be taken on administration of intravenous fluids, as hypovolemia impairs tissue perfusion and overload may worsen oxygenation (Schultz et al., 2017).

Critical cases with acute respiratory distress syndrome (ARDS) are classified to those with mild, moderate, and severe ARDS (Rhodes et al., 2017; Rimensberger et al., 2015).

Mild ARDS cases can be treated with noninvasive oxygenation methods including high-flow nasal oxygen (HFNO), continuous positive airway pressure (CPAP), or bilevel positive airway pressure (BiPAP). That cannot be applied on cases with respiratory failure, major organ failure, disturbed conscious level, or hemodynamic instability. Cases on noninvasive ventilation should be closely monitored with availabilities of immediate intubations if the patient deteriorated or not improved within 1 h of its administration. Airborne precautions should be fulfilled during use of these methods till evaluation of its safety for its potential for aerosolization. The use of HFNO may decrease the need for intubation when compared with the standard oxygen therapy (Rochwerg et al., 2017) and is considered safe in cases with nonworsening hypercapnia (Luo et al., 2015; Lee et al., 2018; Rochwerg et al., 2017).

The use of noninvasive ventilation (NIV) in hypoxemic respiratory failure (except in pandemic viral illness—studied in SARS and influenza (Rochwerg et al., 2017)) is not recommended as it carries the risk of delayed intubation, large tidal volumes, and injurious transpulmonary pressures (Arabi et al., 2014).

Cases with severe ARDS should be intubated and receive invasive mechanical ventilation as respiratory failure in ARDS is caused by intrapulmonary ventilation–perfusion mismatch or shunt (Rhodes et al., 2017).

Pregnant women with ARDS may desaturate rapidly during intubation so preoxygenation with 100% FiO_2 through a face mask with reservoir bag for 5 min is recommended before intubation. A rapid-sequence intubation by an expert provider using airborne precautions is recommended after proper airway assessment (Cheung et al., 2020; Peng et al. 2020; Detsky et al., 2019).

Mechanical ventilation with tidal volumes of 4–8 mL/kg and inspiratory plateau pressure less than 30 cm H_2O are recommended (Rhodes et al., 2017). In nonpregnant cases, prone position is recommended for 12–16 h daily (Guérin et al., 2013; Messerole et al., 2002). The evidence of prone position is little but can be considered in early pregnancy, while lateral decubital position is advised in late trimester.

Titration of positive end-expiratory pressure (PEEP) is individualized by weighting its benefits of improving alveolar recruitment and decreasing atelectrauma against its risks of lung injury and high pulmonary vascular resistance, and the patient should be monitored during titration for its effects. Tables are used for guiding PEEP titration based on the FiO_2 needed to maintain SpO_2 (NIH NHLBI ARDS Clinical Network, 2008).

The use of continuous infusion of neuromuscular blockers should not be routine in cases with moderate and severe ARDS (Papazian et al., 2010).

Disconnection from the ventilator is risky as it causes loss of PEEP and atelectasis with increased risk of infection of the care providers so it is better avoided, and if needed, the use of in-line catheters with clamping of the tube is recommended along with avoidance of manual hyperinflation (Thomas et al., 2020).

Cases with refractory hypoxemia can be transferred to extracorporeal membrane oxygenation if available.

Critically ill patients should receive treatment to prevent complications as thromboembolism. Both arterial and venous thromboembolic complications are reported in severe cases of COVID-19 (Siddamreddy et al., 2020; Violi et al., 2020; Wichmann et al., 2020). Anticoagulant drugs such as low-molecular-weight heparin are used according to local and international standards (NICE Guideline, 2018). Mechanical prophylaxis such as intermittent pneumatic compression device is used if anticoagulant drugs are contraindicated.

Prevention of Other Complications Is Summarized by the WHO in Table 4.1

The RCOG suggested certain criteria for hospital admission for pregnant women with COVID-19 (López et al., 2020). These criteria include the following:

- Persistent fever >38°C despite using paracetamol
- Chest X-ray demonstrating pneumonia
- Pregnant women with other comorbidities such as chronic hypertension, obstructive pulmonary disease, pregestational diabetes, immunosuppression, organ transplant recipients, HIV infection with <350 CD4+ cells, or patients who receive corticosteroids equivalent to >20 mg of prednisone for >2 weeks, use of immunosuppressive drugs, neutropenia, and so on must be carefully evaluated by an infectious disease specialist
- CURB severity scale with a total score >0 (each item gives a score of one point)
  - C: Acute confusion
  - U: Urea >19 mg/dL

TABLE 4.1
Prevention of Other Complications are Summarized by the World Health Organization.

Anticipated Outcome	Interventions
Reduce days of invasive mechanical ventilation	• Use weaning protocols that include daily assessment for readiness to breathe spontaneously • Minimize continuous or intermittent sedation, targeting specific titration endpoints (light sedation unless contraindicated) or with daily interruption of continuous sedative infusions • Early mobilization • Implementation of the above as a bundle of care (may also reduce delirium), such as the awakening and breathing coordination, delirium assessment/management, and Early mobility
Reduce incidence of ventilator-associated pneumonia	• Oral intubation is preferable to nasal intubation in adolescents and adults • Keep patient in semirecumbent position (head of bed elevation 30–45 degrees) • Use a closed suctioning system; periodically drain and discard condensate in tubing • Use a new ventilator circuit for each patient; once patient is ventilated, change circuit if it is soiled or damaged, but not routinely • Change heat moisture exchanger when it malfunctions, when soiled, or every 5–7 days
Reduce incidence of catheter-related bloodstream infection	• Use a checklist with completion verified by a real-time observer as a reminder of each step needed for sterile insertion and as a daily reminder to remove catheter if no longer needed
Reduce incidence of pressure ulcers	• Turn patient every 2 h
Reduce incidence of stress ulcers and gastrointestinal bleeding	• Give early enteral nutrition (within 24–48 h of admission) • Administer histamine-2 receptor blockers or proton-pump inhibitors in patients with risk factors for GI bleeding. Risk factors for GI bleeding include mechanical ventilation for ≥48 h, coagulopathy, renal replacement therapy, liver disease, multiple comorbidities, and higher organ failure score
Reduce the development of antimicrobial resistance Reduce the development of adverse drug effects Promote appropriate antimicrobial prescribing and use during the COVID-19 pandemic	• Utilize deescalation protocols as soon as the patient is clinically stable and there is no evidence of bacterial infection • Expose the patient to empiric antimicrobial therapy for the shortest time possible, to prevent nephrotoxicity, cardiac, and other side-effects from unnecessary antimicrobial use • Do not prescribe antibiotics to suspected or confirmed COVID-19 patients with low suspicion of a bacterial infection, to avoid more short-term side effects of antibiotics in patients and negative long-term consequences of increased antimicrobial resistance

Clinical management of Covid-19 based on WHO recommendations. https://www.who.int/publications/i/item/clinical-management-of-covid-19.

- R (respiratory rate): ≥30 bpm
- B (blood pressure): SBP (systolic) ≤90 mm Hg or DBP (diastolic) ≤60 mm Hg
- Intensive care unit admission criteria
  The admission criteria suggested for intensive care unit (López et al., 2020) (adapted from the American Thoracic Society and Infectious Diseases Society of America) include the presence of one major criterion or three minor criteria of the following:
- Major criteria
  - The need for invasive mechanical ventilation
  - Shock with the need for vasopressors
- Minor criteria
  - Respiratory rate ≥30 bpm
  - PaO_2 (partial pressure of oxygen)/FiO_2 (fraction of inspired oxygen) ratio < 250
  - Multilobar infiltrates
  - Confusion/disorientation
  - Uremia (blood urea nitrogen > 20 mg/dL
  - Leukopenia <4000 cells/mm^3
  - Thrombocytopenia <100,000 platelets/mm^3
  - Hypothermia/central temperature <36°C
  - Hypotension in need of aggressive fluid resuscitation

A suggested delivery room processing is summarized in Fig. 4.1.

A Summary of Guidelines suggested by diffrent organizations for Intrapartum Care of Pregnant Patients During the COVID-19 Pandemic is shown in Table 4.2.

BEFORE HOSPITAL ADMISSION

Full explanation of the process of delivery and care all through labor and postpartum time should be discussed with the parturient woman. She needs to realize that symptomatic pregnant women with COVID-19 are at increased risk of more severe illness compared with nonpregnant peers (Panagiotakopoulos et al., 2020). The CDC now includes pregnant women in its "increased risk" category for COVID-19 illness. Although the absolute risk for severe COVID-19 is low, there is an increased risk of ICU admission, need for mechanical ventilation, and death reported in pregnant women with symptomatic COVID-19 infection, when compared with symptomatic nonpregnant women (Zambrano et al., 2020). Pregnant patients with comorbidities such as obesity and gestational diabetes may be at an even higher risk of severe illness consistent with the general population with similar comorbidities (Panagiotakopoulos et al., 2020; Knight

et al., 2020; Zambrano et al., 2020). Similar to the general population, Black and Hispanic individuals who are pregnant appear to have disproportionate SARS CoV-2 infection and death rates (Zambrano et al., 2020). While data indicate an increased risk of severe illness and maternal death, data also indicate that the majority of pregnant individuals diagnosed with COVID-19 experience relatively mild symptoms; however, symptoms lasting up to 8 weeks have been reported (Yee et al., 2020).

TIMING OF DELIVERY

Timing of delivery, in most cases, should not be affected by maternal COVID-19 infection. For women with suspected or confirmed COVID-19 early in pregnancy who recover, no alteration to the usual timing of delivery is indicated. For women with suspected or confirmed COVID-19 in the third trimester who recover, it is reasonable to attempt to postpone delivery (if no other medical indications arise) until a negative testing result is obtained or quarantine status is lifted in an attempt to avoid transmission to the neonate. In general, COVID-19 infection itself is not an indication for delivery.

Inductions of labor delivery and CD should continue to be performed as indicated. Decisions on

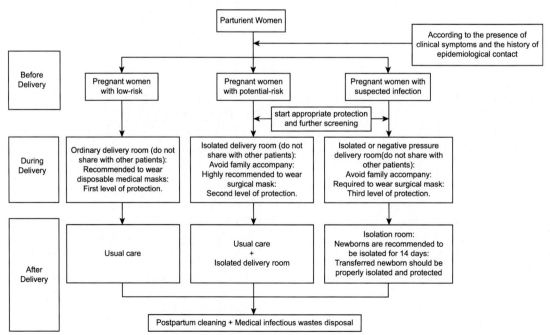

FIG. 4.1 The processing flow of delivery during the COVID-19 outbreak based on the risk level of pregnancy infection. A suggested delivery room processing (Qi et al., 2020).

TABLE 4.2

Summary of Guidelines for Intrapartum Care of Pregnant Patients During the COVID-19 Pandemic.

Title	Professional Society	ISUOG	CNGOF	ACOG	SMFM	RCOG	WHO	CDC	CatSalut	ISS/SIEOG
Intrapartum care	Predelivery Preparation	Social distancing	Social distancing	Notification of the teams of obstetric, pediatric, and anesthesia	Social distancing and stop working 2 weeks before expected delivery time. Screen mother and partner 24 h before admission through phone call.	Social distancing. Minimum staffing. Screen mother and partner at maternity unit. Self-isolation of symptomatic artners within 7 days before admission and prevented from hospital entry. Suspected and confirmed COVID19 mothers better stay at home during latent phase of labor; and admitted when in active labor to dry isolated room with running simulations for elective/emergency procedures.	In suspected and confirmed COVID19 cases, health workers should take all appropriate precautions including hand hygiene and protective clothing such as gloves, gown, and medical mask.	N/A	Screening of symptomatic women at maternity unit. Minimum ataffing. Only one partner allowed.	Screen symptomatic women at maternity unit. Asymptomatic: Managed routinely. Stable symptomatic: Admitted to hospital. Unstable symptomatic: referred to hospitals with ICU.
	Delivery time	Routine obstetric indications. In critically ill cases, early delivery is considered.	Routine obstetric indications if patient recovered from infection in early pregnancy. In critically ill cases, early delivery is considered.	Routine obstetric indications if patient recovered from infection in early pregnancy. Postpone delivery if not contraindicated and if patient recovered from infection in late pregnancy till testing negative or quarantine lifted to minimize neonatal transmission.	Routine obstetric indications. Induction of labor is not contraindicated except in limited beds. Delivery should be considered in terms of COVID-19 patients as symptoms peak after 7–14 days after its onset.	Routine obstetric indications.	N/A	N/A	Use PPE in all cases. Continue fetal electronic monitoring.	Routine obstetric indications.

Continued

TABLE 4.2
Summary of Guidelines for Intrapartum Care of Pregnant Patients During the COVID-19 Pandemic.—cont'd

Title	Professional Society	ISUOG	CNGOF	ACOG	SMFM	RCOG	WHO	CDC	CatSalut	ISS/SIEOG
Delivered location		Designated negative-pressure isolation room.	Designated isolation room for suspected or confirmed cases of COVID-19.	N/A	Designated delivery and operating rooms.	Designated isolation room for suspected or confirmed cases.	N/A	N/A	Designated isolation room for suspected or confirmed cases of COVID-19. Designated negative-pressure isolation room for CS.	Designated isolated room for suspected or confirmed cases.
Mode of delivery		As per routine obstetric indications. COVID-19 is *not* an indication for CS. Expedite CS delivery if fetal distress or maternal deterioration occurs. Avoid water birth.	As per Rroutine obstetric indications. COVID-19 is *not* an indication for CS. Expedite CS delivery if fetal distress or maternal deterioration occurs.	As per routine obstetric indications. No specific recommendations for CS. Operative vaginal delivery is not indicated for suspected or confirmed cases alone but can be used as routinely indicated.	As per routine obstetric indications. COVID-19 is *not* an indication for CS.	As per routine obstetric indications except in demanding maternal respiratory condition. Avoid water birth.	As clinically indicated.	N/A	As per routine obstetric indications. COVID-19 is *not* an indication for CS.	As per routine obstetric indications. COVID-19 is *not* an indication for CS.
Support person		Limit visitors, no clear number specified.	**No visitors**	One asymptomatic support person allowed.	One consistent support person allowed. No attendance below 16–18 years of age.	One consistent asymptomatic support person allowed with restriction to the patient's bedside.	N/A	N/A	One of the relatives is encouraged during all the labor and delivery.	Single accompanying asymptomatic person.
Obstetric analgesia and anesthesia		Both regional and general anesthesia are considered.	Both regional and general anesthesia are considered.	N/A	Avoid nitrous oxide	NO evidence against regional or GA. Epidural analgesia is recommended in suspected or confirmed cases to minimize the need for GA in urgent need for delivery.	N/A	N/A	Epidural analgesia is recommended to women with suspected or confirmed COVID-19 to minimize the need for GA if urgent delivery is needed.	Both regional and general anesthesia are considered.
Second stage of Labor		Consider shortening with operative delivery to	N/A	N/A	Do not delay pushing. Considered aerosolizing. N95 should be	Consider shortening with operative vaginal delivery in symptomatic	N/A	N/A	N/A	N/A

		minimize erosolization and maternal respiratory effort	worn by HCW and patients.	women who become exhausted or hypoxic.				
Third stage of labor	N/A	N/A	Management to reduce blood loss (national blood shortage)	N/A	N/A	N/A	Active management in all cases.	Per routine.
Oxygen supplementation	N/A	N/A	Consider aerosolizing. HCW must wear appropriate PPE. Do not use O_2 for intrauterine resuscitation.	Measurements (in addition to routine maternal—fetal observations) for women with suspected/confirmed COVID-19. Aim to keep O_2 sat >94%.	N/A	N/A	N/A	N/A
Umbilical cord Clamping	Avoid delayed cord clamping in confirmed and suspected cases	No recommendations against delayed clamping of umbilical cord.	Avoid delayed cord clamping	Delayed cord clamping is still recommended in the absence of contraindications.	N/A	N/A	Delayed cord clamping is still recommended in the absence of contraindications.	General rule.
PPE use	N/A	N/A	Asymptomatic or negative patients. Patient and provider wear surgical mask. Aerosolizing procedures—N95. for patient and N95, gown, gloves, and face shield for provider.	Level of PPE should be based on the risk of requiring GA. Aerosolizing procedures—use FFP3 mask.	N/A	Same as general population.	N/A	Symptomatic with stable or unstable condition: mothers, medical staff, and accompanying person must wear all protection devices. Masks should be FFP2/FFP3 type.
Elective cesarean delivery/ induction of labor (IOL)	N/A	N/A	No Ccontraindication to IOL unless there is limited beds.	For suspected/confirmed cases, consider delay of elective CD or IOL if safely feasible.	N/A	N/A	N/A	N/A

CD, cesarean delivery; CS, cesarean section; GA, general anesthesia; (Narang et al., 2020).

how to schedule these procedures in the time of the COVID-19 pandemic are best made at the local facility and systems level, with input from obstetric care professionals and based on healthcare personnel availability, geography, access to readily available local resources, and coordination with other centers (American College of Obstetricians and Gynecologists, 2020).

- Patients undergoing planned induction or CD can have screening and laboratory testing 24–72 h before the planned procedure in an attempt to have results available before admission.
- Symptomatic women should be evaluated to determine whether it is obstetrically feasible to reschedule induction or CD until results of COVID-19 testing are available. This is an individualized decision and requires assessing patient-specific risks of continuing the pregnancy in the setting of an unknown, positive, or negative test result. In particular, if the test result is positive, patients may benefit from delivery since they may become more severely ill over time (symptoms are often more severe in the second week of the illness).
- In asymptomatic patients with a positive SARS-CoV-2 test result, induction of labor delivery or CD performed for appropriate medical/obstetric indications (including week 39 induction of labor) should not be postponed or rescheduled. For those who require cervical ripening, outpatient mechanical ripening with a balloon catheter is an option. For inpatient cervical ripening, using two methods (e.g., mechanical and misoprostol or mechanical and oxytocin) may decrease the time from induction to delivery, compared with using one agent only (UpToDate, Vincenzo Berghella, 2020).

ROUTE OF DELIVERY

- **In asymptomatic or nonsevere** COVID-19 **infection,** the planned route of delivery in these patients should not be altered (American College of Obstetricians and Gynecologists, 2020). CD does not appear to reduce the already low risk for neonatal infection (Walker et al., 2020). Even if vertical transmission at delivery is confirmed as additional data are reported, this would not be an indication for CD since it would increase maternal risk and would be unlikely to improve newborn outcome. Reports of COVID-19 infection in the neonate have generally described mild disease (UpToDate.Vincenzo Berghella, 2020).
- In patients with severe or critical COVID-19, CD is performed for standard obstetric indications including concerns of acute decompensation of intubated and critically ill mothers. Induction can be performed safely in intubated patients (Slayton-Milam et al., 2020; Liu et al., 2019), and in one study of 37 CDs and 41 vaginal deliveries in COVID-19 patients, cesarean birth was associated with an increased risk for clinical deterioration (8/37 [22%] vs. 2/41 [5%]), which remained after adjustment for confounders (adjusted odds ratio 13, 95% CI: 1.5–122.0) (Martínez-Perez et al., 2020). Confounders included maternal age >35 years, body mass index >30 kg/m, maternal comorbidities, need for oxygen supplementation at admission, abnormal chest radiograph at admission, nulliparity, smoking, and preterm delivery. The issue of possible harm from CD should not preclude indicated CD, but it requires further evaluation as the small number of events and bias in case selection for route of delivery could account for the findings (UpToDate.Vincenzo Berghella, 2020).

- Long induction is better avoided in intubated patients and those laboring in an operating room or intensive care unit, because of the specialized equipment and personnel in these sites. CD is often performed in such patients.
- Regardless of the type or site of delivery (e.g., labor and delivery unit, main operating room, intensive care unit), a multidisciplinary care team should be present (e.g., intensivists, maternal–fetal medicine, neonatology, nursing support from obstetrics, pediatrics, and medical disciplines) to care for the severely/critically ill mother and the potentially heavily sedated newborn.

On admission, primary screening of all women (with first-level protective equipment applied) through checking of temperature, the fetal heart rate, and evaluation of symptoms that may suggest COVID-19 infection, such as fever, respiratory symptoms (cough, dyspnea), gastrointestinal symptoms (vomiting, diarrhea), and other symptoms (loss of taste and smell) before allowing women to sit in the maternity waiting area. Ask about COVID-19 exposure through contact with a suspected or confirmed case or traveling to epidemic area within the past 2 weeks. Any positive history of the aforementioned indicates "potential risk" status (Qi et al., 2020).

Pregnant women with potential risk and/or suspected infection should go through further screening (with second-level protective equipment applied) for other respiratory pathogens (such as influenza and parainfluenza virus, *Mycoplasma pneumoniae*, and *Chlamydia pneumoniae*), routine blood tests, and markers of inflammation such as C-reactive protein. Testing for the new coronavirus nucleic acid should be performed. In the presence of any chest manifestations necessitating doing chest CT, it should be done after obtaining an informed written consent. Obstetric management should not be delayed waiting for COVID-19 testing or the results of the performed tests (Qi et al., 2020).

Delivery Room Management

- Pregnant women suspected to be COVID-19 positive should be immediately transferred to an isolated delivery room (avoiding contact with other patients) or negative pressure delivery room and be required to wear surgical mask (National Health Commission of the People's Republic of China, 2020).
- Family members accompanying the parturient women must not be permitted in suspected or confirmed cases.
- Management of delivery should be carried through specific experienced senior medical specialists supplied with third-level protective equipment to avoid cross-infection.
- Low-risk pregnant women without any history of epidemiological exposure or clinical symptoms are transferred to a separate delivery room for delivery and should wear disposable medical masks. Family members with no history of epidemiological contact and clinical symptoms within the past 2 weeks are allowed to attend the childbirth after wearing disposable medical masks (National Health Commission of the People's Republic of China, 2020).
- Person-to-person contact and time in the labor unit and hospital should be limited, as is safely feasible (UpToDate.Vincenzo Berghella, 2020).
- Continuous electronic fetal monitoring in labor is recommended for all women with suspected or confirmed COVID-19 infection as fetal compromise is relatively common in these women (Royal Colleague of Obstetrics and Gynecology, 2020).
- Unnecessary obstetric interventions should be avoided, and cautions should be taken in procedures as episiotomy and instrumental delivery.
- Currently, we do not recommend water deliveries for pregnant women with suspected infection.
- Epidural analgesia and spinal anesthesia are not contraindicated. On the contrary, epidural analgesia should be recommended to pregnant women with suspected or confirmed COVID-19 infection before or in early labor to reduce cardiopulmonary stress from pain and anxiety and, in turn, the chance of viral dissemination (UpToDate.Vincenzo Berghella, 2020) and minimize the need for general anesthesia

(GA) in emergency situations (Royal Colleague of Obstetrics and Gynecology, 2020).

- Delayed cord clamping is still appropriate in the setting of appropriate clinician PPE. Although some experts have recommended against delayed cord clamping, the evidence is based on opinion; a single report later confirmed COVID-19 transmission most likely occurred from the obstetric care clinician to the neonate. Current evidence-based guidelines for delayed cord clamping should continue to be followed until emerging evidence suggests a change in practice.
- Operative vaginal delivery is not indicated for suspected or confirmed COVID-19 alone. Practitioners should follow usual clinical indications for operative vaginal delivery, in the setting of appropriate PPE.
- Labor units may consider suspending use of nitrous oxide for those with suspected or confirmed COVID-19, as there are insufficient data regarding the physiologic safety of its inhalation.
- Oxygen should continue to be considered if maternal hypoxia is noted. Based on limited data, high-flow oxygen use is not considered an aerosol-generating procedure (CDC). Still, there is insufficient evidence about the cleaning and filtering when using oxygen, so it is better to suspend routine use of intrapartum oxygen for indications where benefits of use are not well established (e.g., category II and III fetal heart rate tracings).
- In nonintubated patients with respiratory compromise due to COVID-19, magnesium sulfate for maternal seizure prophylaxis and/or neonatal neuroprotection should be monitored very carefully since high magnesium levels (10−13 mEq/L [12−16 mg/dL or 5.0−6.5 mmol/L]) can cause respiratory paralysis. Consultation with maternal−fetal medicine and pulmonary/critical care specialists is advised (American College of Obstetricians and Gynecologists, 2020).
- The third stage of labor is managed as in non-COVID-19 women, and women who developed postpartum hemorrhage are managed according to standard protocols. However, some clinicians avoid tranexamic acid in COVID-19 patients because its antifibrinolytic properties may increase the risk for thrombosis in those with a hypercoagulable state, such as those with severe or critical disease, and alternative strategies for control of bleeding are available (Ogawa and Asakura, 2020). Some authors suggest avoiding methylergometrine because it has been associated with rare cases of respiratory failure

and severe vasoconstriction in severely ill patients (Donders et al., 2020).

- In patients who develop intrapartum fever, differential diagnosis should include COVID-19 infection especially in the presence of suspected symptoms or hypooxygenation. These women should be tested (or retested) for the virus, along with evaluation for other common causes of intrapartum fever (e.g., chorioamnionitis, epidural fever) (Breslin et al., 2020).
- SARS-CoV-2 has rarely been detected in vaginal secretions or amniotic fluid, so rupture of fetal membranes and internal fetal heart rate monitoring may be performed for usual indications, but data are limited (Schwartz, 2020). It should be noted that labor, and particularly pushing, often causes loss of stools, which can contain the virus and spread the infection (Zhang et al., 2020; Wang et al., 2020).

Management During Cesarean Delivery

- Suspected or confirmed COVID-19 infection is not an indication for CD, unless the patient's respiratory condition requires urgent termination of pregnancy. Otherwise, CD should be done for routine indications.
- MDT consultation regarding premature delivery in suspected or confirmed cases. The team should include expert obstetricians, anesthesiologists, neonatologists, intensivists familiar with obstetric care, and infectious disease physicians.
- CD should be performed in a negative pressure isolation operating room (third-level protective equipment should be applied). The choice of anesthetic mode is determined by the anesthetist, based on the patient's respiratory function.
- For pregnant women with potential infection (potential risk), CD can be done in isolated operating room with second-level protective equipment if properly protected. That could be also applied to low-risk infection women with the use of first-level protective equipment (Qi et al., 2020).

ANESTHESIA IN EMERGENCY CESAREANS FOR PREGNANT WOMEN WITH CORONAVIRUS DISEASE 2019

- A systematic strategy and adequate preparation are mandatory in COVID-19 cases who need to have emergency CD (30-min decision-to-incision interval) to minimize cross-contaminations. The possibilities of suspected/confirmed COVID-19 patients requiring imminent operative deliveries have to be communicated to the operating room team so that they could be conducted in negative-pressure operating rooms (Ti et al., 2020).
- Desaturation COVID-19 parturient (oxygen saturation ≤93%) should receive GA in case of emergency CD. GA is done with rapid sequence induction and tracheal intubation with a cuffed tube. The airway team should wear full PPE and powered air-purifying respirator (PAPR). The use of invasive monitoring (as intraarterial blood pressure and central venous pressure) should be done with cautions in presence of COVID-19 complications as renal failure and disseminated intravascular coagulation (Ashokka et al., 2020).
- The use of regional anesthesia (epidural or spinal) is recommended over GA in COVID-19 parturient when the oxygen saturation is adequate (≥94%) as it minimizes aerosolization and cross-infection during airway management (Royal Colleague of Obstetrics and Gynecology, 2020).
- If the parturient has epidural catheter, administration of a top-up with potent local anesthetics (as 10−15 mL of 1.5% lignocaine, alkalinized with 8.4% sodium bicarbonate) can achieves a rapid onset of 3.5 min surgical anesthetic plane. Rapid sequence spinal anesthesia (Kinsella et al., 2010) is described for emergency CD, wherein patients are placed in a left lateral position with supplemental oxygen, and a single-shot subarachnoid blockade is administered by the most experienced prescrubbed anesthetist. The time required for surgical readiness is comparable with that for GA with better neonatal outcomes (Lim et al., 2018).
- The same precautions used with intubation should be done with extubation after GA (Wax and Christian, 2020). Patients tend be more agitated during emergence from anesthesia and extubation, which could result in increased likelihood of viral dissemination from coughing as compared with the intubation process (Chan et al., 2018). All operating room personnel should wear full PPE until patients are safely extubated and transferred out of the operating room (Wax and Christian, 2020).
- The disposition for patients with COVID-19 after unplanned CD should be decided at the earliest instance. Early transfer to the postanesthesia care unit (PACU) may result in spreading of infection to other postoperative patients. Suspected and confirmed COVID-19 cases may be allowed to recover within the operating rooms where CD was done, if possible. Then the patients can be transferred directly to isolation wards after recovery (Ashokka et al., 2020).

Postpartum Care (UpToDate.Vincenzo Berghella, 2020)

1. Infection control measures:
   a. Mothers with suspected or confirmed SARS-CoV-2 infection should be isolated from other healthy mothers and cared for according to standard infection control guidelines.
   b. After the mother was transferred to the ward, routine cleaning should be undertaken. The surfaces of the equipment (including the obstetric table, ultrasound machine, and neonatal warm bed) in the isolation delivery room and the negative-pressure delivery room need to be wiped and disinfected immediately, preferably with 1000 mg/L chlorine-containing disinfectant; 75% ethanol can be used for the noncorrosion resistance instruments (Royal Colleague of Obstetrics and Gynecology, 2020). Spraying is not a recommended method of disinfecting the equipment, as this can affect the components. Dedicated cleaning tools are required to avoid cross-contamination. The inspection room should be disinfected with ultraviolet light for at least 60 min each time, once or twice a day, with at least 30 min ventilation after irradiation. The ultrasound probe should be protected with a dark cloth during the irradiation. The room should be vacated when ultraviolet lamps are used (Qi et al., 2020).
   c. Medical waste disposal: Protective supplies used by medical personnel and all patient waste should be regarded as infectious medical waste, which requires double-layer sealing, clear labeling, and airtight transport (National Health Commission of the People's Republic of China, 2020). If testing of the placenta and/or amniotic fluid is required, strict sampling and sealing should be carried out to avoid contamination of the surface of the container and the spread of infection. The surface of the container should be disinfected before sample inspection to further avoid infection of any personnel (Qi et al., 2020).

2. Maternal monitoring:
   a. For asymptomatic patients with known or suspected COVID-19, postpartum maternal monitoring is routine (including postpartum vital signs, uterine contractions, maternal mental health, and other conditions of the mother should be monitored, and attention paid to the prevention of postpartum hemorrhage and thrombosis).
   b. For mild COVID-19 patients, monitoring vital signs and fluid balance (intake and output) every 4 h for 24 h after vaginal delivery and 48 h after CD is advised.
   c. For moderate COVID-19 patients, continuous pulse oximetry monitoring for the first 24 h or until clinical improvement occurs, whichever takes longer. Follow-up laboratory investigations and chest imaging are individualized according to the course of the disease.
   d. For patients with severe or critical illness, very close maternal monitoring and care on the labor and delivery unit or intensive care unit are indicated.

3. Venous thromboembolism prophylaxis
   a. Prophylactic-dose anticoagulation is recommended for pregnant/postpartum patients hospitalized for severe COVID-19, if not contraindicated.
   b. Anticoagulation is better discontinued when the patient is discharged to home.
   c. Patients with COVID-19 who do not warrant hospitalization for the infection or who are asymptomatic or mildly symptomatic and hospitalized for reasons other than COVID-19 (e.g., labor and delivery) do not require anticoagulation, unless they have other thrombotic risk factors, such as prior VTE or, in some cases, CD.
   d. Either low-molecular-weight heparin or unfractionated heparin is acceptable, and both are compatible with breastfeeding.

4. **Postpartum analgesia**: Pain management is routine. Acetaminophen is the preferred analgesic agent; however, nonsteroidal antiinflammatory drugs can be used when clinically indicated.

5. Postpartum fever
   a. The differential diagnosis of postpartum fever in patients with COVID-19 includes the infection itself as well as postpartum endometritis, surgical site infection, breast inflammation or infection, influenza, pyelonephritis, other viral or bacterial respiratory infections, and, rarely, pseudomembranous colitis due to *Clostridioides difficile*. The combination of composite symptoms, physical examination, and laboratory tests can usually distinguish among these disorders.
   b. Acetaminophen is the preferred antipyretic agent.

6. **Postpartum patients with new onset of symptoms of COVID-19**—In newly symptomatic patients who previously tested negative for SARS-CoV-2, retesting is appropriate as part of the evaluation of fever or other potential manifestations of COVID-19.

Newborn Evaluation

- The infants of mothers with suspected or confirmed COVID-19 are considered persons under investigation, and they should be tested for SARS-CoV-2 RNA by reverse-transcription polymerase chain reaction (RT-PCR) (Center for Disease Control and Prevention, 2020b).
- The American Academy of Pediatrics (AAP) suggests for diagnosis of newborn infection to test at approximately 24 h of age and, if negative, again at approximately 48 h of age since some infants have had a negative test at 24 h only to have a positive test at a later time. If a healthy, asymptomatic newborn will be discharged prior to 48 h of age, a single test at 24–48 h of age can be performed and to obtain either a single swab of the nasopharynx, a single swab of the throat followed by the nasopharynx, or two separate swabs from each of these sites, and submit for a single test. Some centers have transitioned to swabs of the anterior nares. The specifics of testing depend on the requirements of local testing platforms (American Academy of Pediatrics, 2020b).
- Approximately 2%–3% of infants born to women who test positive for SARS-CoV-2 near the time of delivery have tested positive in the first 24–96 h after birth (Kotlyar et al., 2020; Woodworth et al., 2020).

MotherNewborn Contact in the Hospital

- Separation of the newborn from the suspected or confirmed COVID-19-infected mother is not recommended after birth. The newborn's risk for acquiring SARS-CoV-2 from its mother is low, and data suggest no difference in risk of neonatal SARS-CoV-2 infection whether the neonate is cared for in a separate room or remains in the mother's room (Center for Disease Control and Prevention, 2020b). However, mothers should wear a mask and practice hand hygiene during contact with their infants, and at other times, reasonable physical distancing between the mother and neonate or placing the neonate in an incubator is desirable, when feasible (UpToDate.-Vincenzo Berghella, 2020).
- Factors to consider include the following:
  - Rooming-in helps establish breastfeeding, facilitates bonding and parental education, and promotes family-centered care.
  - Separation may be necessary for mothers who are too ill to care for their infants or who need higher levels of care.
  - Separation may be necessary for neonates who may be at higher risk for severe illness (e.g.,

preterm infants, infants with underlying medical conditions, infants needing higher levels of care).
- Separation to reduce the risk of mother-to-newborn transmission is not useful if the neonate tests positive for SARS-CoV-2, and probably not useful if the mother and newborn will not be able to maintain separation after discharge until they meet criteria for discontinuation of quarantine.
- If separation is implemented, CDC suggest that newborn COVID-19 suspects/confirmed cases should be isolated from other healthy newborns.
- If another healthy family member is providing newborn care (e.g., diapering, bathing, feeding), he or she should use appropriate PPE and procedures (e.g., mask, hand hygiene).

Criteria for Discontinuing Mothernewborn Infection Precautions

- **Symptomatic mothers**—Previously symptomatic mothers with suspected or confirmed COVID-19 are not considered a potential risk of virus transmission to their neonates if they have met the criteria for discontinuing isolation and precautions (American Academy of Pediatrics, 2020b):
  - At least 10 days have passed since symptoms onset (20 days if have severe or critical illness or are severely immunocompromised).
  - At least 24 h have passed since their last fever without the use of antipyretics.
  - Symptoms have improved.
    However, some obstetric services are recommending that 20 days pass since symptoms first appeared for all pregnant women for time-based clearance, given that the viral shedding data from the CDC did not include pregnant women, and there is some concern that they may shed longer (UpToDate.Vincenzo Berghella, 2020).
- **Asymptomatic mothers**—For asymptomatic mothers identified through routine SARS-CoV-2 screening upon hospital admission, at least 10 days should have passed since the positive test before discontinuing mother–newborn infection precautions.
  These are symptom- and time-based strategies for discontinuing transmission precautions. Test-based strategies also exist but are not recommended for most patients because a positive SARS-CoV-2 RT-PCR result can persist for weeks and reflects presence of viral RNA but does not necessarily mean that viable virus is present and can be transmitted (Sethuraman et al., 2020). Data regarding postinfection risk of

transmission and personal immunity are limited (Kirkcaldy et al., 2020).

Breastfeeding and Formula Feeding

- The risk of SARS-CoV-2 transmission from ingestion of breast milk is unclear. Although several small series reported that all samples of breast milk from mothers with COVID-19 tested negative (Elshafeey et al., 2020; Liu et al., 2020), other investigators subsequently reported identifying samples of breast milk positive for SARS-CoV-2 by RT-PCR (Kirtsman et al., 2020; Wu et al., 2020; Groß et al., 2020; Chambers et al., 2020). In a WHO study, breast milk samples from 43 mothers were negative for SARS-CoV-2 by RT-PCR and samples from three mothers tested positive, but specific testing for viable and infective virus was not performed (World Health Organization, 2020a). Samples that are SARS-CoV-2 RT-PCR positive do not necessarily contain viable and transmissible virus (Chambers et al., 2020).
- Breastfeeding should be encouraged because of its many maternal and infant benefits. In the setting of maternal COVID-19 infection, the infant may receive passive antibody protection from the virus since breast milk is a source of antibodies and other antiinfective factors (UpToDate.Vincenzo Berghella, 2020).

Breastfeeding

- The AAP supports breastfeeding in mothers with COVID-19, but mothers should perform hand hygiene before and wear a mask during breastfeeding (American Academy of Pediatrics, 2020b). This approach considers the clear mother–infant benefits of breastfeeding, the low likelihood of passing maternal infection to the newborn when infection precautions are taken, and the nonsevere course of newborn infection when it does occur. This policy was based, in part, on a study from New York City that tested and followed 82 infants of 116 mothers who tested positive for SARS-CoV-2: no infant was positive for SARS-CoV-2 postnatally, although most roomed in with their mothers and were breastfed (Salvatore et al., 2020). The infants were kept in a closed isolette while rooming-in, and the mothers wore surgical masks while handling their infants and followed frequent hand- and breast-washing protocols.

Feeding Pumped Breast Milk

- If mother and baby separation has been implemented, ideally, the infant is fed expressed breast milk by another healthy caregiver until the mother has recovered or has been proven uninfected, provided that the other caregiver is healthy and follows hygiene precautions (Center for Disease Control and Prevention, 2020c). Expressing breast milk is important to support establishment of the maternal milk supply.
- Before pumping, ideally with a dedicated breast pump, mothers should wear a mask and thoroughly clean their hands and breasts with soap and water and clean pump parts, bottles, and artificial nipples (American Academy of Pediatrics, 2020a). If possible, the pumping equipment should be thoroughly cleaned by a healthy person.
- If feeding by a healthy caregiver is not possible, mothers with confirmed COVID-19 or symptomatic mothers with suspected COVID-19 should take precautions to prevent transmission to the infant during breastfeeding (wear a mask, hand and breast hygiene, disinfect shared surfaces that the symptomatic mother has contacted). However, it should be noted that the value of precautions, such as cleansing the breast prior to breastfeeding/milk expression or disinfecting external surfaces of milk collection devices (e.g., bottles, milk bags) for reducing potential transmission of SARS-CoV-2, has not been formally studied (Center for Disease Control and Prevention, 2020a).
- **Pasteurized donor human milk**—Holder pasteurization is commonly used in human milk banks and appears to eliminate replication-competent SARS-CoV-2 virus (Unger et al., 2020).
- **Formula feeding**—Ideally, women who choose to formula-feed should have another healthy caregiver feed the infant. If this is not possible or desired, such women must also take appropriate infection control precautions, as described before, to prevent transmission through close contact when feeding (UpToDate.Vincenzo Berghella, 2020).

Permanent and Reversible Contraception

Permanent contraception (tubal sterilization) does not add significant additional time or risk when performed at an uncomplicated cesarean birth and, thus, should be performed if planned. Permanent contraception after a vaginal birth is more of an elective procedure, so such decisions should be made on a local level, based on available resources. If not performed or if a reversible contraceptive method is desired, an alternative form of contraception should be provided (e.g., immediate postpartum long-acting reversible contraception or depot medroxyprogesterone acetate) as long as the patient desires one of these methods. This avoids additional outpatient postpartum visits (UpToDate. Vincenzo Berghella, 2020).

Discharge from Hospital

Patients without COVID-19—In stable patients, we suggest early discharge postpartum (24 and 48 h after vaginal delivery and two CDs, respectively) to limit their personal risk of acquiring infection in the hospital environment (Boelig et al., 2020). However, this should be considered in the context of the clinical scenario since early discharge may place additional burdens on families to access recommended newborn care and pediatric offices to provide this care (American Academy of Pediatrics, 2020b).

Patients with Known or Suspected COVID-19

The decision to discharge a patient with COVID-19 is generally the same as that for other conditions and depends on the need for hospital-level care and monitoring. Patients are counseled about the warning symptoms that should prompt reevaluation by telehealth visit and in-person visit, including emergency department evaluations. These include new onset of dyspnea, worsening dyspnea, dizziness, and mental status changes, such as confusion. Patients are also counseled about what to expect after recovery. Early discharge will require discussion with the facility's pediatric care team and should be linked to home telehealth visits for the mother and infant (UpToDate.Vincenzo Berghella, 2020).

   Postpartum office visit (UpToDate.Vincenzo Berghella, 2020):
- Postpartum outpatient care is better minimized during the pandemic and is appropriate to decrease the risk of exposure to infection. Early postpartum assessments for wound and blood pressure can be achieved through telemedicine facilities.
- A comprehensive postpartum visit is important after 12 weeks, especially for those having comorbidities and in patients who lose insurance coverage at that time.
- Postpartum depression screening should be done 4–8 weeks after delivery for all patients. The most widely used instrument is the self-reported, 10-item Edinburgh Postnatal Depression Scale, which can be completed in less than 5 minutes.
- The psychological impact of COVID-19, which may include moderate to severe anxiety, should also be recognized and support offered.
- Offer modified postpartum counseling regarding
  - any potential changes to the length of hospital stay and postpartum care;

- how to best communicate with their postpartum care team, especially in the case of an emergency;
- when and how to contact their postpartum care clinician;
- any special considerations for infant feeding;
- checking with their pediatric clinician or family physician regarding newborn visits because pediatric clinicians or family physicians also may be altering their procedures and routine appointments;
- postpartum contraception; ideally, all methods of contraception should be discussed in context of how provision of contraception may change within the limitations of decreased postpartum in-person visits;
- any potential changes to their postpartum care team and support system; most patients will likely have had changes to expected care support resources at home (e.g., family who can no longer travel, childcare providers who are no longer available). To the extent possible, patients should be connected to community support resources.
- It should be noted that it may be necessary to provide these services or enhanced resources by phone or electronically where possible. If telehealth visits are anticipated, patients should be provided with any necessary equipment (e.g., blood pressure cuffs) if available and as appropriate.

Precautions of the Contact Personnel

When caring for patients with confirmed or suspected COVID-19, healthcare workers should use contact and droplet precautions (i.e., gown, gloves, surgical mask, face shield, or goggles). During the patient's second stage of labor, and during episodes of deep respiratory efforts, healthcare workers should also use airborne precautions (i.e., N95 mask or PAPR, where available), in addition to contact and droplet precautions (American College of Obstetricians and Gynecologists, 2020).

Prevention of Transmission of Infection

Prevention of spread of infection is mandatory. Caution when dealing with aerosol-generating procedures (AGPs) and applications of PPE is particularly important.

Aerosol-Generating Procedures

Aerosols generated by medical procedures are one route for the transmission of the SARS-CoV-2. AGPs should be used only if necessary and for the shortest possible duration for suspected and confirmed COVID-19 patients. AGPs should be carried out in a single-room closed door and preferably completed in a negative-pressure side room with the presence of the least needed healthcare staff wearing their full PPE (Physiopedia-Contributors, 2020).

Potentially infectious AGPs (Moses and Consultant Respiratory Physiotherapist, 2020) include the following:

- Intubation, extubation, and related procedures
- Tracheotomy/tracheostomy procedures
- Manual ventilation
- Open suctioning
- Bronchoscopy
- NIV, e.g., BiPAP and CPAP ventilation
- Surgery and postmortem procedures in which high-speed devices are used
- High-frequency oscillating ventilation
- HFNO
- Induction of sputum (*typically involves administration of nebulized saline to moisten and loosen respiratory secretions) (this may be accompanied by chest physiotherapy such as percussion and vibration to induce forceful coughing). This may be required if lower respiratory tract samples are needed*

Certain other procedures/equipment may generate an aerosol from material other than patient secretions but are not considered to represent a significant infectious risk. These include the following:

- *Administration of pressurized humidified oxygen*
- *Administration of medication via nebulization. During nebulization, the aerosol derives from a nonpatient source (the fluid in the nebulizer chamber) and does not carry patient-derived viral particles. If a particle in the aerosol coalesces (combines) with a contaminated mucous membrane, it will cease to be airborne and therefore will not be part of an aerosol. Staff should use appropriate hand hygiene when helping patients to remove nebulizers and oxygen masks.*

Decontamination

Reusable (communal) noninvasive equipment must be decontaminated in the following circumstances:

- between each patient and after patient use;
- after blood and body fluid contamination; and
- at regular intervals as part of equipment cleaning.

An increased frequency of decontamination should be considered for reusable noninvasive care equipment when used in isolation/cohort areas (Moses and Consultant Respiratory Physiotherapist, 2020).

Equipment (Physiopedia-Contributors, 2020)

- Reusable equipment should be avoided if possible; if used, it should be decontaminated according to the manufacturer's instructions before removal from the room. If it is not possible to leave equipment inside a room, then follow IPC Guidelines on Decontamination. This usually involves cleaning with neutral detergent and then a chlorine-based disinfectant, in the form of a solution at a minimum strength of 1000 ppm available chlorine (e.g., "Haz-Tab" or other brands).
- If possible, use dedicated equipment in the isolation room. Avoid storing any extraneous equipment in the patient's room.
- Dispose of single-use equipment as per clinical waste policy inside a room.
- Point-of-care tests, including blood gas analysis, should be avoided unless a local risk assessment has been completed and shows it can be undertaken safely.
- Ventilators and mechanical devices (e.g., cough-assist machines) should be protected with a high-efficiency viral–bacterial filter such as BS EN 13328-1.
- When using mechanical airway clearance, filters should be placed at the machine end and the mask end before any expiratory or exhalation ports. Filters should be changed when visibly soiled or dependent on the filter used either after each use or every 24 h. Complete circuit changes should be undertaken every 72 h.
- Closed-system suction should be used if patients are intubated or have tracheostomies.
- Disconnecting a patient from mechanical ventilation should be avoided at all costs, but if required, the ventilator should be placed on standby.
- Manual hyperinflation (bagging) should be avoided if possible and attempt ventilator recruitment maneuvers where possible and required.
- Water humidification should be avoided, and a heat and moisture exchanger should be used in ventilator circuits.
- Disposable crockery and cutlery may be used in the patient's room as far as possible to minimize the numbers of items which need to be decontaminated.
- Any additional items such as stethoscopes, pulse oximeters, or ultrasound probes that are taken into a room will also need to be disinfected, regardless of whether there has been direct contact with the patient or not. This is due to the risk of environmental contamination of the equipment within the isolation room.

Patients' Rooms (Physiopedia-Contributors, 2020)

- If AGPs are undertaken in the patient's own room, the room should be decontaminated 20 min after the end of the procedure.

- If a different room is used for a procedure, it should be left for 20 min and then cleaned and disinfected before being put back into use.
- Clearance of any aerosols is dependent on the ventilation of the room. In hospitals, rooms commonly have 12 to 15 air changes per hour, and so after about 20 min, there would be less than 1% of the starting level (assuming cessation of aerosol generation).
- If it is known locally that the design or construction of a room may not be typical for a clinical space, or that there are fewer air changes per hour, then the local infection prevention and control team would advise on how long to leave a room before decontamination.

PRECAUTIONS FOR HEALTHCARE PERSONNEL: PERSONAL PROTECTIVE EQUIPMENT

COVID-19 infection is highly contagious, and this must be taken into consideration when planning intrapartum care. PPE recommended by the CDC is listed in the following, and the CDC provides strategies for how to optimize the supply of PPE.

General Considerations

- To protect patients and coworkers, all healthcare personnel should always wear a facemask while they are in a healthcare facility, regardless if patients are wearing a face covering or facemask (Center for Disease Control and Prevention, 2020d). Recent data suggest that universal masking, appropriate use of N95 respirators, and close evaluation of extended use or reuse of N95 respirators in the healthcare setting can play a crucial role in decreasing healthcare-related COVID-19 infection (Chu et al., 2020; Degesys et al., 2020; Seidelman et al., 2020).
- In areas with moderate to substantial community transmission, healthcare personnel should also wear eye protection in addition to their facemask (CDC).
- In areas where universal testing is not employed and adequate PPE is available, universal PPE, including respirators (e.g., N95 respirators), is recommended until the patient's status is known.
- Importantly, all medical staff should be trained in and adhere to proper donning and doffing of PPE.
- Although there is understandable emphasis on facial protection, data from the SARS outbreak suggest that the comprehensive array of recommended PPE (listed below) used alongside hand hygiene and environmental cleaning leads to the optimal decreased risk of transmission of respiratory viruses, and this is likely true for COVID-19.
- During a possible N95 shortage, extended use or limited reuse of N95 masks may be implemented or necessary. If extended use or limited reuse is being implemented, polices regarding extended use or limited reuse should be in accordance with CDC/NIOSH recommendations, considering the actual masks being used. These policies should also be developed in coordination with local occupational health and infection control departments.
- Although limited data have noted subtle physiologic changes (with no known clinical impact) associated with extended wear of N95 masks (Bae et al., 2020), the reduction of infectious risk outweighs any theoretical physiologic concern.

Caring for Individuals with Potential or Confirmed COVID-19

All medical staff caring for potential or confirmed COVID-19 patients should use PPE listed in the following, including respirators (e.g., N95 respirators).
  The CDC recommended the following PPE:

- Respirator or facemask (cloth face coverings are *not* PPE and should not be worn for the care of patients with known or suspected COVID-19 or in other situations where a respirator or facemask is warranted):
  - Put on a respirator or facemask (if a respirator is not available) before entry into the patient's room or care area.
  - N95 respirators or respirators that offer a higher level of protection should be used instead of a facemask when performing or present for an AGP. Disposable respirators and facemasks should be removed and discarded after exiting the patient's room or care area and closing the door. Perform hand hygiene after discarding the respirator or facemask.
  - If reusable respirators (e.g., PAPRs) are used, they must be cleaned and disinfected according to manufacturer's reprocessing instructions before reuse.
  - When the supply chain is restored, facilities with a respiratory protection program should return to use of respirators for patients with known or suspected COVID-19.
- Eye protection:
  - Put on eye protection (i.e., goggles or a disposable face shield that covers the front and sides of the face) upon entry to the patient's room or care area. Personal eyeglasses and contact lenses are *not* considered adequate eye protection.

- Remove eye protection before leaving the patient's room or care area.
- Reusable eye protection (e.g., goggles) must be cleaned and disinfected according to manufacturer's reprocessing instructions before reuse.
- Disposable eye protection should be discarded after use.
- Gloves:
  - Put on clean, nonsterile gloves upon entry into the patient's room or care area.
  - Change gloves if they become torn or heavily contaminated.
  - Remove and discard gloves when leaving the patient's room or care area, and immediately perform hand hygiene.
- Gown:
  - Put on a clean isolation gown upon entry into the patient's room or area. Change the gown if it becomes soiled. Remove and discard the gown in a dedicated container for waste or linen before leaving the patient's room or care area. Disposable gowns should be discarded after use. Cloth gowns should be laundered after each use.
  - If there are shortages of gowns, they should be prioritized for the following:
    - AGPs
    - Care activities where splashes and sprays are anticipated
    - High-contact patient care activities that provide opportunities for transfer of pathogens to the hands and clothing of the healthcare practitioner. Examples include the following:
      - Dressing
      - Bathing/showering
      - Transferring
      - Providing hygiene
      - Changing linens
      - Changing briefs or assisting with toileting
      - Device care or use
      - Wound care

During N95 respirator shortages, facilities might need to prioritize N95 respirator use for AGP* or surgical procedures that involve anatomic regions where viral loads might be higher (e.g., nose and throat, oropharynx, respiratory tract). Even during a shortage, it is important that medical staff use appropriate forms of PPE, including surgical masks. During shortages, facilities are encouraged to take steps that facilitate the protection of medical staff and enable personnel to protect themselves. Finally, although individual physicians, after careful consideration, may opt to provide care without adequate PPE, physicians are not ethically obligated to provide care to high-risk patients without protections in place. ACOG continues to advocate for congressional and regulatory action to increase access to PPE for obstetrician—gynecologists, particularly in labor and delivery units.

REFERENCES

Academy of Breastfeeding Medicine, 2021. Considerations for COVID-19 Vaccination in Lactation. ABM Statement. https://abm.memberclicks.net/abm-statement-considerations-for-covid-19-vaccination-in-lactation.

ACIP, 2020. Advisory Committee on Immunization Practices (ACIP). https://www.cdc.gov/vaccines/acip/recommendations.html.

American Academy of Pediatrics, 2020a. Breastfeeding Guidance Post Hospital Discharge for Mothers or Infants with Suspected or Confirmed SARS-CoV-2 Infection. https://services.aap.org/en/pages/2019-novel-coronavirus-covid-19-infections/breastfeeding-guidance-post-hospital-discharge.

American Academy of Pediatrics, 2020b. Management of Infants Born to COVID-19 Mothers. https://services.aap.org/en/pages/2019-novel-coronavirus-covid-19-infections/clinical-guidance/faqs-management-of-infants-born-to-covid-19-mothers.

American Colleague of Obstetrics and Gynecology, 2021. Vaccinating Pregnant and Lactating Patients Against Summary of Key Information and Recommendations, pp. 1—20.

American College of Obstetricians and Gynecologists, 2020. COVID-19 FAQs for Obstetricians-Gynecologists.

Arabi, Y.M., Arifi, A.A., Balkhy, H.H., Najm, H., Aldawood, A.S., Ghabashi, A., et al., 2014. Clinical course and outcomes of critically ill patients with Middle East respiratory syndrome coronavirus infection. Annu. Intern. Med. 160 (6), 389—397.

Ashokka, B., Loh, M.-H., Tan, C.H., Su, L.L., Young, B.E., Lye, D.C., Biswas, A., E Illanes, S., Choolani, M., 2020. Care of the pregnant woman with coronavirus disease 2019 in labor and delivery: anesthesia, emergency cesarean delivery, differential diagnosis in the acutely ill parturient, care of the newborn, and protection of the healthcare personnel. Am. J. Obstet. Gynecol. 223 (1), 66—74.e3. https://doi.org/10.1016/j.ajog.2020.04.005.

Bae, S., Kim, M.-C., Kim, J.Y., Cha, H.-H., Lim, J.S., Jung, J., Kim, M.-J., et al., 2020. "Effectiveness of surgical and cotton masks in blocking SARS—CoV-2: a controlled comparison in 4 patients. Ann. Intern. Med. 173 (1), W22—W23. https://doi.org/10.7326/M20-1342.

Bhatia, P., Biyani, G., Mohammed, S., Sethi, P., Bihani, P., 2016. Acute respiratory failure and mechanical ventilation in pregnant patient: a narrative review of literature. J. Anaesthesiol. Clin. Pharmacol. 32 (4), 431—439. https://doi.org/10.4103/0970-9185.194779.

Boelig, R.C., Manuck, T., Oliver, E.A., Di Mascio, D., Gabriele Saccone, Federica Bellussi, Berghella, V., 2020. Labor and delivery guidance for COVID-19. Am. J. Obstet. Gynecol. MFM 2 (2), 100110. https://doi.org/10.1016/j.ajogmf.2020.100110.

Breslin, N., Baptiste, C., Gyamfi-Bannerman, C., Miller, R., Martinez, R., Bernstein, K., Ring, L., et al., 2020. Coronavirus disease 2019 infection among asymptomatic and symptomatic pregnant women: two weeks of confirmed presentations to an affiliated pair of New York city hospitals. Am. J. Obstet. Gynecol. MFM 2 (2), 100118. https://doi.org/10.1016/j.ajogmf.2020.100118.

Center for Disease Control and Prevention, 2020a. Care for Breastfeeding Women. Interim Guidance on Breastfeeding and Breast Milk Feeds in the Context of COVID-19. https://www.cdc.gov/coronavirus/2019-ncov/hcp/care-for-breastfeeding-women.html.

Center for Disease Control and Prevention, 2020b. Care for Newborns. https://www.cdc.gov/coronavirus/2019-ncov/hcp/caring-for-newborns.html.

Center for Disease Control and Prevention, 2020c. Coronavirus Disease 2019. Considerations for Inpatient Obstetric Healthcare Settings. https://www.cdc.gov/coronavirus/2019-ncov/hcp/inpatient-obstetric-healthcare-guidance.html.

Center for Disease Control and Prevention, 2020d. Infection Control Guidance for Healthcare Professionals about Coronavirus (COVID-19). https://www.cdc.gov/coronavirus/2019-ncov/hcp/infection-control.html.

Center for Disease Control and Prevention, 2020e. People With Certain Medical Conditions. https://www.cdc.gov/coronavirus/2019-ncov/need-extra-precautions/people-with-medical-conditions.html.

Chambers, C., Paul, K., Bertrand, K., Contreras, D., Tobin, N.H., Bode, L., Aldrovandi, G., 2020. Evaluation for SARS-CoV-2 in breast milk from 18 infected women. J. Am. Med. Assoc. 324 (13), 1347–1348. https://doi.org/10.1001/jama.2020.15580.

Chan, M.T.V., Chow, B.K., Lo, T., Ko, F.W., Ng, S.S., Gin, T., Hui, D.S., 2018. Exhaled air dispersion during bag-mask ventilation and Sputum suctioning - implications for infection control. Sci. Rep. 8 (1), 198. https://doi.org/10.1038/s41598-017-18614-1.

Cheung, J.C.H., Ho, L.T., Cheng, J.V., Cham, E.Y.K., Lam, K.N., 2020. Staff safety during emergency airway management for COVID-19 in Hong Kong. Lancet Respir. Med. 8 (4), e19. https://doi.org/10.1016/S2213-2600(20)30084-9.

Chu, D.K., Akl, E.A., Duda, S., Solo, K., Yaacoub, S., Schünemann, H.J., El-harakeh, A., et al., 2020. Physical distancing, face masks, and eye protection to prevent person-to-person transmission of SARS-CoV-2 and COVID-19: a systematic review and meta-analysis. Lancet 395 (10242), 1973–1987. https://doi.org/10.1016/S0140-6736(20)31142-9.

Cohen, J. (n.d.)a. Incredible Milestone for Science. Pfizer and BioNTech Update Their Promising COVID-19 Vaccine Result. https://www.sciencemag.org/news/2020/11/covid-19-vaccine-trial-complete-pfizer-and-biontech-update-their-promising-result. (Accessed 2 January 2021).

Cohen. (n.d.)b. Russia's Claim of a Successful COVID-19 Vaccine Doesn't Pass the 'Smell Test,' Critics Say. https://www.sciencemag.org/news/2020/11/russia-s-claim-successful-covid-19-vaccine-doesn-t-pass-smell-test-critics-say.

Degesys, N.F., Wang, R.C., Kwan, E., Fahimi, J., Noble, J.A., Raven, M.C., 2020. Correlation between N95 extended use and reuse and fit failure in an emergency department. J. Am. Med. Assoc. 324 (1), 94–96. https://doi.org/10.1001/jama.2020.9843.

Detsky, M.E., Jivraj, N., Adhikari, N.K., Friedrich, J.O., Pinto, R., Simel, D.L., Wijeysundera, D.N., Scales, D.C., 2019. Will this patient Be difficult to intubate? The rational clinical examination systematic review. J. Am. Med. Assoc. 321 (5), 493–503. https://doi.org/10.1001/jama.2018.21413.

Donders, F., Lonnée-Hoffmann, R., Tsiakalos, A., Mendling, W., Martinez de Oliveira, J., Judlin, P., Xue, F., Donders, G.G.G., Isidog Covid-Guideline Workgroup, 2020. ISIDOG recommendations concerning COVID-19 and pregnancy. Diagnostics 10 (4), 243. https://doi.org/10.3390/diagnostics10040243.

Dooling, K., 2020. ACIP COVID-19 Vaccines Work Group Phased Allocation of COVID-19 Vaccines.

Duncan, H., Hutchison, J., Parshuram, C.S., 2006. The pediatric early warning system score: a severity of illness score to predict urgent medical need in hospitalized children. J. Crit. Care 21 (3), 271–278. https://doi.org/10.1016/j.jcrc.2006.06.007.

Elshafeey, F., Rana, M., Hindi, N., Elshebiny, M., Farrag, N., Mahdy, S., Sabbour, M., et al., 2020. A systematic scoping review of COVID-19 during pregnancy and childbirth. Int. J. Gynaecol. Obstet. 150 (1), 47–52. https://doi.org/10.1002/ijgo.13182.

Folegatti, P.M., Ewer, K.J., Aley, P.K., Angus, B., Becker, S., Belij-Rammerstorfer, S., Duncan, B., et al., 2020. Safety and immunogenicity of the ChAdOx1 NCoV-19 vaccine against SARS-CoV-2: a preliminary report of a Phase 1/2, single-blind, randomised controlled trial. Lancet 396 (10249), 467–478. https://doi.org/10.1016/S0140-6736(20)31604-4.

Goossens, H., Ferech, M., Vander Stichele, R., Elseviers, M., 2005. Outpatient Antibiotic use in Europe and association with resistance: a cross-national database study. Lancet 365 (9459), 579–587. https://doi.org/10.1016/s0140-6736(05)17907-0.

Greenhalgh, T., Koh, G.C.H., Car, J., 2020. Covid-19: a remote assessment in primary care. Br. Med. J. 368, 1–5. https://doi.org/10.1136/bmj.m1182.

Groß, R., Conzelmann, C., Müller, J.A., Stenger, S., Steinhart, K., Frank, K., Jan, M., 2020. Detection of SARS-CoV-2 in human breastmilk. Lancet. https://doi.org/10.1016/S0140-6736(20)31181-8.

Guérin, C., Reignier, J., Richard, J.-C., Beuret, P., Gacouin, A., Boulain, T., Mercier, E., et al., 2013. Prone positioning in severe acute respiratory distress syndrome. N. Engl. J. Med. 368 (23), 2159–2168. https://doi.org/10.1056/nejmoa1214103.

Kaur, S.P., Gupta, V., 2020. Since January 2020 Elsevier has Created a COVID-19 Resource Centre With Free Information in English and Mandarin on the Novel Coronavirus COVID- 19. The COVID-19 Resource Centre Is Hosted on

Elsevier Connect, the Company's Public News and Information.

Kinsella, S.M., Girgirah, K., Scrutton, M.J.L., 2010. Rapid sequence spinal anaesthesia for category-1 urgency caesarean section: a case series. Anaesthesia 65 (7), 664–669. https://doi.org/10.1111/j.1365-2044.2010.06368.x.

Kirkcaldy, R.D., King, B.A., Brooks, J.T., 2020. COVID-19 and postinfection immunity: limited evidence, many remaining questions. J. Am. Med. Assoc. 323 (22), 2245–2246. https://doi.org/10.1001/jama.2020.7869.

Kirtsman, M., Diambomba, Y., Poutanen, S.M., Malinowski, A.K., Vlachodimitropoulou, E., Parks, W.T., Erdman, L., Morris, S.K., Shah, P.S., 2020. Probable congenital SARS-CoV-2 infection in a neonate born to a woman with active SARS-CoV-2 infection. Can. Med. Assoc. J. 192 (24), E647–E650. https://doi.org/10.1503/cmaj.200821.

Knight, M., Bunch, K., Vousden, N., Morris, E., Simpson, N., Gale, C., O'Brien, P., Quigley, M., Brocklehurst, P., Kurinczuk, J.J., 2020. Characteristics and outcomes of pregnant women admitted to hospital with confirmed SARS-CoV-2 infection in UK: national population based cohort study. Br. Med. J. 369 (June), m2107. https://doi.org/10.1136/bmj.m2107.

Kotlyar, A.M., Grechukhina, O., Chen, A., Popkhadze, S., Grimshaw, A., Tal, O., Taylor, H.S., Tal, R., 2020. Vertical transmission of coronavirus disease 2019: a systematic review and meta-analysis. Am. J. Obstet. Gynecol. 224 (1), 35–53.e3. https://doi.org/10.1016/j.ajog.2020.07.049.

Lee, M.K., Choi, J., Park, B., Kim, B., Lee, S.J., Kim, S.H., Yong, S.J., Choi, E.H., Lee, W.Y., 2018. High flow nasal cannulae oxygen therapy in acute-moderate hypercapnic respiratory failure. Clin. Respir. J. 12 (6), 2046–2056. https://doi.org/10.1111/crj.12772.

Li, T., Zheng, Q., Yu, H., Wu, D., Xue, W., Xiong, H., Huang, X., et al., 2020. SARS-CoV-2 spike produced in insect cells elicits high neutralization titres in non-human primates. Emerg. Microb. Infect. 9 (1), 2076–2090. https://doi.org/10.1080/22221751.2020.1821583.

Lim, G., Facco, F.L., Nathan, N., Waters, J.H., Wong, C.A., Eltzschig, H.K., 2018. A review of the impact of obstetric anesthesia on maternal and neonatal outcomes. Anesthesiology 129 (1), 192–215. https://doi.org/10.1097/ALN.0000000000002182.

Liu, C., Sun, W., Wang, C., Liu, F., Zhou, M., 2019. Delivery during extracorporeal membrane oxygenation (ECMO) support of pregnant woman with severe respiratory distress syndrome caused by influenza: a case report and review of the literature. J. Matern. Fetal Neonatal Med. 32 (15) https://doi.org/10.1080/14767058.2018.1439471, 2570–74.

Liu, W., Wang, J., Li, W., Zhou, Z., Liu, S., Rong, Z., 2020. Clinical characteristics of 19 neonates born to mothers with COVID-19. Front. Med. 14 (2), 193–198. https://doi.org/10.1007/s11684-020-0772-y.

Llor, C., Bjerrum, L., 2014. Antimicrobial resistance: risk associated with antibiotic overuse and initiatives to reduce the problem. Therapeut. Adv. Drug Saf. 5 (6), 229–241. https://doi.org/10.1177/2042098614554919.

López, M., Gonce, A., Meler, E., Plaza, A., Hernández, S., Raigam, J., Martinez-Portilla, Cobo, T., et al., 2020. Coronavirus disease 2019 in pregnancy: a clinical management protocol and considerations for practice. Fetal Diagn. Ther. 47 (7), 519–528. https://doi.org/10.1159/000508487.

Luo, Y., Ou, R., Ling, Y., Qin, T., 2015. The therapeutic effect of high flow nasal cannula oxygen therapy for the first imported case of Middle East respiratory syndrome to China. Zhonghua Wei Zhong Bing Ji Jiu Yi Xue 27 (10), 841–844.

Martínez-Perez, O., Vouga, M., Melguizo, S.C., Forcen Acebal, L., Panchaud, A., Muñoz-Chápuli, M., Baud, D., 2020. Association between mode of delivery among pregnant women with COVID-19 and maternal and neonatal outcomes in Spain. J. Am. Med. Assoc. 324 (3), 296–299. https://doi.org/10.1001/jama.2020.10125.

McClung, N., Chamberland, M., Kinlaw, K., Matthew, D.B., Wallace, M., Bell, B.P., Lee, G.M., et al., 2020. The advisory committee on immunization practices' ethical principles for allocating initial supplies of COVID-19 vaccine — United States, 2020. Morb. Mortal. Wkly. Rep. 69 (47), 1782–1786. https://doi.org/10.15585/mmwr.mm6947e3.

Messerole, E., Peine, P., Wittkopp, S., Marini, J.J., Albert, R.K., 2002. The pragmatics of prone positioning. Am. J. Respir. Crit. Care Med. 165 (10), 1359–1363. https://doi.org/10.1164/rccm.2107005.

Moses, R., Consultant Respiratory Physiotherapist, 2020. Rachael Moses, Consultant Respiratory Physiotherapist, Lancashire Teaching Hospitals. Version 1 Dated 12th March 2020," No. March: 1–12.

Mulligan, M.J., Lyke, K.E., Kitchin, N., Absalon, J., Gurtman, A., Lockhart, S., Neuzil, K., et al., 2020. Phase I/II study of COVID-19 RNA vaccine BNT162b1 in adults. Nature 586 (7830), 589–593. https://doi.org/10.1038/s41586-020-2639-4.

Narang, K., Enninga, E., Gunaratne, M., Ibirogba, E.R., Trad, A., Elrefaei, A., Theiler, R.N., Ruano, R., Szymanski, L.M., Chakraborty, R., Garovic, D., 2020. SARS-CoV-2 Infection and COVID-19 During Pregnancy: A Multidisciplinary Review. Mayo Clinic proceedings 95 (8), 1750–1765. https://doi.org/10.1016/j.mayocp.2020.05.011.

National Health Commission of the People's Republic of China, 2020. Notice on Issuing Technical Guidelines for the Selection and Instruction of Masks for Prevention of COVID-19 Infection in Different Populations. http://www.nhc.gov.cn/jkj/s7916/202002/.

NICE guideline, 2018. Venous Thromboembolism in over 16s: Reducing the Risk of Hospital-Acquired Deep Vein Thrombosis or Pulmonary Embolism. National Institute for Health and Care Excellence. https://www.nice.org.uk/guidance/ng89.

NIH NHLBI ARDS Clinical Network, 2008. Mechanical ventilation protocol summary. Trial. http://www.ardsnet.org/files/ventilator_protocol_2008-07.pdf.

Ogawa, H., Asakura, H., 2020. Consideration of tranexamic acid administration to COVID-19 patients. Physiol. Rev. 100 (4), 1595–1596. https://doi.org/10.1152/physrev.00023.2020.

Panagiotakopoulos, L., Myers, T.R., Gee, J., Lipkind, H.S., Kharbanda, E.O., Ryan, D.S., Williams, J.T.B., et al., 2020. SARS-CoV-2 infection among hospitalized pregnant women: reasons for admission and pregnancy characteristics — eight U.S. Health care centers, March 1–May 30, 2020. Morb. Mortal. Wkly. Rep. 69 (38), 1355–1359. https://doi.org/10.15585/mmwr.mm6938e2.

Papazian, L., Forel, J.M., Gacouin, A., Penot-Ragon, C., Perrin, G., Loundou, A., et al., 2010. Neuromuscular blockers in early acute respiratory distress syndrome. N. Engl. J. Med. 263 (12), 1107–1116.

Peng, P.W.H., Ho, P.L., Hota, S.S., 2020. Outbreak of a new coronavirus: what anaesthetists should know. Br. J. Anaesth. 124 (5), 497–501. https://doi.org/10.1016/j.bja.2020.02.008.

Physiopedia-Contributors, 2020. Respiratory management of COVID 19. Physiopedia (6), 1–20. https://www.physiopedia.com/index.php?title=Respiratory_Management_of_COVID_19&oldid=240410.

Poon, L.C., Yang, H., Kapur, A., Melamed, N., Dao, B., Divakar, H., David McIntyre, H., et al., 2020. Global interim guidance on coronavirus disease 2019 (COVID-19) during pregnancy and puerperium from FIGO and allied partners: information for healthcare professionals. Int. J. Gynecol. Obstet. 149 (3), 273–286. https://doi.org/10.1002/ijgo.13156.

Qi, H., Chen, M., Luo, X., Liu, X., Shi, Y., Liu, T., Zhang, H., et al., 2020. Management of a delivery suite during the COVID-19 epidemic. Eur. J. Obstet. Gynecol. Reprod. Biol. 250 (July), 250–252. https://doi.org/10.1016/j.ejogrb.2020.05.031.

Rasmussen, S.A., Smulian, J.C., Lednicky, J.A., Wen, T.S., Jamieson, D.J., 2020. Coronavirus disease 2019 (COVID-19) and pregnancy: what obstetricians need to know. Am. J. Obstet. Gynecol. 222 (5), 415–426. https://doi.org/10.1016/j.ajog.2020.02.017.

Rawson, T.M., Moore, L., Zhu, N., Ranganathan, N., Skolimowska, K., Gilchrist, M., Satta, G., Cooke, G., Holmes, A., 2020. Bacterial and fungal coinfection in individuals with coronavirus: A rapid review to support COVID-19 antimicrobial prescribing. Clin. infect Dis. Official Publication Infecti. Dis. Soc. Am. 71 (9), 2459–2468. https://doi.org/10.1093/cid/ciaa530.

Ren, W., Sun, H., Gao, G.F., Chen, J., Sun, S., Zhao, R., Gao, G., et al., 2020. Recombinant SARS-CoV-2 spike S1-fc fusion protein induced high levels of neutralizing responses in nonhuman primates. Vaccine 38 (35), 5653–5658. https://doi.org/10.1016/j.vaccine.2020.06.066.

Rhodes, A., Evans, L.E., Alhazzani, W., Levy, M.M., Antonelli, M., Ferrer, R., Kumar, A., et al., 2017. Surviving Sepsis Campaign: International Guidelines for Management of Sepsis and Septic Shock: 2016. In: Intensive Care Medicine, vol. 43. Springer Berlin Heidelberg. https://doi.org/10.1007/s00134-017-4683-6.

Rimensberger, P.C., Cheifetz, I.M., Jouvet, P., Thomas, N.J., Willson, D.F., Erickson, S., Khemani, R., et al., 2015. Ventilatory support in children with pediatric acute respiratory distress syndrome: proceedings from the pediatric acute lung injury consensus conference. Pediatr. Crit. Care Med. 16 (5), S51–S60. https://doi.org/10.1097/PCC.0000000000000433.

Rochwerg, B., Brochard, L., Elliott, M.W., Hess, D., Hill, N.S., Nava, S., Navalesi, P., et al., 2017. Official ERS/ATS clinical practice guidelines: noninvasive ventilation for acute respiratory failure. Eur. Respir. J. 50 (4) https://doi.org/10.1183/13993003.02426-2016.

Royal Colleague of Obstetrics and Gynecology, 2020. Coronavirus (COVID-19) Infection in Pregnancy, pp. 1–77.

Sahin, U., Alexander, M., Derhovanessian, E., Vogler, I., Kranz, L.M., Vormehr, M., Baum, A., et al., 2020. COVID-19 vaccine BNT162b1 elicits human antibody and TH1 T cell responses. Nature 586 (7830), 594–599. https://doi.org/10.1038/s41586-020-2814-7.

Salvatore, C.M., Han, J.-Y., Acker, K.P., Tiwari, P., Jin, J., Brandler, M., Cangemi, C., et al., 2020. Neonatal management and outcomes during the COVID-19 pandemic: an observation cohort study. Lancet Child Adolesc. Health 4 (10), 721–727. https://doi.org/10.1016/S2352-4642(20)30235-2.

Schultz, M.J., Dunser, M.W., Dondorp, A.M., Adhikari, N.K.J., Iyer, S., Kwizera, A., Lubell, Y., et al., 2017. Current challenges in the management of sepsis in ICUs in resource-poor settings and suggestions for the future. Intensive Care Med. 43 (5), 612–624. https://doi.org/10.1007/s00134-017-4750-z.

Schwartz, D.A., 2020. An analysis of 38 pregnant women with COVID-19, their newborn infants, and maternal-fetal transmission of SARS-CoV-2: maternal coronavirus infections and pregnancy outcomes. Arch. Pathol. Lab Med. https://doi.org/10.5858/arpa.2020-0901-SA.

Seidelman, J., Seidelman, J., Lewis, S., Lewis, S., Advani, S., Advani, S., Akinboyo, I., et al., 2020. Universal masking is an effective strategy to flatten the SARS-2-CoV healthcare worker epidemiologic curve. Infect. Control Hosp. Epidemiol. https://doi.org/10.1017/ice.2020.313.

Sethuraman, N., Jeremiah, S.S., Ryo, A., 2020. Interpreting diagnostic tests for SARS-CoV-2. J. Am. Med. Assoc. 323 (22), 2249–2251. https://doi.org/10.1001/jama.2020.8259.

Siddamreddy, S., Thotakura, R., Dandu, V., Kanuru, S., Meegada, S., 2020. Corona virus disease 2019 (COVID-19) presenting as acute ST elevation myocardial infarction. Cureus 2019 (4). https://doi.org/10.7759/cureus.7782.

Slayton-Milam, S., Sheffels, S., Chan, D., Alkinj, B., 2020. Induction of labor in an intubated patient with coronavirus disease 2019 (COVID-19). Obstet. Gynecol. 136 (5), 962–964. https://doi.org/10.1097/AOG.0000000000004044.

Smith, T.R.F., Patel, A., Ramos, S., Elwood, D., Zhu, X., Yan, J., Gary, E.N., et al., 2020. Immunogenicity of a DNA vaccine candidate for COVID-19. Nat. Commun. 11 (1), 1–13. https://doi.org/10.1038/s41467-020-16505-0.

Thomas, P., Baldwin, C., Bissett, B., Boden, I., Gosselink, R., Granger, C.L., Hodgson, C., et al., 2020. Physiotherapy management for COVID-19 in the acute hospital setting: recommendations to guide clinical practice. J. Physiother. 33 (1), 32–35. https://doi.org/10.1016/j.jphys.2020.03.011.

Ti, L.K., Ang, L.S., Foong, T.W., Ng, B.S.W., 2020. What we do when a COVID-19 patient needs an operation: operating room preparation and guidance. Can. J. Anaesth. https://doi.org/10.1007/s12630-020-01617-4.

Unger, S., Christie-Holmes, N., Guvenc, F., Budylowski, P., Mubareka, S., Gray-Owen, S.D., O'Connor, D.L., 2020. Holder pasteurization of donated human milk is effective in inactivating SARS-CoV-2. Can. Med. Assoc. J. 192 (31), E871–E874. https://doi.org/10.1503/cmaj.201309.

UpToDate, Vincenzo, B., Hughes, B., 2020. Coronavirus Disease 2019 (COVID-19): Labor, Delivery, and Postpartum Issues and Care. https://www.uptodate.com/contents/coronavirus-disease-2019-covid-19-labor-delivery-and-postpartum-issues-and-care?sectionName=Infant evaluation&topicRef=127 535&anchor=H3851809117&source=see_link.

Violi, F., Pastori, D., Cangemi, R., Pignatelli, P., Loffredo, L., 2020. Hypercoagulation and antithrombotic treatment in coronavirus 2019: a new challenge. Thromb. Haemostasis 120 (6), 949–956. https://doi.org/10.1055/s-0040-1710317.

Walker, K.F., O'Donoghue, K., Grace, N., Dorling, J., Comeau, J.L., Li, W., Thornton, J.G., 2020. Maternal transmission of SARS-COV-2 to the neonate, and possible routes for such transmission: a systematic review and critical analysis. BJOG 127 (11), 1324–1336. https://doi.org/10.1111/1471-0528.16362.

Walsh, E.E., Frenck, R., Falsey, A.R., Kitchin, N., Absalon, J., Gurtman, A., Lockhart, S., et al., 2020. RNA-based COVID-19 vaccine BNT162b2 selected for a pivotal efficacy study. medRxiv. https://doi.org/10.1101/2020.08.17.20176651.

Wang, W., Xu, Y., Gao, R., Lu, R., Han, K., Wu, G., Tan, W., 2020. Detection of SARS-CoV-2 in different types of clinical specimens. J. Am. Med. Assoc. 323 (18), 1843–1844. https://doi.org/10.1001/jama.2020.3786.

Wax, R.S., Christian, M.D., 2020. Practical recommendations for critical care and anesthesiology teams caring for novel coronavirus (2019-NCoV) patients. Can. J. Anaesth. 67 (5), 568–576. https://doi.org/10.1007/s12630-020-01591-x.

Wichmann, D., Sperhake, J.P., Lütgehetmann, M., Steurer, S., Edler, C., Heinemann, A., Heinrich, F., et al., 2020. Autopsy findings and venous thromboembolism in patients with COVID-19: a prospective cohort study. Ann. Intern. Med. 173 (4), 268–277. https://doi.org/10.7326/M20-2003.

Woodworth, K.R., Olsen, E.O., Neelam, V., Lewis, E.L., Galang, R.R., Oduyebo, T., Aveni, K., et al., 2020. Birth and infant outcomes following laboratory-confirmed SARS-CoV-2 infection in pregnancy - SET-NET, 16 Jurisdictions, March 29–October 14, 2020. Morb. Mortal. Wkly. Rep. 69 (44), 1635–1640. https://doi.org/10.15585/mmwr.mm6944e2.

World Health Organization, 2016. Guideline: Updates on Paediatric Emergency Triage, Assessment and Treatment: Care of Critically-Ill Children. World Health Organisation, Geneva, Switzerland. https://apps.who.int/iris/bitstream/handle/10665/204463/9789241510219_eng.pdf?sequence=1.

World Health Organization, 2018. Basic Emergency Care WHO. https://www.who.int/publications-detail/basic-emergency-care-approach-to-the-acutely-ill-and-injured.

World Health Organization, 2019. World Health Organization Model List of Essential Medicines. AWARE Classification of Antibiotics. https://apps.who.int/iris/bitstream/handle/10665/325771/WHO-MVP-EMP-IAU-2019.06-eng.pdf?ua=1.

World Health Organization, 2020a. Breastfeeding and Covid-19.Scientific Brief. https://www.who.int/publications/i/item/10665332639. (Accessed 25 June 2020).

World Health Organization, 2020b. Clinical Management of COVID-19. https://www.who.int/publications/i/item/clinical-management-of-covid-19.

World Health Organization, 2020c. The use of non-steroidal anti-inflammatory drugs (NSAIDs) in patients with COVID-19. Ecancermedicalscience. https://doi.org/10.3332/ecancer.2020.1023.

Wu, Y., Liu, C., Dong, L., Zhang, C., Chen, Y., Liu, J., Zhang, C., et al., 2020. Coronavirus disease 2019 among pregnant Chinese women: case series data on the safety of vaginal birth and breastfeeding. BJOG 127 (9), 1109–1115. https://doi.org/10.1111/1471-0528.16276.

Yee, J., Kim, W., Han, J.M., Yoon, H.Y., Lee, N., Lee, K.E., Gwak, H.S., 2020. Clinical manifestations and perinatal outcomes of pregnant women with COVID-19: a systematic review and meta-analysis. Sci. Rep. 10 (1), 18126. https://doi.org/10.1038/s41598-020-75096-4.

Yuan, P., Ai, P., Liu, Y., Ai, Z., Wang, Y., Cao, W., Xia, X., Zheng, J.C., 2020. Safety, tolerability, and immunogenicity of COVID-19 vaccines: a systematic review and meta-analysis. medRxiv. https://doi.org/10.1101/2020.11.03.20224998.

Zambrano, L.D., Ellington, S., Strid, P., Galang, R.R., Oduyebo, T., Van Tong, T., Woodworth, K.R., et al., 2020. Update: characteristics of symptomatic women of reproductive age with laboratory-confirmed SARS-CoV-2 infection by pregnancy status — United States, January 22–October 3, 2020. Morb. Mortal. Wkly. Rep. 69 (44), 1641–1647. https://doi.org/10.15585/mmwr.mm6944e3.

Zhang, W., Du, R.-H., Li, B., Zheng, X.-S., Yang, X.-L., Hu, B., Wang, Y.-Y., et al., 2020. Molecular and serological investigation of 2019-NCoV infected patients: implication of multiple shedding routes. Emerg. Microb. Infect. 9 (1), 386–389. https://doi.org/10.1080/22221751.2020.1729071.

CHAPTER 5

Lines of Treatment of COVID-19 Infection

AHMED M. MAGED EL-GOLY, MD

(Dr. Prof. M Fadel Shaltout. Prof. of Obstetrics and Gynecology, Cairo University, Faculty of Medicine)

Covid-19 Infection and Pregnancy. https://doi.org/10.1016/B978-0-323-90595-4.00002-9

LINES OF TREATMENT

Till the present time, there is no definitive treatment for COVID-19 infection. Various drugs are suggested for managing these cases, but none of them have proven efficacy. Many researches are currently going all over the world to optimize the outcome of the patients with COVID-19 infection.

Different lines for treatment of pregnant women with COVID-19 are summarized in Table 5.1.

ANTIVIRAL DRUGS

As COVID-19 infection is caused by severe acute respiratory syndrome coronavirus 2 (SARS-CoV-2), antiviral drugs were investigated in its management. Antiviral drugs work through many mechanisms. They prevent viral entry inside the target cells through angiotensin-converting enzyme 2 (ACE2) receptor and transmembrane serine protease 2 (TMPRSS2), prevent viral membrane fusion, and prevent 3-chymotrypsin-like protease (3CLpro) and the RNA-dependent RNA polymerase (Sanders et al., 2020).

TABLE 5.1
Lines of Treatment for Covid-19.

1. Antiviral drugs (remdesivir, lopinavir/ritonavir combination, umifenovir, and favipiravir)
2. Antibacterial drugs
3. Antimalarial drugs
4. Antiparasitics
5. Anticoagulants
6. Immune-based therapy
   a. Immunomodulatory therapy
      i. Steroids
      ii. Interleukin inhibitors
      iii. Interferons
   b. Human blood-derived products
      i. Convalescent plasma
      ii. Immunoglobulins
      iii. Mesenchymal stem cells
7. Host-directed therapy
   a. Metformin
   b. Statins
   c. Pioglitazones
8. Other therapeutic agents
   a. Angiotensin-converting enzyme inhibitors
   b. Nonsteroidal antiinflammatory drugs
   c. Vitamin C
   d. Vitamin D
   e. Zinc
   f. lactoferrin
   g. Melatonin
9. Oxygen therapy

These mechanisms suggest that antiviral treatment should start early in the disease to limit viral replications. The role of antiviral drugs is questionable after progression of the disease to the hyperinflammatory state, which characterizes advanced stages of the disease (Siddiqi and Mehra, 2020).

The commonly used antiviral drugs for treatment of COVID-19 include remdesivir, lopinavir/ritonavir combination, umifenovir, and favipiravir.

Remdesivir is an adenosine analog intravenous nucleotide that binds to the viral RNA-dependent RNA polymerase, preventing viral replication through premature termination of RNA transcription. It has a promising antiviral activity against RNA viruses such as SARS/MERS-CoV (Middle East respiratory syndrome coronavirus) and Ebola (Li et al., 2020). In vitro studies confirmed its efficiency in COVID-19 (Wang et al., 2020c), and its early administration after viral inoculation in a rhesus macaque model was associated with lower lungs virus load and less lung damage than the control animals (Williamson et al., 2020).

Side effects of remdesivir include gastrointestinal manifestations, elevated liver transaminases, prolonged prothrombin time, renal toxicity, and allergic reactions. Accordingly, liver function tests including liver enzymes and prothrombin time and concentration should be evaluated before the start of treatment and during it if clinically indicated, and discontinuation of treatment may be indicated if alanine transaminase (ALT) levels reached 10 folds of normal and should be done if ALT increase was associated with manifestations of liver inflammation (Wang et al., 2020c).

Treatment with remdesivir should not be started if glomerular filtration fraction is lower than 30 L/min (Sanders et al., 2020)

Remdesivir was not assigned to any Food and Drug Administration (FDA) category, as clinical trials evaluating its role in COVID-19 excluded pregnant women from participation. However, it was safely used in pregnant women for treatment of Ebola virus (Mulangu et al., 2019).

Remdesivir was approved by the FDA for the treatment of hospitalized COVID-19 cases on supplemental oxygen. Its use in patients on mechanical ventilation is questionable, as there was lack of evidence for its efficiency during these advanced stages (Beigel et al., 2020; Goldman et al., 2020; Spinner et al., 2020; Wang et al., 2020a,b,c,d,e,f).

A randomized controlled trial (RCT) included 1063 severely ill COVID-19 patients from 68 sites worldwide (47 in the United States and 21 in Europe and Asia).

Early reports revealed that remdesivir enhanced recovery time in these patients (from 15 to 11 days). After receiving the initial report, the FDA approved the emergency use of the drug for the treatment of severe hospitalized cases of COVID-19 (FDA, 2020a).

Remdesivir was used in 86 pregnant and postpartum hospitalized women with severe COVID-19, and it was found to be well tolerated, with a low rate of serious adverse events. The NIH recommended that remdesivir should not be withheld from pregnant patients if indicated (Williamson et al., 2020).

The currently recommended dose of remdesivir is 200 mg IV loading dose followed by 100 mg IV infusion daily for 9 days (Favilli et al., 2020).

Drug interactions with remdesivir were not studied in clinical trials. Gilead Sciences reported minimal to no reduction in remdesivir exposure when administered with dexamethasone (written communication, July 2020) and no significant interactions with oseltamivir or baloxavir (written communications, August and September 2020). Chloroquine or hydroxychloroquine (HCQ) may decrease the antiviral activity of remdesivir, so coadministration is not recommended (FDA, 2020b).

Lopinavir/Ritonavir and Other HIV Protease Inhibitors

Lopinavir/ritonavir (LPV/r), used in Chinese treatment schemes against COVID-19, are also known as "anti-HIV drugs" (Lee et al., 2007).

The replication of SARS-CoV-2 depends on two proteases that cleave polyproteins into an RNA-dependent RNA polymerase and a helicase. These proteases are named 3-chymotrypsin-like protease (3CLpro) and papain-like protease (PLpro) (Zumla et al., 2016a).

Lopinavir inhibits the division of HIV Gag-Pol, whereas ritonavir is a protease inhibitor. The combination of the two molecules reduces the replication of HIV by the production of immature particles that block viral replication (Chu et al., 2004).

LPV/r is an inhibitor of SARS-CoV 3CLpro in vitro, and this protease appears to be highly conserved in SARS-CoV-2 (Tahir ul Qamar et al., 2020; Liu and Wang, 2020). LPV/r has a poor selective in vitro activity against SARS-CoV. That may reflect the need for higher drug level than the tolerable levels to achieve the desired effect in vivo (Chen et al., 2004). Lopinavir is excreted in the gastrointestinal tract; therefore, coronavirus-infected enterocytes might be exposed to higher concentrations of the drug (Chu et al., 2004).

Darunavir inhibits the 3CLpro enzyme and possibly the PLpro enzyme. However, in an in vitro study, darunavir did not show activity against SARS-CoV-2 (De Meyer et al., 2020).

Side effects of LPV/r include gastrointestinal manifestations such as anorexia, nausea, vomiting, abdominal pain, diarrhea, QTc prolongation, renal toxicity pancreatitis, cutaneous manifestation, and liver toxicity (Li et al., 2020; Cao et al., 2020). The last side effect is particularly important, as 20%–30% of COVID-19 patients have elevated levels of transaminases (Wu et al., 2020).

LPV/r is not assigned to any FDA category. However, ritonavir alone was assigned as FDA class B (Favilli et al., 2020).

There is extensive experience with the use of LPV/r in pregnant women with HIV, and the drug has a good safety profile with no evidence of teratogenicity. A RCT (Koss et al., 2014) conducted on 356 HIV-infected pregnant women showed no significant risk of preterm labor, even if Berghella reported that it crosses the transplacental barrier and may increase the risk of preterm delivery, but not the risk of teratogenic effects (Vincenzo and Hughes, 2020). That was confirmed in a study where 955 women with exposure to LPV/r during pregnancy were analyzed (Roberts et al., 2009).

For relative safety, LPV/r was suggested for treatment of COVID-19 in pregnant women. A treatment protocol could involve an oral administration of LPV/r 200 mg/50 mg, two capsules every 12 h with interferon-alpha (IFN-α) 5 million IU in 2 mL of nebulized physiologic solution (Liang and Acharya, 2020). Kim and colleagues recommend to avoid the nebulization of solutions for the risk of aerosolization of SARS-CoV-2 and, when possible, to administer inhaled medications by metered dose inhaler (Arthur and Kim, 2020). If the nebulized therapy is necessary, it is important to use some precautions during the nebulization, such as the positioning of the patient in an airborne infection isolation room, the use of adequate PPE, and not to reenter the room for 2–3 h after the therapy (George, 2020).

LPV/r plasma concentrations achieved using typical doses are far below the levels that may be needed to inhibit SARS-CoV-2 replication (Schoergenhofer et al., 2020). A moderately sized randomized trial failed to find a virologic or clinical benefit of LPV/r over standard of care (Cao et al., 2020). Results from a small RCT showed that darunavir/cobicistat was not effective for the treatment of COVID-19 (Chen et al., 2020c). There are no data from clinical trials that support using other HIV protease inhibitors to treat COVID-19.

DRUG–DRUG INTERACTIONS

LPV/r is a potent inhibitor of cytochrome P450 3A. Coadministration of LPV/r with medications that are metabolized by this enzyme may increase the concentrations of those medications, resulting in concentration-related toxicities.

LPV/r oral solution contains 42.4% alcohol and 15.3% propylene glycol and is not recommended for use during pregnancy, and the same is applied for the use of once-daily dosing.

The COVID-19 Treatment Guidelines Panel recommends against using LPV/r (AI) or other HIV protease inhibitors (AIII) for the treatment of COVID-19, except in a clinical trial, as it did not show efficacy in RCT with moderate sample size and the pharmacodynamics of the drug raise the doubts about the possibility of reaching enough blood concentration able to inhibit the virus proteases.

ANTIBACTERIAL DRUGS

Many reports found that most of the hospitalized cases with COVID-19 have received broad-spectrum antibiotics with unknown efficacy (Ding et al., 2020; Du et al., 2020; Zhou et al., 2020a,b; Chen et al., 2020a; Guan et al., 2020). Unnecessary antibiotics upon hospitalization may increase the individual risk of subsequent hospital-acquired pneumonia caused by resistant bacteria (Kalil et al., 2016; Stevens et al., 2011). On a population level, antibiotic administration for all hospitalized COVID-19 patients increase the antibiotic use during a pandemic and consequently an increase in antimicrobial resistance rates (Bell et al., 2014). Bacterial coinfection occurred in 3.5% of COVID-19 cases (Sieswerda et al., 2020).

The WHO recommends against the use of antibiotic therapy or prophylaxis in mild suspected or confirmed COVID-19. In moderate suspected or confirmed cases, antibiotics should not be prescribed unless bacterial infection is clinically suspected. In severe cases, the use of empiric antimicrobials to treat all likely pathogens, based on clinical judgment, patient host factors, and local epidemiology, should start as early as possible, ideally after withdrawal of blood cultures. Antimicrobial therapy should be assessed daily for deescalation. The choice of antibiotics with the least ecologic impact is based on data and guidance from local institution, region, or country (e.g., of the Access group of the AWaRe classification). The AWaRe classification categorizes antibiotics into three different groups (Access, Watch, and Reserve) based on their indication for common infectious syndromes, their spectrum of activity, and their potential for increasing antibiotic resistance (World Health Organization, 2019).

Treatment of other coinfections may be based on a laboratory-confirmed diagnosis or epidemiological criteria. Empiric antibiotic therapy should be deescalated on the basis of microbiology results and clinical judgment, and the duration of empiric antibiotic therapy should be shortened as much as possible, generally 5–7 days. Cautions should be taken, as the increase in the use of antibiotics during pandemics can enhance certain infections as Clostridioides difficile with clinical disease ranging from diarrhea and fever to colitis (Aldeyab et al., 2012).

Table 5.2 summarized evidence-based recommendations for antibacterial therapy in adults with COVID-19.

ANTIMALARIAL DRUGS

Chloroquine is an antimalarial drug that was developed in 1934. In 1946, its analog HCQ was developed. Chloroquine/chloroquine phosphate/HCQ are antimalarial drugs that have both antiviral and immunomodulatory activities. The three drugs differ in chemical structure but have the same clinical effects. However, HCQ has the least side effects (Favilli et al., 2020).

Chloroquine has a proven inhibitory effect on many viruses including HIV, MERS-CoV, and SARS-CoV. The needed dose of drug for treatment of a viral infection is lower than that used for treatment of malaria with subsequent lower side effects (Keyaerts et al., 2009).

Both chloroquine and HCQ increase the endosomal pH, inhibiting fusion of SARS-CoV-2 and the host cell membranes (Wang et al., 2020c). Chloroquine inhibits glycosylation of the cellular ACE2 receptor, which may interfere with binding of the virus to cell receptor (Vincent et al., 2005). Both chloroquine and HCQ prevent release of the viral genome through blocking of the transport of SARS-CoV-2 from early endosomes to endolysosomes (Liu et al., 2020a). However, administration of HCQ—alone or when combined with azithromycin—neither reduced upper or lower respiratory tract viral loads nor demonstrated clinical efficacy in a rhesus macaque model (Maisonnasse et al., 2020).

On March 28, 2020, the FDA authorized the emergency use of chloroquine and HCQ supplied from the Strategic National Stockpile to treat hospitalized adults and adolescents >50 kg body weight with COVID-19 for whom a clinical trial is not available, or participation is not feasible. On April 13, 2020, the Division of Anti-infective (DAI) products opened a priority Tracked Safety Issue (TSI) 2150 to assess the risk of cardiac toxicity with HCQ and chloroquine with or without azithromycin when used for the treatment of COVID-19.

TABLE 5.2

Summarized Evidence-Based Recommendations for Antibacterial Therapy in Adults With COVID-19 (Sieswerda et al., 2020).

Recommendation	Strength	Quality of Evidence
Antibacterial drugs should be restricted in suspected or confirmed COVID-19 patients. This especially applies for mild and moderated cases.	Weak	Very low
Exceptions for the restrictive use of antibacterial drugs can be made for suspected or confirmed COVID-19 patients presenting with radiological findings and/or inflammatory markers compatible with bacterial coinfection, immunocompromised cases,[a] and those with severe illness.	Weak	GPS
Blood, sputum, and pneumococcal urinary antigen testing is better done upon admission before the start of empirical antibiotic therapy in suspected or confirmed COVID-19 patients.	Strong	GPS
In case of suspected bacterial coinfection, empirical antibiotic treatment covering atypical pathogens is better avoided in suspected or confirmed COVID-19 patients hospitalized at the general ward. Legionella urinary antigen testing should be performed according to local and/or national guidelines for CAP.	Weak	Very low
The empirical antibiotic regimens in case of suspected bacterial coinfection depend on the severity of disease and according to local and/or national guidelines. For those fulfilling criteria of mild and moderate severe CAP, following local and/or national guideline recommendations on antibacterial treatment in CAP is recommended.	Weak	Very low
Following local and/or national guideline recommendations on antibacterial treatment for patients with COVID-19 and suspected bacterial secondary infection is recommended.	Strong	GPS
Stopping antibiotics is suggested when blood, sputum, and urinary antigen tests taken before start of empirical antibiotic therapy in patients with suspected or confirmed COVID-19 show no bacterial pathogens after 48 h of incubation.	Weak	GPS
Antibiotic treatment for 5 days is suggested in patients with COVID-19 and suspected bacterial infection upon improvement of signs, symptoms, and inflammatory markers.	Weak	GPS

[a] Immunocompromised is defined as the use of chemotherapy for cancer, bone marrow or organ transplantation, immune deficiencies, poorly controlled HIV or AIDS, or prolonged use of corticosteroids or other immunosuppressive medications. *GPS*, good practice statement.

On April 24, 2020, the FDA issued a Drug Safety Communication (DSC) cautioning against the use of HCQ or chloroquine for COVID-19 outside of the hospital setting or a clinical trial due to risk of arrhythmias. The DSC described reports of serious cardiac events, including QT prolongation, in patients receiving HCQ or chloroquine, often in combination with azithromycin and other QT prolonging medicines, for the prevention or treatment of COVID-19.

The COVID-19 Treatment Guidelines Panel recommends against the use of chloroquine or HCQ with or without azithromycin for the treatment of COVID-19 patients except in a clinical trial (AI).

In a large RCT of hospitalized patients in the United Kingdom, the use of HCQ was associated with increased risk of intubation and death in patients on noninvasive mechanical ventilation and longer hospital stay without reduction of 28-day mortality when compared with the usual standard of care (Horby et al., 2020).

Further RCT failed to prove the efficacy of HCQ in treatment of mild to moderate (Cavalcanti et al., 2020) or severe COVID-19 cases (Furtado et al., 2020).

In addition to these randomized trials, data from large retrospective observational studies do not consistently show evidence of a benefit for HCQ with or without azithromycin in hospitalized patients with COVID-19 (Geleris et al., 2020; Rosenberg et al., 2020).

Conversely, a large retrospective cohort study reported a survival benefit among hospitalized patients who received HCQ with or without azithromycin, compared with those who received neither drug (Arshad et al., 2020a,b). However, cases who did not receive HCQ had a lower rate of intensive care unit (ICU) admission, which suggests that these patients may have received less aggressive care. Furthermore, a substantially higher percentage of patients in the HCQ arms also received corticosteroids (77.1% of patients in the HCQ arms vs. 36.5% of patients in the control arm), which is proved to improve the survival rate of patients with COVID-19 (Recovery Collaborative Group, 2020).

HCQ is administered orally as 400 mg every 12 h for 5 d or 400 mg twice a day for the first day and then 200 mg twice a day for 4 days (Li et al., 2020a). Chloroquine is used as 1 g for the first day of treatment and then 500 mg daily for 4–7 d depending on clinical response (Kim and Gandhi, 2020; Colson et al., 2020; Wei, 2020).

The COVID-19 Treatment Guidelines Panel recommends against using high-dose chloroquine to treat COVID-19 (AI). High-dose chloroquine (600 mg twice daily for 10 days) has been associated with more severe toxicities than lower-dose chloroquine (450 mg twice daily for 1 day, followed by 450 mg once daily for 4 days). An RCT compared high-dose and low-dose chloroquine in hospitalized severe COVID-19 patients. In addition, all participants received azithromycin, and 89% of the participants received oseltamivir. The study was terminated early because of the high rate of mortality and QTc prolongation in the high-dose group (Borba et al., 2020).

Several randomized trials have not shown a clinical benefit for HCQ in nonhospitalized patients with COVID-19. However, other clinical trials are still ongoing (Mitjà et al., 2020; Skipper et al., 2020).

Both chloroquine and HCQ have a similar toxicity profile, although HCQ is better tolerated and has a lower incidence of toxicity than chloroquine.

Adverse effects are mainly cardiac in the form of QTc prolongation, Torsade de Pointes, ventricular arrhythmia, and cardiac deaths (Nguyen et al., 2020). Other side effects included hypoglycemia, rash, nausea, and retinopathy. Bone marrow suppression may occur with long-term use.

Patients receiving chloroquine or HCQ should be monitored for adverse events, especially prolonged QTc interval (AIII). Baseline and follow-up electrocardiograms are recommended when there are potential drug interactions with concomitant medications (e.g., azithromycin) or underlying cardiac diseases. The risk–benefit ratio should be assessed for patients with cardiac disease, a history of ventricular arrhythmia, bradycardia (<50 bpm), or uncorrected hypokalemia and/or hypomagnesemia (American College of Cardiology, 2020).

DRUG–DRUG INTERACTIONS

Chloroquine and HCQ are moderate inhibitors of cytochrome P450 (CYP) 2D6, and these drugs are also P-glycoprotein (P-gp) inhibitors. Use caution when administering these drugs with medications that are metabolized by CYP2D6 (e.g., certain antipsychotics, beta-blockers, selective serotonin reuptake inhibitors, methadone) or transported by P-gp (e.g., certain direct-acting oral anticoagulants, digoxin) (University of Liverpool, 2020). Chloroquine and HCQ may decrease the antiviral activity of remdesivir; coadministration of these drugs is not recommended (Food and Drug Administration, 2020e).

Concomitant medications that pose a moderate-to-high risk for QTc prolongation (e.g., antiarrhythmics, antipsychotics, antifungals, macrolides [including azithromycin] (Nguyen et al., 2020), fluoroquinolone antibiotics) (CredibleMeds, 2020) should be used only if necessary. Consider using doxycycline rather than azithromycin as empiric therapy for atypical pneumonia. Multiple studies have demonstrated that concomitant use of HCQ and azithromycin can prolong the QTc interval (Bessière et al., 2020; Chorin et al., 2020; Mercuro al., 2020); in an observational study, the use of HCQ plus azithromycin was associated with increased odds of cardiac arrest (Rosenberg et al., 2020). The use of this combination warrants careful monitoring.

CONSIDERATIONS IN PREGNANCY

Both chloroquine and HCQ are not assigned to any FDA category, but they have mild effects when administered during pregnancy with no evidence of increased risk of preterm birth or fetal damage. Chloroquine is widely used in malaria areas. Klumpp et al. report that, in 20 years of utilization, about 1 billion people have used chloroquine, including pregnant women reporting no fetal damages or adverse effects on pregnancy, labor, and newborns (Theodore, 1965). HCQ is commonly used for the treatment of malaria in pregnant women and malaria in pregnant women (Vincenzo and Hughes, 2020), and it is reported that it passes the placental barrier and accumulates in fetal ocular tissues, but no toxicity or ocular damages have been found in human species.

The NIH stated that antirheumatic doses of chloroquine and HCQ have been used safely in pregnant women with SLE, and no changing of dose is necessary for chloroquine or HCQ during pregnancy.

ANTIPARASITICS

Ivermectin

Ivermectin is an FDA-approved broad-spectrum anti-parasitic drug (Canga et al., 2008) that has antiviral activity (Lundberg et al., 2013; Tay et al., 2013; Veronika et al., 2016) in vitro. It inhibits integrase protein (IN) nuclear import and HIV-1 replication (Kylie et al., 2012) and nuclear import of host and viral proteins, including simian virus SV 40 large tumor antigen (T-ag) and dengue virus (DENV) nonstructural protein 5 (Kylie et al., 2012). It also inhibits importin (IMP) $\alpha/\beta1$ heterodimer responsible for IN nuclear import through which many RNA viruses can infect the host cells. Accordingly, it limits many RNA viruses infection as DENV 1−4, West Nile Virus, Venezuelan equine encephalitis virus, and influenza (Leon et al., 2020). Ivermectin has some inhibitory activity on some DNA viruses as pseudorabies virus.

Ivermectin inhibits the replication SARS-CoV-2 in cell cultures (Leon et al., 2020). However, pharmacokinetic and pharmacodynamic studies suggest that achieving the plasma concentrations necessary for the antiviral efficacy detected in vitro would require administration of doses up to 100-fold higher than those approved for use in humans (Guzzo et al., 2002; Chaccour et al., 2020). Even though ivermectin appears to accumulate in the lung tissue, predicted systemic plasma and lung tissue concentrations are much lower than 2 µM, the half-maximal inhibitory concentration (IC_{50}) against SARS-CoV-2 in vitro (Arshad et al., 2020a,b; Bray et al., 2020).

Ivermectin is not approved for the treatment of any viral infection, including SARS-CoV-2 infection. The FDA issued a warning in April 2020 that ivermectin intended for use in animals should not be used to treat COVID-19 in humans.

The available clinical data on the use of ivermectin to treat COVID-19 are limited.

A retrospective analysis of confirmed COVID-19 patients (27% of them had the severe form) who were admitted to four Florida hospitals compared patients who received at least one dose of ivermectin ($n = 173$) to those who received "usual care" ($n = 103$). Ivermectin was administered as a single dose of 200 µg/kg, to be repeated after 7 days if the patient was still hospitalized (13 patients received a second dose). In addition, 90% of the ivermectin cases and 97% of the usual care group received HCQ (in the majority in conjunction with azithromycin).

They found that all-cause mortality was significantly lower in the ivermectin group when compared with the usual care group (OR [odds ratio] 0.27; $P = .03$). That improved mortality appeared to be limited to severe cases. There was no difference between the groups for the median length of hospital stay (7 days in both groups) or the proportion of mechanically ventilated patients who were successfully extubated (36% and 15% in the ivermectin and the usual care groups respectively; $P = .07$). However, this study has many limitations. It is a retrospective one with no enough data given about oxygen saturation or radiographic findings or other therapeutic interventions (types and timing). The analyses of the durations of ventilation and hospitalization do not appear to account for death as a competing risk, and no virologic assessments were performed (Juliana et al., 2020).

A recent metaanalysis was conducted to assess the value of ivermectin for the treatment of COVID-19. A total of 629 (397 of them received ivermectin along with usual therapy) confirmed COVID-19 patients were included in four studies. The overall pooled OR to be 0.53 (95% CI [confidence interval]: 0.29 to 0.96, $P = .04$) for all-cause mortality. Ivermectin-received patients had a significant clinical improvement compared with usual therapy (OR = 1.98, 95% CI: 1.11 to 3.53, $P = .02$). However, the quality of evidence is very low (Padhy et al., 2020).

The COVID-19 Treatment Guidelines Panel recommends against the use of ivermectin for the treatment of COVID-19, except in a clinical trial (AIII).

ANTICOAGULANTS

Association Between COVID-19 and Thromboembolism

COVID-19 infection is associated with inflammation and a prothrombotic state, with increases in fibrin, fibrin degradation products, fibrinogen, and D-dimers (Driggin et al., 2020; Han et al., 2020). The increased levels of these markers were reported to be associated with poor clinical outcomes (Guan et al., 2020; Tang et al., 2020).

Venous thromboembolism (VTE) was reported in 14.1% of hospitalized COVID-19 patients (95% CI: 11.6−16.9) (Nopp et al., 2020). The prevalence was higher when ultrasound screening was used (40.3%; 95% CI: 27.0−54.3 vs. 9.5%; 95% CI: 7.5−11.7). The incidence of VTE in non−COVID-19-hospitalized patients on VTE prophylaxis ranged from 0.3% to 1% for symptomatic VTE and from 2.8% to 5.6% for VTE overall (Samama et al., 1999; Leizorovicz et al., 2004). That incidence increased to 5%−16% and reached 37% in critically ill non−COVID-19 patients who received prophylactic dose of anticoagulants and critically ill septic patients, respectively (Fraisse et al., 2000; Shorr and Williams, 2009; Protect Investigators, 2011). VTE guidelines for non–COVID-19 patients have recommended against routine ultrasound screening in critically ill patients due to lack of evidence that this strategy reduces the rate of complications (Kahn et al., 2012).

In American Society of Hematology Guidelines Panel metaanalysis, there was no difference in overall VTE and related mortality when compared patients treated with prophylactic dose or higher doses of anticoagulation (American Society of Hematology, 2020). In critically ill patients, intermediate or therapeutic dose anticoagulation was associated with a lower odd of pulmonary embolism (OR 0.09; 95% CI: 0.02–0.57) but a higher odd of major bleeding (OR 3.84; 95% CI: 1.44–10.21). Incidences of symptomatic VTE between 0% and 0.6% at 30–42 days post–hospital discharge have been reported in patients with COVID-19 (Patell et al., 2020; Roberts et al., 2020). Epidemiologic studies that control for clinical characteristics, underlying comorbidities, prophylactic anticoagulation, and COVID-19-related therapies are needed.

There are limited prospective data demonstrating the safety and efficacy of using therapeutic doses of anticoagulants in patients with COVID-19 to prevent VTE. A single-center retrospective analysis of 2773 hospitalized COVID-19 patients from a single reported in-hospital mortality in 22.5% and 22.8% of patients who received therapeutic anticoagulation and those who did not receive anticoagulation, respectively. Subgroup analysis of 395 mechanically ventilated patients reported mortality rate of 29.1% and 62.7% in those who received anticoagulation and who did not respectively. However, this study has many limitations: data about patient characteristics, indications for anticoagulant initiation, and other therapies were lacking. In addition, the authors did not discuss the potential impact of survival bias on the study results. For these reasons, the data are not sufficient to influence standard of care, and this study further emphasizes the need for prospective trials to define the risks and potential benefits of therapeutic anticoagulation in patients with COVID-19 (Paranjpe et al., 2020).

A single-center, randomized trial of 20 mechanically ventilated patients with D-dimers >1000 μg/L (as measured by the VIDAS D-dimer Exclusion II assay) found that patients treated with therapeutic anticoagulation showed improvement in PaO_2:FiO_2 ratio, higher number of ventilator-free days (15 days [IQR 6–16] vs. 0 days [IQR 0–11]; $P = .028$) when compared with those who received the prophylactic anticoagulation. However, there was no between-group difference in in-hospital or 28-day mortality. Two patients had minor bleeding in the therapeutic anticoagulation group, and two patients in each group experienced thrombosis (Lemos et al., 2020).

Guidelines about coagulopathy and prevention and management of VTE in COVID-19 have been released by multiple organizations, including the Anticoagulation Forum (Barnes et al., 2020), the American College of Chest Physicians(Moores et al., 2020), the American Society of Hematology (Adam et al., 2020), the International Society of Thrombosis and Haemostasis (ISTH) (Thachil et al., 2020), the Italian Society on Thrombosis and Haemostasis (Marietta et al., 2020), and the Royal College of Physicians. In addition, a paper that outlines issues related to thrombotic disease with implications for prevention and therapy has been endorsed by the ISTH, the North American Thrombosis Forum, the European Society of Vascular Medicine, and the International Union of Angiology (Bikdeli et al., 2020).

All guidelines agree that hospitalized patients with COVID-19 should receive prophylactic-dose anticoagulation for VTE. Some guidelines note that intermediate dose anticoagulation can be considered for critically ill patients (Barnes et al., 2020) (Spyropoulos et al., 2020a). Given the variation in VTE incidence and the unknown risk of bleeding in critically ill patients with COVID-19, the COVID-19 Treatment Guidelines Panel and the American Society of Hematology and the American College of Chest Physician Guidelines Panels recommend treating all hospitalized patients with COVID-19, including critically ill patients, with prophylactic dose anticoagulation (Moores et al., 2020). Participation in clinical trials is suggested to understand the safety and efficacy of different anticoagulant doses in patients with COVID-19.

MONITORING COAGULATION MARKERS IN PATIENTS WITH COVID-19

The NIH recommends against the routine testing of markers of coagulopathy, such as D-dimer level, prothrombin time, fibrinogen level, and platelet count (AIII) in nonhospitalized COVID-19 patients, as there is lack of evidence that these tests can predict the risk of VTE in asymptomatic and mild cases of COVID-19 infection. Although hematologic and coagulation parameters are commonly measured in hospitalized cases, there are currently insufficient data to recommend either for or against using such data to guide management decisions (National Institutes of Health, 2020a).

Fig. 5.1 described the suggested approach to hospitalized COVID-19 cases.

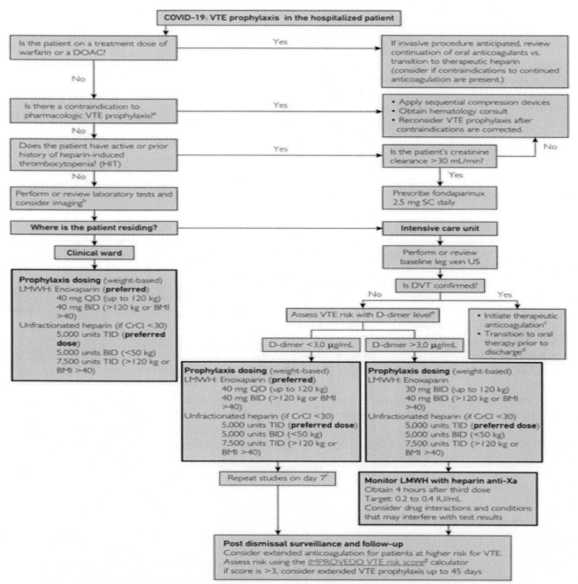

FIG. 5.1 Suggested approach to patients requiring hospitalization for coronavirus disease 2019 (COVID-19)-related complications. [a]Active bleeding, platelet count <30 × 10^9/L, or congenital bleeding disorder including von Willebrand disease or hemophilia. [b]Laboratory tests: complete blood cell count and differential, prothrombin time (PT), activated partial thromboplastin time (aPTT), fibrinogen, D-dimer. If PT and/or aPTT are prolonged, consider a special coagulation profile, which includes a lupus anticoagulant screen. Imaging: for patients presenting with a prolonged illness or those who have had a long hospital stay, consider obtaining bilateral lower extremity venous ultrasonography. [c]Initiate therapeutic anticoagulation therapy as follows: unfractionated heparin infusion is preferred. In a patient with a history of heparin-induced thrombocytopenia, use argatroban or bivalirudin (see direct thrombin inhibitors order set). [d]Continue oral anticoagulation for a minimum of 3 months with clinical reassessment thereafter. A direct oral anticoagulant (DOAC) is preferred unless the patient has another indication for the use of a vitamin K antagonist or low-molecular-weight heparin (LMWH). [e]Assess venous thromboembolism (VTE) risk using the D-dimer level as follows: low risk, <3.0 μg/mL; high risk, 3.0 μg/mL. This recommendation reflects a sixfold increase above the upper limit of normal. Precise cutoff requires external validation. [f]On day 7 of therapy (or earlier if clinical deterioration occurs), repeat the following studies: bilateral lower extremity venous ultrasonography; laboratory tests (complete blood cell count with differential, D-dimer, and fibrinogen). Consider alternating ultrasonography and laboratory tests every 3—4 days (McBane 2nd et al., 2020).

A suggested protocol for anticoagulation therapy in COVID-19 patients includes the following:

- Patients with suspected/confirmed thromboembolic events without possible ischemia or infarction should receive full dose of anticoagulation according to institutional protocols and those with possible ischemic events as myocardial infarction or strokes may be considered for thrombolytic therapy after consultation of pulmonary embolism response or stroke teams and have to receive full dose of anticoagulation according to institutional protocols if thrombolytic therapy is not available or recommended.
- Patients not suspected to have thromboembolic events and using anticoagulation for any other indication as atrial fibrillation (AF) should continue their current therapy if they do not need hospitalization or switch to short acting parenteral agent in those who need hospitalization. Those not using such therapy are not candidate for anticoagulation as long as they do not need hospitalization (prophylactic doses can be used in high-risk cases as patients with previous thromboembolic events, recent surgery or trauma, immobilization, or obese patients). In patients needing hospital admission for routine medical, surgical, or obstetric care, prophylactic dose of anticoagulation preferably low-molecular-weight heparin (LMWH) is used while those needing ICU admission should start anticoagulation. There are controversies of the used dose in these patients due to lack of sufficient data about their safety and efficacy and the risk of thromboembolic events. However, following the institutional protocol is recommended.

NIH Recommendations for Venous Thromboembolism Prophylaxis and Screening

Anticoagulants and antiplatelet therapy should not be administered in nonhospitalized patients for the prevention of venous or arterial thrombosis unless indicated for other reasons or as a part of a clinical trial (AIII) while prophylactic dose anticoagulation (AIII) should be given to all hospitalized nonpregnant adults with COVID-19. Anticoagulant or antiplatelet therapy should not be used to prevent arterial thrombosis outside of the usual standard of care for patients without COVID-19 (AIII).

There are currently insufficient data to recommend either for or against the use of thrombolytics or higher than the prophylactic dose of anticoagulation for VTE prophylaxis in hospitalized COVID-19 patients outside of a clinical trial.

Hospitalized patients with COVID-19 should not routinely be discharged from the hospital while on VTE prophylaxis (AIII) unless being at high risk for VTE and low risk for bleeding. In these cases, continuing anticoagulation with an FDA-approved regimen for extended VTE prophylaxis after hospital discharge can be considered (BI).

There are currently insufficient data to recommend either for or against routine screening for VTE in COVID-19 patients without clinical manifestations of VTE, regardless of the status of their coagulation markers.

The possibility of VTE should be considered in hospitalized COVID-19 patients who experience rapid deterioration of respiratory, cardiac, and/or neurological function, or of sudden, localized loss of peripheral perfusion (AIII) (Piazza and Morrow, 2020).

The American College of Chest Physicians recommended the use of prophylactic dose of LMWH in critically ill patients and prophylactic-dose LMWH or fondaparinux in other patients and does *not* recommend routine prophylaxis in nonhospitalized cases or extended prophylaxis after hospital discharge. The International Society on Thrombosis and Hemostasis recommended the use of prophylactic-dose LMWH in critically ill and noncritically ill patients, half-therapeutic dose LMWH in high-risk critically ill patients, and LMWH or direct oral anticoagulant in patients with high risk of thrombosis and low risk of bleeding for 30 days after hospital discharge and does *not* recommend routine prophylaxis in nonhospitalized cases (Piazza and Morrow, 2020).

NIH Recommendations for Venous Thromboembolism Treatment

- When diagnostic imaging is not possible, patients with COVID-19 who experience an incident thromboembolic event or who are highly suspected to have thromboembolic disease should be managed with therapeutic doses of anticoagulant therapy (AIII).
- Patients with COVID-19 who require extracorporeal membrane oxygenation (ECMO) or continuous renal replacement therapy or who have thrombosis of catheters or extracorporeal filters should be treated with antithrombotic therapy as per the standard institutional protocols for those without COVID-19 (AIII).

Special Considerations During Pregnancy and Lactation

- Anticoagulants treatment that was administered during pregnancy before diagnosis of COVID-19 infection should be continued (AIII).
- For pregnant patients hospitalized for severe COVID-19, prophylactic-dose anticoagulation is recommended unless contraindicated (BIII).

- As for nonpregnant patients, VTE prophylaxis after hospital discharge is not recommended for pregnant patients (AIII). Decisions to continue VTE prophylaxis in the pregnant or postpartum patient after discharge should be individualized, considering concomitant VTE and bleeding risk factors.
- Anticoagulation therapy use during labor and delivery requires specialized care and planning. It should be managed in pregnant patients with COVID-19 in a similar way as in pregnant patients with other conditions that require anticoagulation in pregnancy (AIII).
- Unfractionated heparin, LMWH, and warfarin do not accumulate in breast milk and do not induce an anticoagulant effect in the newborn; therefore, they can be used in breastfeeding individuals with or without COVID-19 who require VTE prophylaxis or treatment (AIII). In contrast, direct-acting oral anticoagulants are not routinely recommended due to lack of safety data (AIII) (Bates et al., 2018)

MANAGING ANTITHROMBOTIC THERAPY IN PATIENTS WITH COVID-19

- In hospitalized, critically ill patients, LMWH or unfractionated heparin is preferred over oral anticoagulants because of their shorter half-lives, ability to be administered intravenously or subcutaneously, and fewer drug–drug interactions (AIII). Potential drug–drug interactions with other concomitant drugs should be considered (AIII).
- COVID-19 outpatients receiving warfarin who are in isolation and thus unable to get international normalized ratio monitoring may be candidates for switching to direct oral anticoagulant therapy. Patients with mechanical heart valves, ventricular assist devices, valvular AF, or antiphospholipid antibody syndrome or patients who are lactating should continue treatment with warfarin (AIII).
- Hospitalized patients with COVID-19 who are taking anticoagulant or antiplatelet therapy for underlying medical conditions should continue their treatment unless significant bleeding develops, or other contraindications are present (AIII).
- For hospitalized patients with COVID-19, prophylactic-dose anticoagulation should be prescribed unless contraindicated (e.g., a patient has active hemorrhage or severe thrombocytopenia) (AIII). Although data supporting this recommendation are limited, a retrospective study showed reduced mortality in patients who received prophylactic anticoagulation, particularly if the patient had a sepsis-induced coagulopathy score ≥4 (Tang et al., 2020). For those without COVID-19, anticoagulant or antiplatelet therapy should not be used to prevent arterial thrombosis outside of the standard of care (AIII). Anticoagulation is routinely used to prevent arterial thromboembolism in patients with heart arrhythmias. Although there are reports of strokes and myocardial infarction in patients with COVID-19, the incidence of these events is unknown.
- When imaging is not possible, patients with COVID-19 who experience an incident thromboembolic event or who are highly suspected to have thromboembolic disease should be managed with therapeutic doses of anticoagulant therapy as per the standard of care for patients without COVID-19 (AIII).
- There are currently insufficient data to recommend either for or against using therapeutic doses of antithrombotic or thrombolytic agents for COVID-19 in patients who are hospitalized. Although there is evidence that multiorgan failure is more likely in patients with sepsis if they develop coagulopathy (Iba et al., 2017), there is no convincing evidence to show that any specific antithrombotic treatment will influence outcomes in those with or without COVID-19.
- After hospital discharge, VTE prophylaxis is not recommended for patients with COVID-19 (AIII) except in high-risk patients. The FDA approved the use of rivaroxaban 10 mg daily for 31–39 days in patients (Cohen et al., 2016; Spyropoulos et al., 2020b) with Modified International Medical Prevention Registry on Venous Thromboembolism (IMPROVE) VTE risk score ≥4 or Modified IMPROVE VTE risk score ≥2 and D-dimer level >2 times the upper limit of normal (Spyropoulos et al., 2020b).
- Any decision to use postdischarge VTE prophylaxis for patients with COVID-19 should consider the individual patient's risk factors for VTE, including reduced mobility, bleeding risks, and feasibility. Participation in clinical trials is encouraged.

SPECIAL CONSIDERATIONS DURING PREGNANCY AND LACTATION

Pregnant and parturient women are at higher risk of VTE when compared with nonpregnant women as pregnancy is a well-known hypercoagulable state (Heit et al., 2005). It is not yet known whether COVID-19 increases this risk or not. In several cohort studies of pregnant women with COVID-19, VTE was not reported as a complication even among women with severe disease, although the receipt of prophylactic or therapeutic anticoagulation varied across the studies (Breslin et al., 2020; Delahoy et al., 2020; Knight et al., 2020). The American College of Obstetricians and Gynecologists (ACOG) advises that although there are no data for or against thromboprophylaxis in the setting of COVID-19 in pregnancy, VTE prophylaxis can reasonably be considered for pregnant women hospitalized with COVID-19, particularly for those who have severe disease (American College of Obstetricians and Gynecologists, 2020). If there are no contraindications to use, the Society of Maternal Fetal Medicine recommends prophylactic heparin or LMWH in critically ill or mechanically ventilated pregnant patients (Society for Maternal Fetal Medicine, 2020). Several professional societies, including the American Society of Hematology and ACOG, have guidelines that specifically address the management of VTE in the context of pregnancy (ACOG Practice Bulletin No. 196 Summary: Thromboembolism in Pregnancy, 2018; Bates et al., 2018). If delivery is threatened, or if there are other risks for bleeding, the risk of bleeding may outweigh the potential benefit of VTE prophylaxis in pregnancy.

There are no data on the use of scoring systems to predict VTE risk in pregnant individuals. Additionally, during pregnancy, the D-dimer level may not be a reliable predictor of VTE because there is a physiologic increase of D-dimer levels throughout gestation (Hu et al., 2020; Réger et al., 2013; Wang et al., 2013).

In general, the preferred anticoagulants during pregnancy are heparin compounds. Because of its reliability and ease of administration, LMWH is recommended rather than unfractionated heparin for the prevention and treatment of VTE in pregnancy (ACOG Practice Bulletin No. 196 Summary: Thromboembolism in Pregnancy, 2018).

Direct-acting anticoagulants are not routinely used during pregnancy due to the lack of safety data in pregnant individuals (Bates et al., 2018). The use of warfarin to prevent or treat VTE should be avoided in pregnant individuals, regardless of their COVID-19 status, and especially during the first trimester due to the concern for teratogenicity.

IMMUNE-BASED THERAPY

Hyperactive inflammatory response is responsible for many manifestations and complications in COVID-19 infection. Agents modulating the immune response were suggested as an adjunctive therapy in moderate, severe, and critical COVID-19 (Zhong et al., 2020). These agents include immunomodulatory and human blood-derived therapy.

Immunomodulatory agents include generalized antiinflammatory drugs as steroid therapy (Horby et al., 2020) and more targeted antiinflammatory therapy as interleukin (IL) inhibitors (Xu et al., 2020; Shakoory et al., 2016), IFNs (Zhou et al., 2020b), kinase inhibitors (Cao et al., 2020b), and others.

Human blood-derived products are obtained from recovered COVID-19-infected individuals (Mair-Jenkins et al., 2015; Wang et al., 2020d). These products can work through direct antiviral actions (as convalescent plasma [CP]) and/or immunomodulatory effects (as mesenchymal stem cells [MSCs]) (Shetty, 2020). Additionally, neutralizing monoclonal antibodies directed against the virus are currently investigated in clinical trials (Marovich et al., 2020).

STEROIDS

In severe COVID-19 infection, lung injury and multisystem organ dysfunction can complicate the condition as a result of exaggerated systemic inflammatory response. As corticosteroids have potent antiinflammatory effects, these drugs were evaluated to prevent or ameliorate these deleterious effects.

RATIONALE FOR USE OF CORTICOSTEROIDS IN PATIENTS WITH COVID-19

The use of corticosteroids in patients suffering from pulmonary infections has been studied and proved to have both beneficial and deleterious clinical outcomes. In patients with *Pneumocystis jirovecii* pneumonia and hypoxia, prednisone therapy reduced the risk of death (Bozzette et al., 1990). However, in outbreaks of previous coronavirus infections (as MERS and SARS), corticosteroid treatment led to delayed virus clearance (Stockman et al., 2006; Arabi et al., 2018). In severe influenza viruses' pneumonia, it was associated with poor clinical outcomes, including secondary bacterial infection and death (Rodrigo et al., 2016).

A metaanalysis of seven RCTs that included 851 patients with acute respiratory distress syndrome (ARDS) found that corticosteroid therapy reduced the risk of all-cause mortality (risk ratio 0.75; 95% CI: 0.59−0.95) and duration of mechanical ventilation (mean difference, −4.93 days; 95% CI: −7.81 to −2.06 days) (Mammen et al., 2020).

In January 2020, the WHO issued against the routine use of steroids for treatment of SARS-CoV-2. That was based on the previous experience in treatment of influenza, MERS, and SARS-CoV. They described a possible role for steroids in ARDS based on their ability to suppress lung tissue inflammation but with a risk of delayed virus clearance (Zhang et al., 2020c).

The RCOG described that steroids administration for SARS-CoV-2 during pregnancy has no potential harms (RCOG Guideline, 2020).

As SARS-CoV-2 infection in pregnancy is associated with increased risk of preterm birth, administration of betamethasone was advised by many studies (Liang and Acharya, 2020; Dashraath et al., 2020; Poon et al., 2020).

RECOVERY trial randomized 2104 patients with SARS-CoV-2 to single dose of oral or intravenous dexamethasone 6 mg daily for 10 days and compared them with 4321 patients received usual care without steroids. The mortality rate was reduced in the dexamethasone ventilated patients by one-third (rate ratio 0.65 [95% CI 0.48 to 0.88]; $P = .0003$) and in dexamethasone oxygen receiving patients by one-fifth (0.80 [0.67 to 0.96]; $P = .0021$). There was no benefit among those patients who did not require respiratory support (1.22 [0.86 to 1.75]; $P = .14$). They concluded that dexamethasone could prevent 1 death if administered in eight ventilated patients or 25 patients requiring oxygen alone (Recovery Collaborative Group, 2020).

CORTICOSTEROIDS OTHER THAN DEXAMETHASONE

- Prednisone, methylprednisolone, or hydrocortisone can replace dexamethasone if not available. The dose should be titrated as 6 mg of dexamethasone is equivalent to 40 mg prednisone, 32 mg methylprednisolone, and 160 mg hydrocortisone (Czock et al., 2005).
- As the half-lives of different forms vary, the frequency of administration varies with it. Dexamethasone is a long-acting corticosteroid that has a half-life of 36−72 h, so administered once daily. Prednisone and methylprednisolone are intermediate-acting corticosteroids with a half-life of 12−36 h, once or twice daily. Hydrocortisone is a short-acting corticosteroid with a half-life of 8−12 h, so administered in two to four doses daily.
- Hydrocortisone is the commonest form of steroids used in management of septic shock in COVID-19 patients. Dexamethasone has the advantage over other corticosteroids studied in patients with ARDS of lacking mineralocorticoid activity and thus has minimal effect on fluid and electrolyte balance (Villar et al., 2020).

Monitoring, Adverse Effects, and Drug−Drug Interactions

- COVID-19 patients receiving dexamethasone should be closely monitor for adverse effects (e.g., hyperglycemia, secondary infections, psychiatric effects, avascular necrosis).
- Prolonged use of systemic corticosteroids may increase the risk of reactivation of latent infections (e.g., hepatitis B virus, herpesvirus infections, strongyloidiasis, tuberculosis).
- The risk of reactivation of latent infections for a 10-day course of dexamethasone (6 mg once daily) is not well defined. When initiating dexamethasone, appropriate screening and treatment to reduce the risk of *Strongyloides* hyperinfection in patients at high risk of strongyloidiasis (e.g., patients from tropical, subtropical, or warm, temperate regions or those engaged in agricultural activities) (Lier et al., 2020; Stauffer et al., 2020) or fulminant reactivations of HBV (Liu et al., 2020b) should be considered.
- Dexamethasone may reduce the concentration and efficacy of drugs that are CYP3A4 substrates being a moderate cytochrome P450 (CYP) 3A4 inducer. Clinicians should review a patient's medication regimen to assess potential interactions.
- Coadministration of remdesivir and dexamethasone has not been formally studied, but a clinically significant pharmacokinetic interaction is not predicted.
- Dexamethasone should be continued for up to 10 days or until hospital discharge, whichever comes first.

CONSIDERATIONS IN PREGNANCY

Steroids are used commonly during pregnancy for enhancement of fetal lung maturation a long time ago without hazards (Dashraath et al., 2020).

There is risk of potential deterioration of the clinical condition of already sick patients. So, 12-mg betamethasone single dose is recommended to minimize hazardous effects on maternal blood sugar and her clinical condition (Kakoulidis et al., 2020).

Betamethasone is not assigned to any FDA category. The use of steroids in management of SARS-CoV-2-infected pregnant women should be individualized according to clinical condition of the woman (Poon et al., 2020) with close monitoring of infected ones (Kakoulidis et al., 2020).

Given the potential benefit of decreased maternal mortality and the low risk of fetal adverse effects for a short course of dexamethasone therapy, the Panel recommends using dexamethasone in hospitalized pregnant women with COVID-19 who are mechanically ventilated (AIII) or who require supplemental oxygen but who are not mechanically ventilated (BIII).

Other steroids such as methylprednisolone are not recommended by the WHO as they delay virus clearance and cause maternal hyperglycemia without adding benefits on maternal survival (Li et al., 2020a).

The WHO recommended the use of systemic corticosteroids for the treatment of patients with severe and critical COVID-19 (strong recommendation, based on moderate certainty evidence) and suggested not to use corticosteroids in the treatment of patients with nonsevere COVID-19 (conditional recommendation, based on low certainty evidence) (World Health Organization, 2020b).

One essential feature of corticosteroid treatment is its duration, especially in COVID-19 cases with persistent ground-glass lung opacities. At the present time, extended corticosteroids therapy beyond 10 days is considered only in selected patients of severe COVID-19 (Villar et al., 2020a,b).

Corticosteroid is a double-edged sword in the treatment of COVID-19 patients. Corticosteroids use and its duration should consider the risk–benefit ratio (Mishra and Mulani, 2020).

Long duration of treatment prevents postdisease fibrosis in COVID-19 patients but may be associated with poor outcomes. The steroid effects in the procoagulant environment of COVID-19 patients are to be considered. A hypercoagulable state with profound endothelial injury may be responsible for thrombosis that occurs in severe COVID-19 infection (Maiese et al., 2020).

Dexamethasone (6 mg per day) increases the clotting factors especially fibrinogen so it can precipitate clinical thrombosis (Brotman et al., 2006).

A protracted corticosteroid course contributes to the so-called long COVID syndrome that presents with fatigue and psychological symptoms, in which steroid-related adverse drug reactions such as myopathy, neuromuscular weakness, and psychiatric symptoms might have a part to play (Warrington and Bostwick, 2006; Steinberg et al., 2006).

INTERLEUKIN-6 INHIBITORS

Interleukin (IL)-6 is a pleiotropic, proinflammatory cytokine produced by a variety of cell types, including lymphocytes, monocytes, and fibroblasts. COVID-19 infection induces IL-6 production from bronchial epithelial cells in a dose-dependent pattern (Yoshikawa et al., 2009). Increased cytokine release in COVID-19 patients with systemic inflammatory response and hypoxic respiratory failure can be detected through elevated blood levels of IL-6, C-reactive protein (CRP), D-dimer, and ferritin (Wang et al., 2020f; Zhou et al., 2020a). Controlling the elevated levels of IL-6 may alter the aggressive course of the disease.

The FDA-approved IL-6 inhibitors include anti-IL-6 receptor monoclonal antibodies (e.g., sarilumab, tocilizumab) and anti-IL-6 monoclonal antibodies (e.g., siltuximab).

ANTI–INTERLEUKIN-6 RECEPTOR MONOCLONAL ANTIBODIES
Sarilumab

Sarilumab is an FDA-approved recombinant humanized anti-IL-6 receptor monoclonal antibody used in treatment of rheumatoid arthritis. The available form is subcutaneous injection that was not approved for the treatment of cytokine release syndrome (CRS).

The safety and efficacy of 400 and 200 mg IV sarilumab was compared with placebo in hospitalized COVID-19 patients in an adaptive Phase 2 and 3, randomized (2:2:1), double-blind, placebo-controlled trial (ClinicalTrials.gov Identifier NCT04315298). Randomization was stratified by the severity of disease (i.e., severe, critical, multisystem organ dysfunction) and use of systemic corticosteroids for COVID-19. The Phase 2 component of the trial reported a reduced level of CRP in both doses of sarilumab. The primary outcome for Phase 3 of the trial was change on a seven-point ordinal scale, and this phase was modified to focus on the dose of sarilumab 400 mg among the patients in the critically ill group. During the conduct of the trial, there were numerous amendments that increased the sample size and modified the dosing strategies being studied, and multiple interim analyses were performed. Ultimately, the trial findings to date do not support a clinical benefit of sarilumab for any of the disease severity subgroups or dosing strategies studied. Additional detail (as would be included in a published manuscript) is required to fully evaluate the implications of these study findings (Sanofi and Regeneron, 2020).

Adverse Effects

Side effects of sarilumab include dose-dependent transient and/or reversible liver enzymes elevations, rarely neutropenia and thrombocytopenia. Serious bacterial and/or fungal infections and intestinal perforations were reported with prolonged use of the drug.

Considerations in Pregnancy

There are insufficient data about the risk of teratogenicity or miscarriage. Monoclonal antibodies actively cross the placenta (with highest transfer during the third trimester), which may affect immune responses in utero in the exposed fetus.

TOCILIZUMAB

Tocilizumab is FDA-recombinant humanized anti-IL-6 receptor monoclonal antibody that is used in patients with rheumatologic disorders and CRS induced by chimeric antigen receptor T cell therapy. It is available in both IV and SQ injection. However, IV route should be used in treating CRS (Le et al., 2018).

Clinical Data for COVID-19

In the industry-sponsored Phase 3 COVACTA trial (ClinicalTrials.gov Identifier NCT04320615), 450 adults hospitalized with severe COVID-19-related pneumonia were randomized to receive tocilizumab or placebo. No statistically significant difference was found between the two study groups regarding improved clinical status, which was measured using a seven-point ordinal scale to assess clinical status based on the need for intensive care and/or ventilator use and the requirement for supplemental oxygen over a 4-week period (OR 1.19; 95% CI: 0.81−1.76; $P = .36$). At week 4, mortality rates did not differ between the tocilizumab and placebo groups (19.7% vs. 19.4%; difference of 0.3%; 95% CI: −7.6%−8.2%; $P = .94$). The difference in median number of ventilator-free days between the tocilizumab and placebo groups did not reach statistical significance (22 days for tocilizumab group vs. 16.5 days for placebo group; difference of 5.5 days; 95% CI: −2.8 to 13.0 days; $P = .32$). Infection rates at week 4 were 38.3% in the tocilizumab group and 40.6% in the placebo group; serious infection rates were 21.0% and 25.9% in the tocilizumab and placebo groups, respectively (Roche, 2020).

Adverse Effects

Tocilizumab has similar side effects such as dose-dependent liver enzymes elevations, uncommonly neutropenia and thrombocytopenia. Serious bacterial and/or fungal infections and intestinal perforations were reported with prolonged use of the drug.

Considerations in Pregnancy

Exactly as with sarilumab, there are insufficient data about the risk of teratogenicity or miscarriage.

Monoclonal antibodies actively cross the placenta (with highest transfer during the third trimester), which may affect immune responses in utero in the exposed fetus.

ANTI−INTERLEUKIN-6 MONOCLONAL ANTIBODY

Siltuximab

Siltuximab is an FDA-approved recombinant human−mouse chimeric monoclonal antibody that binds IL-6 used in treatment of Castleman's disease. It inhibits the binding of IL-6 to its receptors (both soluble and membrane-bound receptors). Siltuximab is dosed as an IV infusion.

Clinical Data for COVID-19

There are limited, unpublished data describing the efficacy of siltuximab in patients with COVID-19 (Gritti et al., 2020). There are no data describing clinical experiences using siltuximab for patients with other novel coronavirus infections (SARS or MERS).

Adverse Effects

The primary adverse effects reported for siltuximab have been related to rash. Additional adverse effects (e.g., serious bacterial infections) have been reported only with long-term dosing of siltuximab once every 3 weeks.

Considerations in Pregnancy

There are insufficient data about the risk of teratogenicity or miscarriage. Monoclonal antibodies actively cross the placenta (with highest transfer during the third trimester), which may affect immune responses in utero in the exposed fetus.

The Panel recommends against the use of anti-IL-6 receptor monoclonal antibodies (e.g., sarilumab, tocilizumab) or anti-IL-6 monoclonal antibody (siltuximab) for the treatment of COVID-19, except in a clinical trial (BI) as preliminary, unpublished data from RCTs failed to demonstrate safety or efficacy of sarilumab or tocilizumab in COVID-19 patients and only limited, unpublished data describing the efficacy of siltuximab in patients with COVID-19 (Gritti et al., 2020).

INTERFERON

IFNs are a family of cytokines with antiviral properties. They have been suggested as a potential treatment for COVID-19 especially, type I (IFN-I) because of their in vitro and in vivo antiviral properties.

The IFN-I family includes IFN-α, IFN-β, and IFN-ω. These molecules provide innate immune initial rapid antiviral response. IFN-I proteins are induced when a cell detects viral RNA through protein sensors, e.g., TLR3, TLR7, and TLR8 that are found within the endosomes. The IFN-I molecules then bind to the cell surface receptor IFNAR (types 1 and 2) with transcription of many genes (Schoggins et al., 2011) that block viral replication (Meffre and Iwasaki, 2020).

CLINICAL DATA FOR COVID-19

In a double-blind, placebo-controlled trial conducted in the United Kingdom on hospitalized nonventilated COVID-19 patients. Compared with the patients receiving placebo (n = 50), the 48 patients receiving inhaled IFN-β1a (once daily for up to 14 days) were more likely to recover to ambulation without restrictions (HR [hazard ratio] 2.19; 95% CI: 1.03–4.69; $P = .04$), had decreased odds of developing severe disease (OR 0.21; 95% CI: 0.04–0.97; $P = .046$), and had less breathlessness when compared with the 50 placebo patients (Synairgen plc, 2020).

Another single center open-label, randomized trial conducted on severe COVID-19 patients found no difference between 42 patients who received subcutaneous IFN-β1a (three times weekly for 2 weeks) and 39 control patients regarding time to clinical response, overall length of hospital stay, length of ICU stay, or duration of mechanical ventilation. They reported a lower 28-day mortality in IFN group; however, four patients in this group were excluded from the analysis as they died before receiving the fourth dose of IFN-β1a, which may reflect biased results (Davoudi-Monfared et al., 2020).

INTERFERON-ALPHA-2B

A Chinese retrospective cohort study of 77 moderate COVID-19 adults patients compared treatment with nebulized IFN-α2b, umifenovir only versus combination of the two drugs. The time to viral clearance in the upper respiratory tract and reduction in systemic inflammation was shorter in the IFN groups. However, bias in these results as participants in the combined therapy group were younger (mean age 40 vs. 65 years old) and had less morbidity events (15 vs. 54%) when compared with participants in the umifenovir only group. The nebulized IFN-α2b formulation is not approved by the FDA for use in the United States (Zhou et al., 2020b).

A prospective observational study was conducted on 814 patients suffering from severe COVID-19 infection in Cuba. They were subjected to combination of oral antivirals (lopinavir/ritonavir and chloroquine) with intramuscular administration of IFN-α2b (761 patients) or the approved COVID protocol without IFN treatment (53 patients). The proportion of patients discharged from hospital (without clinical and radiological symptoms and nondetectable virus by real-time polymerase chain reaction) was higher in the IFN-treated compared with the non–IFN-treated group (95.4% vs. 26.1%, $P < .01$). The case fatality rate (CFR) for all patients was 2.95%, and for those patients who received IFN-α2b, the CFR was reduced to 0.92. Intensive care was required for 82 patients (10.1%), and 42 (5.5%) had been treated with IFN. The authors claimed that the use of Heberon Alpha R may be contributed to complete recovery of patients. The study has the limitation of unmatched demographic characteristics between treatment groups with huge difference in the number of participants between the two arms (Pereda et al., 2020).

CLINICAL DATA FOR SARS AND MERS

IFN-β used alone and in combination with ribavirin in patients with SARS and MERS has failed to show a significant positive effect on clinical outcomes (Al-Tawfiq et al., 2014; Shalhoub et al., 2015; Omrani et al., 2014; Chu et al., 2004).

ADVERSE EFFECTS

The most frequent adverse effects of IFN-α include flu-like symptoms, nausea, fatigue, weight loss, hematological toxicities, elevated transaminases, and psychiatric problems (e.g., depression and suicidal ideation). IFN-β is better tolerated than IFN-α (Food and Drug Administration, 2018, 2019c).

DRUG–DRUG INTERACTIONS

The most serious drug–drug interactions with IFN are the potential for added toxicity with concomitant use of other immunomodulators and chemotherapeutic agents (Food and Drug Administration, 2018, 2019c).

CONSIDERATIONS IN PREGNANCY

No proven adverse birth outcomes as congenital malformations or miscarriage was associated with exposure to IFN-β1b whether in the preconception period or during pregnancy (Hellwig et al., 2020; Sandberg-Wollheim et al., 2011), and IFN exposure did not affect neonatal morphological characteristics such as birth weight, neonatal height, or head circumference (Burkill et al., 2019).

RECOMMENDATION

The COVID-19 Treatment Guidelines Panel recommends against the use of IFN for the treatment of severe or critical COVID-19 patients, except in a clinical trial (AIII). There are insufficient data to recommend either for or against the use of IFN-β for the treatment of early (within 1 week from onset of symptoms) mild and moderate COVID-19 cases.

CONVALESCENT PLASMA

Plasma collected from people recovering from infection, especially after severe form, may contain high levels of polyclonal, pathogen-specific antibodies. These antibodies can provide a rapid passive immunity to recipients and, in viral infection, can neutralize viral particles (Casadevall and Pirofski, 2020). The use of CP to passively transfer antibodies aiming to prevent or treat infections dates back almost 100 years and before the advent of vaccines, including during the influenza pandemic of 1918 (Mair-Jenkins et al., 2015). Hyperimmune globulin is still used for postexposure prophylaxis against various viral infections, including hepatitis B, varicella zoster, and rabies (Estcourt and Roberts, 2020).

During MERS and SARS coronavirus outbreaks, CP was used safely and efficiently to treat infected cases (although evidence was obtained from small case series). The efficacy may be related to faster viral clearance, particularly when administered early in the disease course (Casadevall and Pirofski, 2020). Most patients who recover from COVID-19 infection develop circulating antibodies to various SARS-CoV-2 proteins 14–21 days following infection, which are detectable by enzyme-linked immunosorbent assay or other quantitative assays and often correlate with the presence of neutralizing antibodies. These antibodies appear to be protective, based on several primate studies showing animals could not be reinfected with SARS-CoV-2 weeks to months later.

On August 23, 2020, the US FDA granted emergency use authorization (EUA) of CP in hospitalized COVID-19 patients. They suggested using high-titer CP during early stages of the disease. Titer was measured by specific antiviral Ig testing and titer threshold criteria (Food and Drug Administration, 2020c).

Both high-titer (i.e., Ortho VITROS SARS-CoV-2 IgG tested with signal-to-cutoff ratio ≥12) and low-titer COVID-19 CPs are authorized for use (Food and Drug Administration, 2020c).

The process of CP preparation is complex and requires cooperation between recovered patients, plasma collection centers, treating clinicians, receiving patients, and healthcare administrators. Plasmapheresis is used to collect large volumes of plasma. Clinical assays that measure the level of antibodies reacting against various SARS-CoV-2 protein are widely available and may correlate with neutralizing antibody titers, and thus might be used to predict the potency of CP units, although data on this relationship continue to evolve (Li et al., 2020b).

Potential donors must have had confirmed COVID-19 infection (documented through positive nasopharyngeal swab or positive serological tests), free of symptoms for at least 14 days, and have standard blood donor eligibility requirements. At the present time, patients treated with CP are not allowed to donate blood or its products including plasma for 3 months. Donations are accepted as frequently as weekly for several months following recovery before the decline in antibody titers (Li et al., 2020b).

ADVERSE EFFECTS

Before administering CP to patients with a history of severe allergic or anaphylactic transfusion reactions, the Panel recommends consulting a transfusion medicine specialist who is associated with the hospital blood bank.

Serious adverse reactions following the administration of CP are uncommon and similar to the usual risks associated with plasma infusions for any indications. These risks include blood transmitted diseases (e.g., HIV, hepatitis B, hepatitis C), allergic reactions, anaphylactic reactions, febrile nonhemolytic reactions, transfusion-related acute lung injury (TRALI), circulatory overload (TACO), and hemolytic reactions. Hypothermia, metabolic complications, and posttransfusion purpura have also been described (Food and Drug Administration, 2020a).

Additional risks include a theoretical risk of antibody-dependent enhancement and a theoretical risk of suppressed long-term immunity.

CONSIDERATIONS IN PREGNANCY

The safety and effectiveness of COVID-19 CP during pregnancy have not been evaluated.

An open-label randomized trial of CP use in patients with severe or critical COVID-19 was conducted in Wuhan, China, from February 14 to April 1, 2020. Only plasma units with a SARS-CoV-2 viral spike receptor–binding domain-specific IgG titer of ≥1:640 were transfused. There was no significant difference between the treatment and control groups in time to clinical improvement within 28 days (HR 1.40; 95% CI: 0.79–2.49; $P = .26$). 91% versus 68% in severe cases (difference of 23%; OR 1.34; 95% CI: 0.98–1.83; $P = .07$) and 21% versus 24% in life-threatening disease in CP and control groups, respectively. There was no significant difference in mortality (21% vs. 24% in CP and control groups, respectively; ($P = .30$). At 24 h, the rates of negative SARS-CoV-2 viral polymerase chain reaction were significantly higher in the CP group (45%) than in the control group (15%; $P = .003$), and differences persisted at 72 h. The nonblinding and early termination of the trial are the main limitations of the study (Li et al., 2020b).

PLACID Trial is an open-label, randomized clinical trial of CP versus standard of care for hospitalized patients with COVID-19 conducted in 39 tertiary care centers in India between April 22 and July 14, 2020. Confirmed severe COVID-19 patients with hypoxia were eligible if matched donor plasma was available at the time of enrollment. Critically ill patients (those with a ratio of arterial partial pressure of oxygen to fraction of inspired oxygen [PaO$_2$/FiO$_2$] <200 mmHg or shock) were excluded. Participants in the intervention arm received two doses of 200 mL plasma, transfused 24 h apart. Antibody titer of transfused plasma was not assessed.

Among 454 participants, 235 were randomized into the CP arm and 229 were randomized into the standard of care arm. The arms were well balanced regarding age (median of 52 years in both arms) and days from symptom onset to enrollment (median of 8 days in both arms). There was no difference in the primary outcome (time to disease progression and 28-day mortality) across the trial arms. The composite outcome occurred in 44 patients (18.7%) in the CP arm and 41 (17.9%) in the control arm. 34 (14.5%) in the CP arm and 31 patients (13.6%) in the control arm died. In each arm, 17 participants progressed to severe disease (7.2% in the CP arm vs. 7.4% in the standard of care arm) (Agarwal et al., 2020). The nonblinding and non-testing of the antibody titer in the donated plasma were the main limitations of the study.

The Expanded Access to Convalescent Plasma for the Treatment of Patients with COVID-19 program was an open-label, nonrandomized that was primarily designed to provide adult patients with severe or critical COVID-19 infection with access to CP. Secondary objectives were to obtain data on the safety of the intervention. Exploratory objectives included assessment of 7-day and 28-day mortality. The program was sponsored by the Mayo Clinic and included a diverse range

of clinical sites. SARS-CoV-2 antibody testing of plasma donors and assessment of SARS-CoV-2 neutralization potential were not mandated. Patients were transfused with 1 or 2 units (200–500 mL) of CP. The main outcomes for the safety analysis were serious adverse events (including death) reported after 4 h and 7 days of transfusion (Joyner et al., 2020a,b).

A peer-reviewed publication described the safety outcomes for the first 20,000 EAP plasma recipients, enrolled between April 3 and June 2, 2020. One-third of the participants were aged $\geq$70 years, 60% were men, and 71% had severe or life-threatening COVID-19. 20% of the participants were African American, 35% were Hispanic/Latino, and 5% were Asian. Thirteen deaths were assessed as possibly or probably related to the CP treatment. The 83 nonfatal SAEs that were assessed as possibly or probably related to the CP treatment included 37 TACO events, 20 TRALI events, and 26 severe allergic reactions. The life-threatening events that were reported up to 7 days after transfusion included 87 thrombotic/thromboembolic complications, 406 sustained hypotension events, and 643 cardiac events. The overall mortality rate was 8.6% at 7 days (Joyner et al., 2020a,b).

Both the FDA and the Mayo Clinic performed retrospective, indirect evaluations of the efficacy of COVID-19 CP by using subsets of EAP data, hypothesizing that patients who received plasma units with higher titers of neutralizing antibodies would have better clinical outcomes than those who received plasma units with lower titers of antibodies.

The FDA analysis included 4330 patients, and donor-neutralizing antibody titers were measured by the Broad Institute using a pseudovirus assay. The analysis revealed no difference in 7-day mortality between the patients who received high-titer plasma and those who received low-titer plasma, in the patient population overall, or in the subset of patients who were intubated. However, among nonintubated patients (approximately two-thirds of those analyzed), mortality within 7 days of transfusion was 11% for those who received high-titer plasma and 14% for those who received low-titer plasma ($P = .03$). In a post hoc analysis of patients aged <80 years who were not intubated and who were treated within 72 h of COVID-19 diagnosis, 7-day mortality was lower among the patients who received high-titer plasma than among those who received low-titer plasma (6.3% vs. 11.3%, respectively; $P = .0008$) (FDA, 2020a).

A similar efficacy analysis by the Mayo Clinic, which has not been peer reviewed, included 3082 participants who received a single unit of plasma out of the 35,322 participants who had received plasma through the EAP by July 4, 2020. Antibody titers were measured by using the Ortho Clinical Diagnostics COVID-19 IgG assay, and outcomes in patients transfused with low- (lowest 18%), medium-, and high- (highest 17%) titer plasma were compared. After adjusting for baseline characteristics, the 30-day mortality in the low-titer group was 29% and 25% in the high-titer group. This difference did not reach statistical significance. Similar to the FDA analyses, post hoc subgroup analyses suggested a benefit of high-titer plasma in patients aged <80 years who received plasma within 3 days of COVID-19 diagnosis and who were not intubated (Joyner et al., 2020a,b).

Limitations of the Study

- The lack of an untreated control arm limits interpretation of the safety and efficacy data. For example, the possibility that differences in outcomes are attributable to harm from low-titer plasma rather than benefit from high-titer plasma cannot be excluded.
- The EAP data may be subject to multiple confounders, including regional differences and temporal trends in the management of COVID-19.
- There is no widely available and generally agreed-upon best test for measuring neutralizing antibodies, and the antibody titers in CP from patients who have recovered from COVID-19 are highly variable.
- The efficacy analyses rely on a subset of EAP patients who only represent a fraction of the patients who received CP through the EAP.
- The subgroup that demonstrated the largest estimated effect between high-titer and low-titer CP— patients aged <80 years who were not intubated and who were transfused within 3 days of COVID-19 diagnosis—was selected post hoc by combining several subset rules, which favored subgroups that showed a trend toward benefit of high-titer plasma. This approach tends to overestimate the treatment effect.
- The FDA analysis relied on 7-day mortality, which may not be clinically meaningful in the context of the prolonged disease course of COVID-19. Because participants in this observational study were not rigorously followed after they were discharged from the hospital, the 30-day mortality estimates are uncertain.

A randomized trial was conducted in hospitalized adult patients with severe COVID-19 pneumonia in a 2:1 ratio to receive CP or placebo. A total of 228 patients were assigned to receive CP and 105 to receive placebo. The median titer of the infused CP was 1:3200 of total SARS-CoV-2 antibodies (interquartile range, 1:800 to 1:3200]. After 30 days, no significant difference was detected between the two groups regarding the distribution of clinical outcomes according to the ordinal scale (OR 0.83; 95% CI: 0.52 to 1.35; $P = .46$). Overall mortality was 10.96% and 11.43% in CP and placebo group, respectively, for a risk difference of -0.46 percentage points (95% CI: -7.8 to 6.8). Total SARS-CoV-2 antibody titers tended to be higher in the CP group at day 2 after the intervention. Adverse events and serious adverse events were similar in the two groups (Simonovich et al., 2020).

RECOMMENDATION

There are insufficient data for the COVID-19 Treatment Guidelines Panel (the Panel) to recommend either for or against the use of COVID-19 CP for the treatment of COVID-19.

Currently, there are insufficient data from well-controlled, adequately powered, randomized clinical trials to evaluate the efficacy and safety of CP for the treatment of COVID-19. However, >70,000 patients in the United States have received COVID-19 CP through the Mayo Clinic's Expanded Access Program (EAP), which was designed primarily to provide broad access to investigational CP and thus did not include an untreated control arm. Both the FDA and the Mayo Clinic performed retrospective, indirect evaluations of efficacy by using the Mayo Clinic EAP data, hypothesizing that patients who received plasma units with higher titers of SARS-CoV-2 neutralizing antibodies would have better clinical outcomes than those who received plasma units with lower antibody titers. The results of their analyses suggest that CP with high antibody titers may be more beneficial than low-titer plasma in nonintubated patients, particularly when administered within 72 h of COVID-19 diagnosis.

The FDA determined that these findings—along with additional data from small randomized and nonrandomized studies, observational cohorts, and animal experiments—met the criteria for EUA issuance (Food and Drug Administration, 2020d). Despite meeting the "may be effective" criterion for EUA issuance, the EAP analyses are not sufficient to establish the efficacy or safety of CP due to the lack of a randomized, untreated control group and potential confounding. There is no widely available and generally agreed-upon best test for measuring neutralizing antibodies, and the antibody titers of plasma from patients who have recovered from COVID-19 are highly variable. Furthermore, hospitalized patients with COVID-19 may already have SARS-CoV-2 neutralizing antibody titers that are comparable with those of plasma donors, potentially limiting the benefit of CP in this patient population (Gharbharan et al., 2020; Agarwal et al., 2020). Several randomized, placebo-controlled trials of COVID-19 CP are ongoing.

The Panel's assessment of the EAP data is consistent with the FDA statements in the CP EUA documents (FDA, 2020a).

IMMUNOGLOBULINS: SARS-COV-2 SPECIFIC

Intravenous immunoglobulins (IVIGs) are commonly used immunotherapeutic agents for treatment of various autoimmune and inflammatory diseases. These are normal IgG isolated from the pooled plasma of several thousand healthy donors (Perez et al., 2017). The therapeutic dose of IVIG is 2 g/kg infused over one, two, or five consecutive days. IVIGs are used to treat many diseases including Kawasaki disease, immune thrombocytopenic purpura, inflammatory myopathies, Guillain–Barré syndrome, graft-vs-host disease, and blistering diseases. IVIG exerts its therapeutic effects through targeting the inflammatory immune response mediators (both soluble and cellular mediators). That can be achieved through complement scavenging, autoantibodies neutralization; enhancement of autoantibodies degradation by neonatal Fc receptor saturation; inhibition of activation of various innate immune cells, including dendritic cells, monocytes, macrophages, and neutrophils, and secretion of inflammatory mediators; suppression of effector T helper cells Th1 and Th17, and reciprocal enhancement of immunoprotective regulatory T cells (Tregs); and blockade of B cell activation (Galeotti et al., 2017).

The possible mechanisms of action of IVIGs in COVID-19 are through reduction in the inflammatory mediators. IVIGs target cytokine storm in severe and critically ill COVID-19 patients (Fig. 5.2). Although IVIGs could neutralize seasonal coronavirus, they could not provide cross-neutralizing antibodies against SARS-CoV-2 (Schwaiger et al., 2020). Therefore, the benefits of IVIGs cannot be explained by passive virus neutralization. SARS-CoV-2 encodes a superantigen motif near its S1/S2 cleavage site that might trigger cytokine storm (Cheng et al., 2020). As IVIG contains antibodies reacting against SARS-CoV-2 antigens (Díez et al., 2020), IVIG might inhibit superantigen-mediated T cell activation and cytokine release.

Potential risks may include transfusion reactions. Theoretical risks may include antibody-dependent enhancement of infection.

CONSIDERATIONS IN PREGNANCY

Pathogen-specific immunoglobulins are used clinically during pregnancy to prevent varicella zoster virus and rabies and have also been used in clinical trials of therapies for congenital CMV infection.

RECOMMENDATION

There are insufficient data for the COVID-19 Treatment Guidelines Panel to recommend either for or against SARS-CoV-2 IG for the treatment of COVID-19. Trials evaluating SARS-CoV-2 immunoglobulins are in development but not yet active and enrolling participants. Similarly, there are no clinical data on use of specific immunoglobulin or hyperimmunoglobulin products in patients with SARS or MERS.

IMMUNOGLOBULINS: NON-SARS-COV-2 SPECIFIC

Recommendation

The COVID-19 Treatment Guidelines Panel recommends against the use of non-SARS-CoV-2)-specific IVIG for the treatment of COVID-19, except in a clinical trial (AIII). This recommendation should not prevent the use of IVIG when otherwise indicated for the treatment of COVID-19 complications.

RATIONALE FOR RECOMMENDATION

It is unknown whether products derived from the plasma of donors without confirmed SARS-CoV-2 infection contain high titer of SARS-CoV-2 neutralizing antibodies. Furthermore, although other blood components in IVIG may have general immunomodulatory effects, it is unclear whether these theoretical effects will benefit patients with COVID-19.

CLINICAL DATA FOR COVID-19

This study has not been peer reviewed.

A retrospective, non–peer-reviewed nonrandomized cohort study of IVIG for the treatment of

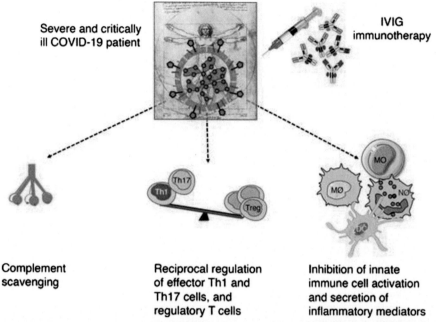

Severe and critically
ill COVID-19 patient

IVIG
immunotherapy

Complement
scavenging

Reciprocal regulation
of effector Th1 and
Th17 cells, and
regulatory T cells

Inhibition of innate
immune cell activation
and secretion of
inflammatory mediators

FIG. 5.2 Possible mechanisms of action of intravenous immunoglobulin (IVIG) in coronavirus disease-19 (COVID-19). It is not yet clear how IVIG benefits severe and critically ill COVID-19 patients. Several studies have reported that IVIG reduces IL-6 and C-reactive protein in the COVID-19 patients suggesting that IVIG targets inflammatory process. Based on the known mechanisms of IVIG demonstrated in autoimmune and inflammatory diseases, and considering the pathophysiology of COVID-19, we propose that IVIG might target cytokine storm in COVID-19 patients by complement scavenging; reciprocal regulation of effector Th1 and Th17 cells, and regulatory T cells (Treg); and inhibition of innate immune cell activation and secretion of inflammatory mediators. *DC*, dendritic cell; *MO*, monocyte; *MØ*, macrophage; *NØ*, neutrophil (Galeotti et al., 2020).

COVID-19 was conducted across eight treatment centers in China between December 2019 and March 2020. The study showed no difference in 28-day or 60-day mortality between 174 patients who received IVIG and 151 patients who did not receive IVIG. More patients in the IVIG group had critical disease at study entry (41% vs. 21% in IVIG and control groups respectively). The median hospital stay was longer in the IVIG group (24 vs.16 days), and the median duration of disease was also longer in the IVIG group (31 vs. 23 days). A subgroup analysis that was limited to the critically ill patients suggested a mortality benefit at 28 days, which was no longer significant at 60 days (Shao et al., 2020). Interpretations of these results are difficult as it was nonrandomized, unbalanced arms as patients in the IVIG group were older with higher proportion of them had more severe disease when compared with non-IVIG arm and nonbalancing of concomitant therapeutic drug administration among the two groups.

CONSIDERATIONS IN PREGNANCY

IVIG is commonly used in pregnancy for other indications such as immune thrombocytopenia with an acceptable safety profile (ACOG Practice Bulletin No. 207: Thrombocytopenia in Pregnancy, 2019; Neunert et al., 2011).

MESENCHYMAL STEM CELLS

Stem cell–based therapies, especially MSCs, are currently used to treat many diseases (Wang et al., 2012; Murphy et al., 2013). MSC-based therapy is used for management of cases of ARDS based on its property of secretion of antiinflammatory, antifibrosis, and antiapoptosis cytokines, which eventually reduce the cytokine storm (Sadeghi et al., 2020a; Leng et al., 2020).

MSCs can inhibit and control the cytokine storm through their immunomodulatory properties through both cellular actions and release of soluble factors (Leng et al., 2020).

These two mechanisms modulate the proliferation and activation of T cells and induce the polarization of the mononuclear cells to an antiinflammatory phenotype (Lee et al., 2009). MSCs can inhibit T cell activation through several immunomodulatory factors (e.g., transforming growth factor-beta 1 (TGF-β1), prostaglandin E2 (PGE2), and HLA-G5) and membrane-bounded molecules (e.g., PD-L1, VCAM-1, and Gal-1) (Wang and Ma, 2008).

MSCs also increase regulatory T cells (TReg) and antiinflammatory TH2 cells. MSCs release NO and IDO that suppress the T cell cytokine production (Pittenger et al., 2019). MSCs suppress NK cell cytotoxicity with a decrease in IFN-γ expression and prevent dendritic cells maturation (by downregulating the surface expression of CD80, CD86, and MHC class II molecules), thus retaining the dendritic cells in a tolerogenic phenotype and also inducing antiinflammatory M2-macrophage polarization with the increased levels of PGE2, TSG-6, and IL-1RA (Lee et al., 2019).

Unlike other therapeutic options used as immunomodulators, MSCs create an immunomodulated environment via the secretion of many immunomodulatory factors. The unique immunomodulatory property of MSCs is reflected in many clinical trials; it has been shown that MSCs reduce the inflammatory responses and defend the host against cytokine storm with lowered mortality, without serious side effects (Lee et al., 2019; Pittenger et al., 2019). As a result, MSC therapy has emerged as an attractive strategy through several favorable changes in the management of respiratory models such as H7N9-induced ARDS (Trounson and McDonald, 2015; Squillaro et al., 2016; Han et al., 2019). Since H7N9 and COVID-19 have similar complications, MSC therapy could be an alternative approach in COVID-19 treatment (Chen et al., 2020a,b,c).

Furthermore, MSCs lack the ACE2 receptor that SARS-CoV-2 uses for viral entry into cells; therefore, MSCs are resistant to infection (Shetty, 2020; Lukomska et al., 2019). The protective effects of MSCs in COVID-19 are shown in Fig. 5.3.

Clinical Data

Data supporting the use of MSCs in patients with viral infections, including COVID-19 infection, are limited to case reports and small, open-label studies.

A pilot study of intravenous MSC transplantation in China enrolled 10 patients with confirmed COVID-19. Seven of them (one has critical, four have severe, and two have common form). Seven patients received MSCs, and the other three (all have critical form) received placebo. All patients in the MSCs group

recovered. Among the three control patients, one died, one developed ARDS, and one remained stable with severe disease (Leng et al., 2020).

A small clinical trial was conducted on 41 patients with severe COVID-19 infection, 12 of them received human umbilical cord mesenchymal stem cell (hUC-MSC) infusion, and 29 received standard of care therapies only. The groups were well balanced with regard to demographic characteristics, laboratory test results, and disease severity. All participants in the hUC-MSC group recovered without requiring mechanical ventilation, whereas four patients in the standard care group progressed to critical illness requiring mechanical ventilation (three of them died). These results are statistically nonsignificant, and interpretation of the study is limited by nonrandomization and small sample size (Shu et al., 2020).

A case report of a 65-year-old female with critical COVID-19 infection mentioned that she had severe pneumonia, respiratory, and multiorgan failure. She received allogeneic human umbilical cord blood-derived MSCs for three times (5×10^7 cells each time). After the second dose, the patient was off the ventilator and discharged from the ICU 2 days after the third dose and testing for SARS-CoV-2 turned negative (Liang et al., 2020).

Adverse effects of MSC transfusion are uncommon. These include failure to produce the desired effects and its potential to multiply or change into inappropriate cell types, product contamination, growth of tumors, infections, thrombus formation, and administration site reactions (Center for Disease Control and Prevention, 2019).

Considerations in Pregnancy

There are insufficient data to assess the risk of MSC use during pregnancy (Sadeghi et al., 2020b).

Recommendation

The COVID-19 Treatment Guidelines Panel recommends against the use of MSCs for the treatment of COVID-19, except in a clinical trial (AII). No MSCs are approved by the FDA for the treatment of COVID-19. There are insufficient data to assess the role of MSCs for the treatment of COVID-19.

The FDA has recently issued several warnings about patients being potentially vulnerable to stem cell treatments that are illegal and potentially harmful (Food and Drug Administration, 2019b). Several cord blood-derived products are currently licensed by the FDA for indications such as the treatment of cancer (e.g., stem cell transplant) or rare genetic diseases, and as scaffolding for cartilage defects and wound beds. None of

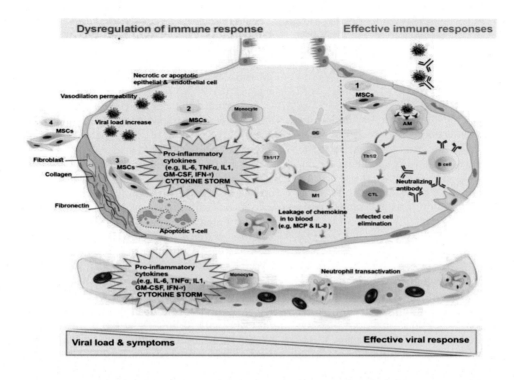

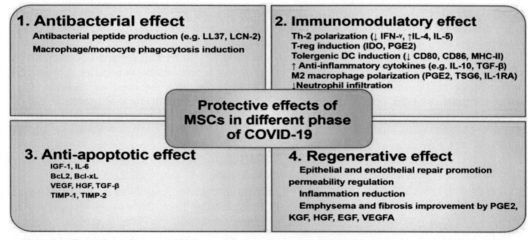

FIG. 5.3 Protective effects of MSCs in different phases of COVID-19 (Sadeghi et al., 2020b). *MSCs,* mesenchymal stem cells.

these products are approved for the treatment of COVID-19 or any other viral disease (Food and Drug Administration, 2019a). In the United States, MSCs should not be used for the treatment of COVID-19 outside of an FDA-approved clinical trial, expanded access programs, or an Emergency Investigational New Drug application (AII).

For the use of immunotherapy in treatment of COVID-19 infection, the Treatment Guidelines Panel (the Panel) does not have recommendations either for or against the use of COVID-19 CP or SARS-CoV-2 immunoglobulins as the available data are insufficient, but the Panel recommends against the use MSCs (AII) or non-SARS-CoV-2-specific IVIG (AIII) outside the clinical trials. This recommendation should not preclude the use of IVIG when it is otherwise indicated for the treatment of complications that arise during the course of COVID-19.

HOST DIRECTED THERAPY

Host-directed therapy (HDT) is a term used to describe therapeutic agents that do not act directly against the virus itself but modulate the host immune system to minimize tissue damage resulted from intense inflammatory reactions (Zhao et al., 2020). HDT use in acute inflammation is safe and efficient and includes treatment with metformin, statins, and glitazone (Zumla et al., 2016).

METFORMIN

Metformin was originally introduced as an antiinfluenza therapy, and its glucose-lowering ability was described as side effect of treatment (Amin et al., 2019). Some scientists described metformin as the aspirin of the 21st century because of its many pleiotropic effects and its widespread utility in medicine (Romero et al., 2017).

Metformin works through activation of AMP-activated protein kinase (AMPK) in hepatocytes by causing its phosphorylation, which results in favorable effects on glucose and lipid metabolism (Zhou et al., 2001).

AMPK activation caused by metformin leads to phosphorylation of ACE2 (Liu et al., 2019), which results in conformational and functional changes in the ACE2 receptor (Plattner and Bibb, 2012). This could lead to decreased its binding with SARS-CoV-2.

Entry of the SARS-CoV-2 inside the host cells causes downregulation of ACE2 receptors with the resultant imbalance in the renin—angiotensin—aldosterone system (RAS) promoting the harmful effects of its proinflammatory and profibrotic arm, further giving rise to the lethal cardiopulmonary complications (Wang et al., 2020a,b,c,d,e,f). By upregulating ACE2, the imbalance in RAS is corrected. So, metformin is not just working through prevention of viral cell entry but also prevents the damaging sequelae by causing activation of ACE2 (Sharma et al., 2020).

A well-known risk factor for development of the severe form of COVID-19 and its complications includes diabetes and obesity. High levels of tumor necrosis factor α (TNFα), which contribute to insulin resistance, were found in the lungs of COVID-19 patients; IL-6 is linked in COVID-19 complications (Blüher et al., 2005).

Metformin decreases TNFα and IL-6, raises IL-10 (an antiinflammatory cytokine), and has been found to cause these beneficial effects significantly more in females than males in both animal and human studies (Matsiukevich et al., 2017; Park et al., 2017; Quan et al., 2016).

Metformin enhances the production of mitochondrial reactive oxygen species and the macrophages autophagy (Beigel et al., 2019), lowering the lung damage in murine models (Zmijewski et al., 2008).

In a retrospective cohort analysis of 6256 hospitalized confirmed COVID-19 patients who have type 2 diabetes or obesity, Metformin use was not associated with significantly decreased mortality in the overall sample of men and women by either Cox proportional hazards stratified model (HR 0·887; 95% CI: 0·782−1·008) or propensity matching (OR 0·912; 95% CI: 0·777−1·071; $P = 0·15$). Metformin was associated with decreased mortality in women by Cox proportional hazards (HR 0·785; 95% CI: 0·650−0·951) and propensity matching (OR 0·759; 95% CI: 0·601−0·960; $P = 0·021$). There was no significant reduction in mortality among men (HR 0·957; 95% CI: 0·82−1·14; $P = 0·689$ by Cox proportional hazards) (Bramante et al., 2020).

Gilbert and colleagues studied the use of metformin during the first trimester of pregnancy (Gilbert et al., 2006) and seemed safe as regarding congenital malformations. Li et al. found that the use of metformin in treatment of women with gestational diabetes reduced the risk of complications as gestational hypertension, hyperglycemia, and the need of neonatal ICU (Li et al., 2015). Metformin is not assigned to any FDA category (Favilli et al., 2020).

HMG-COA REDUCTASE INHIBITORS (STATINS)

Statins are commonly used to treat hyperlipidemia, and they have the ability to decrease cytokines in numerous noninfective conditions (Fang et al., 2005). Long-term statin therapy improves the outcome in patients with bacterial pneumonia (Novack et al., 2009; Mortensen et al., 2005) and influenza (Vandermeer et al., 2012).

There is controversy about the effect of statins on the course of COVID-19 infection. Dysregulation of the myeloid differentiation primary response protein (MYD) 88 pathway with resultant overwhelming inflammation is associated with poor prognosis in previously studied coronaviruses but is still not proved for SARS-CoV-2 (Yuan, 2015)

Statins are known inhibitors of MYD88 and its level in the presence of external stressors, so they can protect COVID-19 patients from the development of hyperinflammatory reaction (Totura et al., 2015).

On the other hand, statins cause deficiency of endogenous cellular cholesterol with upregulation of low-density lipoprotein receptors with subsequent increase of exogenous cholesterol transfer through the cell membrane and formation of multiple lipid bundles increasing the accessibility for coronaviruses (Shrestha, 2020).

Statins might promote the development of a more severe course of COVID-19 due to activation of the inflammasome pathway in ARDS leading to increased proinflammatory interleukin-18 (IL-18) levels and subsequent cytokine storm (Goldstein et al., 2020)

A retrospective cohort study analyzed 717 patients admitted to a tertiary center in Singapore for COVID-19 infection. Clinical outcomes of interest were the need for supplemental oxygen (PO$_2$ ≤ 94%), ICU admission, invasive mechanical ventilation, and death. Patients were considered to have dyslipidemia if they were receiving dyslipidemia medications (statins, fibrates, or ezetimibe) for a long time. One hundred fifty-six (21.8%) patients had dyslipidemia, and 97% of these were on statins. ICU admission rate was lower in statin when compared with nonstatin users (ATET: Coeff [risk difference]: − 0.12 [−0.23, − 0.01]; P = .028). There are no significant differences between statin and nonstatin users regarding other clinical outcomes. The authors described that these findings support to continue statins prescription for COVID-19 patients (Tan et al., 2020).

A metaanalysis involved four studies with a total of 8990 COVID-19 patients: three of them were of good quality. The pooled analysis revealed a significantly reduced hazard for fatal or severe disease with the use of statins pooled (HR 0.70; 95% CI: 0.53−0.94) compared with nonuse of statins in COVID-19 patients. They suggested that the use of statins was associated with 30% reduction in fatal or severe disease (Kow and Hasan, 2020).

Considerations in Pregnancy

Data about the use of statins during pregnancy are limited. Its use was linked to major congenital malformations (Pollack et al., 2005). They are currently assigned, as FDA class thus contraindicated in pregnancy based on animal studies in which the dose administered was much higher than that normally used in human. A systematic review reported the need of evidence for safety of statins use during pregnancy. It should be avoided during the first trimester, as they carry a major teratogenic risk (Karalis et al., 2016).

Both metformin and statins can be used with antiviral drugs as adjuvants to reduce the needed dose of antiviral and consequently its side effects (Zumla et al., 2015). However, no data are available regarding their clinical use with antiviral purpose in pregnant women (Favilli et al., 2020).

Recommendations

- Persons on statin therapy for the prevention or treatment of cardiovascular disease should continue their medications if they become infected with COVID-19 (AIII).
- The Panel recommends against the use of statins for the treatment of COVID-19, except in a clinical trial (AIII)

Pioglitazone

Pioglitazone belongs to the family of thiazolidinediones (TZDs). These drugs are used in treatment of insulin resistance (Lebovitz, 2019). Insulin resistance is associated with numerous cardiovascular risk factors; increases in C-reactive protein, IL-6, and TNF-α (Liu et al., 2016a); and produces a procoagulant state with increased fibrinogen and plasminogen activator inhibitor (PAI-1) (King et al., 2016). These effects ma of insulin resistance raise many concerns about the response of patients with type 2 diabetes to COVID-19 infection (Pfützner et al., 2010). Pioglitazone inhibits the secretion of proinflammatory cytokines (e.g., IL-1b, IL-6, and IL-8) and enhances the secretion of antiinflammatory ones (e.g., IL-4 and IL-10) in astrocytes stimulated with lipopolysaccharide (Qiu and Li, 2015). Pioglitazone decreases ferritin in angiotensin II−induced hypertension in a rat model (Sakamoto et al., 2012).

Administration of pioglitazone at a dose of 30−45 mg/day for 3 months can significantly reduce IL-6 and TNFα (Xie et al., 2017), and a 4-month course of 45 mg/day reduces the monocyte gene and protein expression of IL-1b, IL-6, and IL-8 and lymphocyte IL-2, IL-6, and IL-8 (Zhang et al., 2008).

It was found that pioglitazone decreases lung injury when modulating adipose inflammation in a cecal ligation puncture (CLP) model in mice (Kutsukake et al., 2014). It has a direct effect on lung inflammation and fibrosis (Aoki et al., 2009) and can reduce the lung fibrotic reaction to silica-exposed rats (Barbarin et al., 2005).

The safety of glitazones in pregnancy was not studied, and most data came from case reports of accidental use or therapeutic use for PCOS, which ended in normal outcome (Yaris et al., 2004) or spontaneous abortion (Ota et al., 2008). They were not assigned to any FDA category.

OTHER THERAPEUTIC AGENTS

Concomitant drugs used include ACE inhibitors, angiotensin receptor blockers, nonsteroidal antiinflammatory drugs (NSAIDs), vitamin C, vitamin D, zinc, lactoferrin, and melatonin.

ANGIOTENSIN-CONVERTING ENZYME INHIBITORS AND ANGIOTENSIN RECEPTOR BLOCKERS

Upregulation of ACE2 expression has been shown in several animal models but a limited number of human studies showing mixed results on plasma ACE2 levels (Vaduganathan et al., 2020; Mehta et al., 2020). These findings suggested that the use of angiotensin-converting enzyme inhibitor (ACEI)/angiotensin receptor blocker (ARB) may enhance SARS-CoV-2 cell entry and its replication (Fang et al., 2020). Conversely, ACE2 expression is downregulated following SARS infection. ACE2 serves as the key enzyme for balance between two pathways: one is the classic ACE/angiotensin II, whereas the other is the angiotensin1 7/Mas/"anti—renin angiotensin system pathway (RAS)." Downregulation of ACE2 then leads to overactivation of the classic angiotensin II pathway, which is normally counteracted by the anti-RAS pathway, which can result in excessive RAS activation and lung damage, vessel leakage, inflammation, and fibrosis. Therefore, ACEI/ARB administration may block ACE2 downregulation-induced hyperactivation of RAS with the resultant acute lung injury and risk of ARDS (Sarzani et al., 2020; Zhang et al., 2020a).

A recent metaanalysis that included 21 studies was done to investigate the impact of ACEI/ARB on COVID-19 disease severity and mortality. For mortality with ACEI/ARB use, the pooled OR was 1.29 [0.89—1.87] $P = .18$ with heterogeneity of 91%, while the pooled OR for COVID-19 severity was 0.94 [0.59—1.50] $P = .81$ with heterogeneity of 89%. In combining both mortality and severe disease outcomes, the pooled OR was 1.09 [0.80—1.48] $P = .58$ but with heterogeneity of 92%. The authors concluded that the use of ACEI/ARB was not associated with increased mortality or severe COVID-19 (Lo et al., 2020).

RECOMMENDATIONS

- Persons receiving ACEI and/or ARB for cardiovascular disease (or other indications) should continue these medications if they got infected with COVID-19 (AIII).

- The COVID-19 Treatment Guidelines Panel (the Panel) recommends against the use of ACE inhibitors or ARBs for the treatment of COVID-19, except in a clinical trial (AIII).

It is unclear whether these medications are helpful, harmful, or neutral in the pathogenesis of SARS-CoV-2 infection.

NONSTEROIDAL ANTIINFLAMMATORY DRUGS

NSAIDs are one of the most common medications used as antipyretics and analgesics (Cleveland Clinic, 2020). NSAIDs work by inhibiting cyclooxygenase (COX), an enzyme-converting arachidonic acid to prostaglandins, which play an important role in mediating the inflammatory response (Kirkby et al., 2016).

The use of NSAID in patients with COVID-19 is controversial. On March 14, 2020, the French Minister of Health recommended the use of acetaminophen instead of ibuprofen based on a trial hypothesizing the worsening effects of ibuprofen on COVID-19 infection (Picheta, 2020). Rinott et al. conducted a large retrospective cohort study on 403 confirmed COVID-19 patients who received ibuprofen versus acetaminophen versus no antipyretic medication and found no significant differences in mortality rates or need for respiratory support between the groups (Rinott et al., 2020).

One study conducted by Fu et al. showed an upregulation of ACE2 receptors with the use of NSAIDs, which may facilitate SARS-CoV-2 entry inside the cells (Fu et al., 2020). On the contrary, several NSAIDs as indomethacin and naproxen have antiviral activity (Russell et al., 2020; Lejal et al., 2013). Furthermore, the antiplatelet and anti-inflammatory properties of select NSAIDs may be instrumental in symptomatically treating patients with COVID-19 and reducing the risk of morbidity and mortality (Liu et al., 2016b; Toner et al., 2015).

The common side effects of COX-2 inhibition are related to a decrease in prostaglandin 2 production in the gastrointestinal tract, leading to mucosal injury (Tai and McAlindon, 2018) and nephrotoxicity (Little, 2020). The reduced production of other prostaglandins, particularly prostacyclin (PGI2), is linked to an increased risk of adverse cardiovascular events in patients taking NSAIDs (Grosser et al., 2017). The long-term use of NSAIDs, such as selective COX-2 inhibitors and diclofenac, has also been shown to increase the risk of major vascular events, whereas other NSAIDs such as naproxen did not increase the risk of vascular events (Bhala et al., 2013).

RECOMMENDATIONS

- Persons taking NSAIDs for a comorbid condition should continue therapy as previously directed by their physician if they get infected with COVID-19 (AIII).
- The Panel recommends that there be no difference in the use of antipyretic strategies (e.g., with acetaminophen or NSAIDs) between patients with or without COVID-19 (AIII).

The FDA stated that there is no evidence linking the use of NSAIDs with worsening of COVID-19 and advised patients to use NSAIDs as directed (Food and Drug Administration, 2020b).

VITAMIN C

Vitamin C (ascorbic acid) is a water-soluble vitamin that has beneficial effects in patients with severe and critical illnesses. It has an antioxidant property and acts as a free radical scavenger that has antiinflammatory properties, influences cellular immunity and vascular integrity, and serves as a cofactor in the generation of endogenous catecholamines (Fisher et al., 2011; Wei et al., 2020).

More than 100 animal studies have indicated that a daily dose of a few grams of vitamin C may alleviate or prevent infections (Hemilä, 2017). Already during the outbreak of SARS-CoV-1 in 2003, vitamin C was used as a nonspecific treatment for severe cases (Hemilä, 1997, 2003). Vitamin C supports cellular functions of both the innate and adaptive immune systems with modification of susceptibility viral infections and manipulation of inflammatory process (Ang et al., 2018; Carr and Maggini, 2017). Vitamin C was proved to increase resistance to infection by a coronavirus in chick embryo tracheal organ cultures (Atherton et al., 1978). It also restores the stress response and improves the survival of stressed humans (Marik, 2020).

Because serious COVID-19 may cause sepsis and ARDS, the potential role of high doses of vitamin C in ameliorating inflammation and vascular injury in patients with COVID-19 is being studied.

CLINICAL DATA ON VITAMIN C IN CRITICALLY ILL PATIENTS WITHOUT COVID-19

A small, three-arm pilot study compared two regimens of intravenous (IV) vitamin C to placebo in 24 critically ill patients with sepsis. Over the 4-day study period, patients who received vitamin C 200 mg/kg per day and those who received vitamin C 50 mg/kg per day had lower sequential organ failure assessment scores and levels of proinflammatory markers than patients who received placebo (Fowler et al., 2014).

In an RCT in 167 critically ill patients with sepsis induced ARDS, patients who received IV vitamin C 200 mg/kg per day for 4 days had sequential organ failure assessment (SOFA) scores and levels of inflammatory markers that were similar to those observed in patients who received placebo. However, 28-day mortality was lower in the treatment group (29.8% vs. 46.3%; $P = .03$), coinciding with more days alive and free of the hospital and the ICU (Fowler et al., 2019). A post hoc analysis of the study data reported a difference in median SOFA scores between the treatment group and placebo group at 96 h; however, this difference was not present at baseline or 48 h (Fowler et al., 2020).

Two small studies that used historic controls reported favorable clinical outcomes in patients with sepsis or severe pneumonia who received a combination of vitamin C, thiamine, and hydrocortisone. These outcomes include lowered mortality, lesser risk of progression to organ failure, and improved radiographic findings (Kim et al., 2018; Marik et al., 2017).

In three RCTs in patients with septic shock treated with vitamin C and thiamine (with or without hydrocortisone), no survival benefit was reported. In two trials reduction in organ dysfunction (as measured by a SOFA score at Day 3) (Chang et al., 2020; Fujii et al., 2020) was observed, and the third reported reduction was observed in the duration of shock (Iglesias et al., 2020) without an effect on clinical outcomes. Two other trials found no differences in any physiologic or outcome measure between the treatment and placebo groups (Moskowitz et al., 2020; Hwang et al., 2020).

RECOMMENDATION FOR NONCRITICALLY ILL PATIENTS WITH COVID-19

There are insufficient data for the COVID-19 Treatment Guidelines Panel (the Panel) to recommend either for or against the use of vitamin C for the treatment of COVID-19 in noncritically ill patients, as these patients are less likely to experience oxidative stress or severe inflammation so the role of vitamin C in this setting is unknown.

There are insufficient data for the Panel to recommend either for or against the use of vitamin C for the treatment of COVID-19 in critically ill patients, as there are only sparse and inconclusive observational trials with no completed controlled trials of vitamin C in patients with COVID-19, and the studies of vitamin C in sepsis and ARDS patients have reported variable efficacy and few safety concerns.

VITAMIN D

Vitamin D is a fat-soluble vitamin required for bone and mineral metabolism. It reduces the risk of microbial infection through three main mechanisms: physical barriers, cellular natural immunity, and adaptive immunity (Rondanelli et al., 2018). Vitamin D enhances innate cellular immunity through the induction of antimicrobial peptides, including the human cathelicidin LL-37 and by 1,25-dihdroxyvitamin D and defensins, while maintaining tight junctions, gap junctions, and adherens junctions (Schwalfenberg, 2011; Laaksi, 2012). Cathelicidins also have direct antimicrobial effects against various microbes including bacteria (both gram-positive and gram-negative ones), viruses (both enveloped and nonenveloped viruses), and fungi (Herr et al., 2007). Cathelicidins also induce a variety of proinflammatory cytokines, stimulation of the chemotaxis of neutrophils, monocytes, macrophages, and T lymphocytes into the site of infection, and promotion of the clearance of respiratory pathogens by inducing apoptosis and autophagy of infected epithelial cells (Greiller and Martineau, 2015). Furthermore, 1,25(OH)2D—vitamin D receptor complex acts on the cathelicidin gene promoter vitamin D response elements to enhance transcription of cathelicidin (Wang et al., 2004). Vitamin D reduces the production of proinflammatory T-helper (Th)1 cytokines (TNF-α and IFN-γ) and increases the expression of antiinflammatory cytokines by macrophages (Gombart et al., 2020). It increases cytokine production by Th2 lymphocytes causing suppression of Th1 cells (Cantorna et al., 2015). It also favors induction of the T regulatory (Treg) cells, thereby inhibiting inflammatory processes (Murdaca et al., 2019). With advancement of age, serum vitamin D concentrations tend to decrease as a result of lower levels of 7-dehydrocholesterol in the skin and less sun exposure (Vásárhelyi et al., 2011).

Vitamin D enhances expression of genes glutathione reductase and the glutamate—cysteine ligase modifier subunit linked to antioxidation (Lei et al., 2017), which in turn spares the use of vitamin C enhancing its antimicrobial activities (Mousavi et al., 2019; Colunga Biancatelli et al., 2020).

Drug Interactions

Vitamin D serum levels are reduced by antiepileptics, antineoplastics, antibiotics, antiinflammatory agents, antihypertensives, antiretrovirals, endocrine drugs, and some herbal medicines, through the activation of the pregnane-X receptor (Gröber and Kisters, 2012).

Adverse Effects

Most people do not commonly experience side effects with vitamin D, unless too much is taken. High levels of vitamin D may cause weakness, fatigue, sleepiness, headache, loss of appetite, dry mouth, metallic taste, nausea, vomiting, hypercalcemia, and nephrocalcinosis (Ross et al., 2011)

The role of vitamin D supplementation in the prevention or treatment of COVID-19 is not known. The rationale for using vitamin D is based largely on immunomodulatory effects that could potentially protect against COVID-19 infection or decrease the severity of illness. Some investigational trials on the use of vitamin D in people with COVID-19 are using vitamin D alone or in combination with other agents to participants with and without vitamin D deficiency.

Vitamin D deficiency is more common in old age, obese, and hypertensive patients; these factors were confirmed to be associated with poor outcomes in COVID-19 patients. In observational studies, low vitamin D levels have been associated with an increased risk of community-acquired pneumonia in older adults (Lu et al., 2018) and children (Science et al., 2013).

In a metaanalysis of RCTs, vitamin D supplementation was proven to be protective against acute respiratory tract infection (Martineau et al., 2017). However, in two RCTs that were double-blind and placebo-controlled, high-dose vitamin D administration to critically ill (non- Covid-19) patients with vitamin D deficiency (but not COVID-19) did not reduce the length of the hospital stay or the mortality rate when compared with placebo (Ginde et al., 2019).

Considerations in Pregnancy

Vitamin D is *likely safe* during pregnancy and breastfeeding when used in daily amounts below 4000 units (100 mcg). Higher doses are better avoided unless instructed by healthcare provider. Vitamin D is *possibly safe* when used in higher amounts during pregnancy or while breast-feeding. Using higher doses might cause serious harm to the infant (Webmed, 2020a).

RECOMMENDATION

- There are insufficient data to recommend either for or against the use of vitamin D for the prevention or treatment of COVID-19.

ZINC SUPPLEMENTATION AND COVID-19

Zinc can reduce the risk of viral respiratory tract infections, including COVID-19 and shorten the duration and severity of these infections. Zinc inhibits the enzymatic activity, replication of SARS-CoV RNA polymerase, and ACE2 activity as shown in in vitro studies (Velthuis et al., 2010; Skalny et al., 2020).

Zinc can modify the host's response to infections as it is an essential cofactor element for many functions in the body. Most of the beneficial effects of zinc occur at the cell membrane. Zinc (Zn^{2+}) reduces the cell membrane permeability and alters the capillary epithelium, thus inhibiting transcapillary movement of plasma protein that in turn may reduce local edema, inflammation, exudation, and mucus secretion (Novick et al., 1996).

Zinc enhances cytotoxicity and induces apoptosis when used in vitro with a zinc ionophore like chloroquine. Chloroquine, in turn, enhances intracellular zinc uptake in vitro (Xue et al., 2014). The relationship between zinc and COVID-19, including how zinc deficiency affects the severity of COVID-19 and whether zinc supplements can improve clinical outcomes, is currently under investigation (Calder et al., 2020). Zinc levels are difficult to measure accurately, as it is a component of various proteins and nucleic acids (Hambridge, 2007).

The optimal dose of zinc for the treatment of COVID-19 is not established. The recommended dietary allowance for elemental zinc is 11 mg daily for men and 8 mg for nonpregnant women. There are variability of zinc doses used in registered clinical trials for treatment of COVID-19 with a maximum dose of zinc sulfate 220 mg (50 mg of elemental zinc) twice daily (National Institutes of Health, 2020b).

Side Effects

Routine zinc supplementation is not recommended without healthcare professional advice. In some people, zinc might cause nausea, vomiting, diarrhea, metallic taste, kidney and stomach damage, and other side effects. Zinc is *possibly safe* when taken by mouth in doses greater than 40 mg daily, especially when these doses are taken only for a short period of time (Webmed, 2020a).

Long-term zinc supplementation can cause copper deficiency with subsequent reversible hematologic defects such as anemia, leukopenia, and potentially irreversible neurologic manifestations such as myelopathy, paresthesia, ataxia, and spasticity (Myint et al., 2018). Zinc supplementation for a duration as short as 10 months has been associated with copper deficiency (Hoffman et al., 1988). In addition, oral zinc can decrease the absorption of medications that bind with polyvalent cations (National Institutes of Health, 2020b).

Because zinc has not been shown to have clinical benefit and may be harmful, the Panel recommends against using zinc supplementation above the recommended dietary allowance for the prevention of COVID-19, except in a clinical trial (BIII).

CLINICAL DATA

A non–peer-reviewed retrospective observational study was conducted on 932 hospitalized COVID-19 patients to compare zinc supplementation (n = 411) with no zinc supplementation (n = 521) in patients who received HCQ and azithromycin. Patients who received zinc had higher absolute lymphocyte count and lower troponin and procalcitonin levels at baseline than those who did not receive zinc. In univariate analysis, no differences were observed between the two groups regarding duration of hospital stay, duration of mechanical ventilation, maximum oxygen flow rate, average oxygen flow rate, or average FiO_2 while in bivariate logistic regression analysis, zinc supplementation was associated with a decreased mortality rate; however, the association with a decreased mortality rate was no longer significant when analysis was limited to patients who were treated in the ICU. Limitations of the study included the following: it is retrospective nonrandomized one, the statistical methods used do not account for confounding variables or patient differences, and no details on the timing of zinc initiation or patients' condition at the start of treatment were clarified (Carlucci et al., 2020). Given the nature of the study design and its limitations, the authors do not recommend using this study to guide clinical practice.

Considerations in Pregnancy

Zinc is *likely safe* for most pregnant and breastfeeding women when used in the recommended daily amounts (RDA) but possibly unsafe and likely unsafe if used in high doses by breastfeeding and pregnant women, respectively. Pregnant and breastfeeding women aged 14 to 18 and over 18 should not take more than 34 and 40 mg per day, respectively (Webmed, 2020b).

RECOMMENDATIONS

- There are insufficient data to recommend either for or against the use of zinc for the treatment of COVID-19.
- The COVID-19 Treatment Guidelines Panel (the Panel) recommends against using zinc supplementation above the recommended dietary allowance for the prevention of COVID-19, except in a clinical trial (BIII).

LACTOFERRIN

Lactoferrin (LF) is an iron-binding glycoprotein with a molecular weight in the range of 70–80 kDa, which transports iron in the blood and serum (Wang et al.,

2019). LF has two symmetrical lobes of a polypeptide chain; each lobe contains two domains, which can bind a metal atom (Baveye et al., 1999). The antiviral effect of LF is mediated by binding iron and is not affected by unsaturated iron levels (Lönnerdal and Iyer, 1995). LF is produced by mucosal epithelial cells in several different mammalian and fish species and is found in mucosal secretions, body fluids, and secondary neutrophil granules (González-Chávez et al., 2009). LF has antifungal, antiviral, and antiinflammatory activities, along with its effects on the immune response (Hao et al., 2019). These activities are mediated through the capacity of LF to bind iron and to interact with components of the host and the pathogens (González-Chávez et al., 2009). LF is positively charged in vivo and can bind large molecules with negative charges, such as lipopolysaccharides and glycosaminoglycans, which is one of the key mechanisms underlying its antiviral activity (Elass-Rochard et al., 1998). The protective effects of LF were first confirmed in 1987 in mice infected with the polycythemia-inducing strain of friend virus complex (Lu et al., 1987). LF has been identified to be effective against several viruses including influenza A virus, adenovirus, SARS-CoV, dengue virus, and others (Pietrantoni et al., 2015; Chen et al., 2017; Oda et al., 2020; Ishikawa et al., 2013).

Two promising in vitro studies, one conducted in 2011 on SARS-CoV and the other conducted in 2020 on SARS-CoV-2, have shown that lactoferrin can inhibit viral infection in the early stages and is effective against SARS-CoV-2 in the postinfection phase (Mirabelli et al., 2020).

A preprint study was conducted by teams at university hospitals in Rome to evaluate the role of oral and intranasal administration of lactoferrin in 32 COVID-19 patients with mild to moderate symptoms, as well as an asymptomatic form of the disease. The study also used a control group of 32 healthy volunteers. A dose of 1 g of liposomal apolactoferrin in 10 capsules per day was administered orally for 30 days, in addition to the same form administered nasally three times per day.

All patients demonstrated improvement in every symptom except fatigue, which continued in about a third of the group. Other very promising data have emerged from these studies, such as a drop in D-dimer concentration, which is crucial in prognosis for the disease, as well as regulation of IL-6, one of the three proinflammatory cytokines (Campione et al., 2020).

Lactoferrin is *likely safe* for pregnant and breastfeeding women when taken in food amounts. Lactoferrin is *possibly safe* in doses of 250 mg daily in women who are in the second or third trimester of pregnancy.

Lactoferrin can cause diarrhea. In very high doses, skin rash, loss of appetite, fatigue, chills, and constipation have been reported (Webmed, 2020c).

MELATONIN

Melatonin, the main hormone secreted by the pineal gland, has various properties. These include antioxidant, antiinflammatory, antiexcitatory, sleep initiation, and immunoregulation effects (Juybari et al., 2019). Melatonin affects mitochondrial function through its protection against free radicals, modulating its permeability transition pore, changing its electron flux, and influencing energy metabolism (Mehrzadi et al., 2020). Melatonin is effective therapy in sleep disturbances, cardiovascular diseases, and eye diseases and acts as a complementary therapy in neonatal care, in vitro fertilization hemodialysis, and anesthesia (Sánchez-Barceló et al., 2010).

The antioxidative and antiinflammatory properties of melatonin counter acute lung injury (ALI) and ARDS induced by viral and bacterial infections. In critically ill patients, melatonin reduces vessel permeability, induces sedation, decreases agitation, and increases sleep quality. These beneficial effects support the hypothesis that melatonin may exert further clinical outcomes for COVID-19 patients (Zhang et al., 2020b).

The innate resistance of bats to viral disease is poorly understood. Melatonin production level in humans is significantly lower than in bats, particularly in the elderly ones (Tresguerres et al., 2006). Given that the elderly people were excessively affected by SARS-CoV-2 than people under the age of 20; besides other factors, it could be hypothesized that high levels of melatonin exert protective properties in bats against the severity of SARS-CoV-2 (Bahrampour Juybari et al., 2020).

Due to a positive correlation between immune dysfunction and disease severity in patients with COVID-19, it is necessary to consider this condition for preparing the optimal vaccine. The safety of melatonin profile has been broadly examined in different preclinical and clinical studies on wide-range doses. Because of the lack of an available vaccine or effective antiviral treatment for COVID-19, the use of melatonin as an adjuvant might be worth consideration. Although the direct protective action of melatonin against COVID-19 is unknown, its extensive application in animal studies and human clinical trials has repeatedly verified its efficacy and safety in a broad range of disorders. Therefore, melatonin practical usage in the current COVID-19 outbreak is suggested to be beneficial (Bahrampour Juybari et al., 2020).

Information regarding safety and efficacy in pregnancy and lactation is lacking.

Possible adverse effects include dizziness, enuresis, excessive daytime somnolence, headache, nausea, insomnia, nightmares, and transient depression. Drowsiness may be experienced within 30 min after taking melatonin and may persist for approximately 1 h; as a result, melatonin may affect driving ability. Use of the animal tissue-derived product is discouraged because of the risk of contamination or viral transmission (Drugs.com, 2020).

OXYGENATION AND VENTILATION

Oxygen therapy targets vary according to the clinical condition of the patient with COVID-19. The aim of oxygen therapy in severe COVID-19 patients with respiratory distress, hypoxemia, or shock is to keep $SpO_2 > 94\%$ (World Health Organization, 2020a). After initial stabilization, SpO_2 should be stabilized above 90% and 92%–95% in nonpregnant (World Health Organization, 2020a) and pregnant (The Queensland Health, 2020) adults, respectively, but not above 96% (Alhazzani et al., 2020).

NONMECHANICALLY VENTILATED ADULTS WITH HYPOXEMIC RESPIRATORY FAILURE

Although the optimal oxygen saturation in adults with COVID-19 is not exactly known, uncertain. However, it is known that an SpO_2 below 92% or above 96% may be harmful.

An RCT was conducted on ARDS patients without COVID-19 compared a conservative oxygen strategy with a target SpO_2 of 88%–92% or to a liberal oxygen strategy (target $SpO_2 \geq 96\%$). It was terminated after enrollment of 205 patients, as there was an increased mortality at 90 days (risk difference was 14%; 95% CI: 0.7%–27%) and at 28 days (risk difference was 8%; 95% CI: −5%–21%) in the conservative oxygen group (Barrot et al., 2020).

A metaanalysis of 25 randomized trials involving patients without COVID-19 found that a liberal oxygen strategy (median SpO_2 of 96%) was associated with an increased risk of in-hospital mortality compared with a lower SpO_2 comparator (relative risk 1.21; 95% CI: 1.03–1.43) (Chu et al., 2018).

Acute Hypoxemic Respiratory Failure

Conventional oxygen therapy may be insufficient to reach the target oxygenation in adults with acute hypoxemic respiratory failure and COVID-19 infection.

Proper respiratory support can be provided through high-flow nasal cannula (HFNC), noninvasive positive-pressure ventilation (NIPPV), intubation and invasive mechanical ventilation, or ECMO.

High-Flow Nasal Cannula and Noninvasive Positive-Pressure Ventilation

Data from clinical trial in non-COVID-19 patients with acute hypoxemic respiratory failure suggested that HFNC is preferred over NIPPV. The study participants of this unblind study were randomized to HFNC, conventional oxygen therapy, or NIPPV. The ventilator-free days were 24, 22, and 19 days ($P = .02$) in the HFNC, conventional oxygen therapy, and NIPPV groups, respectively. The 90-day mortality was lower in the HFNC group compared with both the conventional oxygen therapy group (HR 2.01; 95% CI: 1.01–3.99) and the NIPPV group (HR 2.50; 95% CI: 1.31–4.78). In more severely hypoxemic patients with PaO_2/FiO_2 mm Hg ≤ 200, the intubation rate was lower for HFNC when compared with the conventional oxygen therapy or NIPPV patients (HR 2.07 and 2.57, respectively) (Frat et al., 2015).

A metaanalysis of eight trials that involved 1084 patients conducted to assess the efficacy of oxygenation strategies prior to intubation. HFNC reduced the rate of intubation (OR 0.48; 95% CI: 0.31–0.73) and ICU mortality (OR 0.36; 95% CI: 0.20–0.63) when compared with NIPPV (Ni et al., 2018).

NIPPV carries a high risk for aerosol spread of SARS-CoV-2 (Yu et al., 2007; Tran et al., 2012). The risk of aerosol spread for HFNC is undetermined.

Prone Positioning for Nonintubated Patients

In ARDS patients receiving mechanical ventilation, prone position improves oxygenation and clinical outcomes in patients with moderate-to-severe ARDS (Fan et al., 2017). The benefit of prone position in awake patients on supplemental oxygen without mechanical ventilation is less evident. A case series of patients with COVID-19 requiring oxygen or NIPPV have reported that awake prone positioning is well tolerated and improves oxygenation (Elharrar et al., 2020; Sartini et al., 2020), with some series also reporting low intubation rates after proning (Sartini et al., 2020; Sun et al., 2020).

A single-center prospective study of awake prone positioning in 56 COVID-19 patients on HFNC or NIPPV reported that prone positioning for ≤ 3 h was feasible in 84% of the patients, and it was associated with significant improvement in oxygenation (PaO_2/FiO_2 286 and181 mm Hg in prone and supine position, respectively). However, when compared with baseline oxygenation before initiation of prone positioning, this improvement in oxygenation was not sustained (PaO_2/FiO_2 of 181 and 192 mm Hg at baseline and 1 h after resupination, respectively). Among patients put in the prone position, there was no difference in intubation rate between patients who maintained improved oxygenation and others (Caputo et al., 2020).

A prospective, multicenter observational cohort study was conducted on 199 COVID-19 patients with acute respiratory failure receiving HFNC to evaluate the effect of prone positioning on the rate of intubation. Although the time to intubation was 1 day (IQR 1.0–2.5) in patients receiving HFNC and prone positioning (55 patients) versus 2 days [IQR 1.0–3.0] in patients receiving only HFNC ($P = .055$), the use of awake prone positioning did not reduce the risk of intubation (RR 0.87; 95% CI: 0.53–1.43; $P = .60$) (Ferrando et al., 2020).

It is unclear to which patients and for how long should prone position be applied in nonintubated patients with COVID-19 pneumonia. Appropriate candidates for awake prone positioning are those who can adjust their position independently and tolerate lying prone. Awake prone positioning is contraindicated in patients with respiratory distress, as they require immediate intubation, in hemodynamically unstable patients, patients with recent abdominal surgery, and those having an unstable spine (Bamford et al., 2020). Awake prone positioning is acceptable and feasible for pregnant patients and can be performed in the left lateral decubitus position (especially in late pregnancy) or the fully prone position (Society for Maternal Fetal Medicine, 2020).

In patients with COVID-19, there is a possibility of rapid deterioration of hypoxia with increased needs for intubation and invasive mechanical ventilation, so close monitoring is advised (Meng et al., 2020).

Recommendations

- The Panel recommends HFNC oxygen over NIPPV for adults with COVID-19 and acute hypoxemic respiratory failure despite conventional oxygen therapy (BI).
- In the absence of an indication for endotracheal intubation and HFNC, the Panel recommends a closely monitored trial of NIPPV for adults with COVID-19 and acute hypoxemic respiratory failure (BIII).
- For patients with persistent hypoxemia despite increasing supplemental oxygen requirements in whom endotracheal intubation is not otherwise indicated, the Panel recommends considering a trial

of awake prone positioning to improve oxygenation (CIII).
- The Panel recommends against using awake prone positioning as a rescue therapy for refractory hypoxemia to avoid intubation in patients who otherwise meet the indications for intubation and mechanical ventilation (AIII).
- If intubation becomes necessary, the procedure should be performed by an experienced practitioner in a controlled setting due to the enhanced risk of SARS-CoV-2 exposure to healthcare practitioners during intubation (AII).

MECHANICALLY VENTILATED ADULTS
Recommendations
For mechanically ventilated adults with COVID-19 and ARDS:
- The Panel recommends using low tidal volume (VT) ventilation (VT 4–8 mL/kg of predicted body weight) over higher VT ventilation (VT > 8 mL/kg) (AI).
- The Panel recommends targeting plateau pressures of <30 cm H_2O (AII).
- The Panel recommends using a conservative fluid strategy over a liberal fluid strategy (BII).
- The Panel recommends against the routine use of inhaled nitric oxide (AI).

The rationale behind these recommendations is related to the absence of **evidence that ventilator management of patients with hypoxemic respiratory failure due to COVID-19 should differ from ventilator management of patients with hypoxemic respiratory failure due to other causes.**

Recommendations
For mechanically ventilated adults with COVID-19 and moderate-to-severe ARDS:
- The Panel recommends using a higher positive end-expiratory pressure (PEEP) strategy over a lower PEEP strategy (BII).
- For mechanically ventilated adults with COVID-19 and refractory hypoxemia despite optimized ventilation, the Panel recommends prone ventilation for 12–16 h per day over no prone ventilation (BII).
- The Panel recommends using, as needed, intermittent boluses of neuromuscular blocking agents (NMBAs) or continuous NMBA infusion to facilitate protective lung ventilation (BIII).
- In the event of persistent patient–ventilator dyssynchrony, or in cases where a patient requires ongoing deep sedation, prone ventilation, or persistently high

plateau pressures, the Panel recommends using a continuous NMBA infusion for up to 48 h as long as patient anxiety and pain can be adequately monitored and controlled (BIII).

The beneficial effects of PEEP in patients with ARDS include prevention of alveolar collapse, improvement of oxygenation, and reduction of atelectotrauma, a source of ventilator-induced lung injury. A metaanalysis of three large trials in patients without COVID-19 found lower rates of ICU and in-hospital mortality with higher PEEP in those with moderate (PaO_2/FiO_2 100−200 mm Hg) and severe ARDS (PaO_2/FiO_2 <100 mm Hg) (Briel et al., 2010).

High level of PEEP is not definitely defined but usually considered if > 10 cm H_2O (Alhazzani et al., 2020). Recent reports have suggested that, in contrast to non-COVID-19 ARDS patients, some COVID-19 patients with moderate or severe ARDS have normal static lung compliance so that higher PEEP levels may cause harm by compromising hemodynamics and cardiovascular performance (Tsolaki et al., 2020; Marini and Gattinoni, 2020). Other studies reported that COVID-19 patients with moderate-to-severe ARDS had low compliance as patients with non-COVID-19 ARDS (Ziehr et al., 2020; Schenck et al., 2020; Bhatraju et al., 2020; Cummings et al., 2020). These seemingly contradictory observations suggest that COVID-19 patients with ARDS are a heterogeneous population, and assessment for responsiveness to higher PEEP should be individualized based on oxygenation and lung compliance. Clinicians should monitor patients for known side effects of higher PEEP, such as barotrauma and hypotension.

The recommendation for intermittent boluses of NMBA or continuous infusion of NMBA to facilitate lung protection may require a healthcare provider to enter the patient's room frequently for close clinical monitoring. Therefore, in some situations, the risks of SARS-CoV-2 exposure and the need to use personal protective equipment for each entry into a patient's room may outweigh the benefit of NMBA treatment.

Recommendations

For mechanically ventilated adults with COVID-19, severe ARDS, and hypoxemia despite optimized ventilation, and other rescue strategies:

- The Panel recommends using recruitment maneuvers rather than not using recruitment maneuvers (CII).
- If recruitment maneuvers are used, the Panel recommends against using staircase (incremental PEEP) recruitment maneuvers (AII).

- The Panel recommends using an inhaled pulmonary vasodilator as a rescue therapy; if no rapid improvement in oxygenation is observed, the treatment should be tapered off (CIII).

There are no studies to assess the effects of recruitment maneuvers on oxygenation in COVID-19 patients with severe ARDS. A metaanalysis of six studies in non-COVID-19 patients with ARDS found that recruitment maneuvers reduced mortality, improved oxygenation 24 h after the maneuver, and decreased the need for rescue therapy (Goligher et al., 2017). Because recruitment maneuvers can cause barotrauma or hypotension, patients should be closely monitored during recruitment maneuvers. If a patient decompensates during recruitment maneuvers, the maneuver should be stopped immediately. The importance of properly performing recruitment maneuvers was illustrated by an analysis of 8 RCTs in 2544 non-COVID-19 patients, which found that recruitment maneuvers did not reduce hospital mortality (RR 0.90; 95% CI: 0.78−1.04). Subgroup analysis found that traditional recruitment maneuvers significantly reduced hospital mortality (RR 0.85; 95% CI: 0.75−0.97), whereas incremental PEEP titration recruitment maneuvers increased mortality (RR 1.06; 95% CI: 0.97−1.17) (Alhazzani et al., 2020).

Although there are no published studies of inhaled nitric oxide in patients with COVID-19, a Cochrane review of 13 trials of inhaled nitric oxide use in patients with ARDS found no mortality benefit (Gebistorf et al., 2016). Because the review showed a transient benefit in oxygenation, it is reasonable to attempt inhaled nitric oxide as a rescue therapy in COVID patients with severe ARDS after other options have failed. However, if there is no benefit in oxygenation with inhaled nitric oxide, it should be tapered quickly to avoid rebound pulmonary vasoconstriction that may occur with discontinuation after prolonged use.

Oxygen nasal cannula

Nasal oxygen cannula is easy, cheap, and easily used oxygen supplying system. It is the most initially used method for oxygen therapy in patients with mild hypoxia. It has a minimal aerosol generation and a low risk of spreading the virus in COVID-19 patients. However, it can only provide up to 40% inspired fraction of oxygen (FiO_2) and requires humidification when oxygen flow is above 6 L per minute. Therefore, nasal oxygen cannula typically cannot provide efficient oxygen therapy in a patient with severe hypoxia due to significant lung damage (Auriant et al., 2001).

Oxygen face mask

Oxygen face masks, especially nonrebreathing face masks, can provide high FiO_2 oxygen therapy, but does not increase oral pharyngeal pressure, and is therefore not efficient enough to treat hypoxia due to severe lung damage and significant alveolar collapse (Jiang and Wei, 2020).

High-flow nasal oxygenation

High-flow nasal oxygenation (HFNO) therapy is used increasingly before invasive ventilation in adults with acute respiratory failure (Weingart and Levitan, 2012; Levy et al., 2016; Koga et al., 2020). It supplies warm, humidified oxygen through the pliable nasal cannula with a fraction of inspired oxygen (FiO_2) up to 1.0 and maximum flow rate up to 70 L/min. At the beginning of the COVID-19 pandemic, the lack of resources named invasive mechanical ventilators, critical care providers, and the easiness of use of HFNO resulted in its use in some COVID-19 patients for oxygen therapy (Xie et al., 2020). A one study 85% of the survivors and 50% of the nonsurvivors received HFNO. Additionally, 14% of patients were treated with HFNO before intubation, and 34.5% of patients who died of COVID-19 received HFNO (Xie et al., 2020).

HFNO use has a higher efficiency than conventional oxygen therapy provided through nasal cannula or oxygen face mask (Zhu et al., 2019) with better tolerability and comfortability (Stéphan et al., 2015). The risks of treatment failure and 30-day mortality were not significantly different between HFNO and noninvasive ventilation (NIV) as first-line therapy in respiratory failure (Koga et al., 2020). HFNO was recommended over NIV, as it has less side effects (Leone et al., 2020) and more reduction in intubation rates in acute respiratory failure (Huang et al., 2018). HFNO was associated with lower 90-day mortality (Frat et al., 2015) and 30-day mortality (Koga et al., 2020) in patients with pneumonia or patients without hypercapnia.

The use of HFNO is associated with less risk of lung injury, as it has a clearing effect on the upper airway without increasing tidal volume (Mauri et al., 2017). This clearing effect makes it more suitable for patients with excessive secretion (Hasani et al., 2008) (28%−34% of COVID-19 cases and 35%−42% in ICU COVID-19 cases) (Guan et al., 2020). However, most protocols for airway management for patients with COVID-19 now consider HFNO a relative contraindication (Brewster et al., 2020) for fear of increase risk of virus aerosol spread. A systematic review of aerosol-generating procedures in SARS patients suggested that HFNO did not increase the risk of SARS transmission significantly (Tran et al., 2012). Besides, it was reported that HFNO with good interface fitting was associated with a low risk of airborne transmission (Hui et al., 2019).

HFNO could be applied in mild and moderate nonhypercapnia cases, but patients should be assessed for respiratory failure. It is also suggested that if there is no improvement within 1 or 2 h, endotracheal intubation and mechanical ventilation should be considered (Li, 2020).

Noninvasive ventilation

Traditional NIV is primarily composed of continuous positive airway pressure or bilevel positive airway pressure ventilation (García-de-Acilu et al., 2019).

NIV has been used in oxygen/ventilation therapy in SARS- and H1N1-infected patients. In recent studies (Yao et al., 2020; Wang et al., 2020a), NIV was used up to 70% in COVID-19 patients before tracheal intubation for invasive mechanical ventilation. However, it seemed that the mortality in these patients was high. It has also been reported that NIV may delay the intubation in patients with severe respiratory failure and is not recommended (Meng et al., 2020).

Recent international expert recommendations suggested that HFNO should be used before NIV in critically ill COVID-19 patients. If NIV is used, it should be limited to short periods with close monitoring of pulmonary failure and decision for early tracheal intubation for invasive ventilation (Alhazzani et al., 2020).

Helmet ventilation

Helmet ventilation is an alternative mode of NIV, with a helmet to replace a commonly used face mask (Lucchini et al., 2020). Its main advantages over the face mask in COVID-19 patients include the reduction of air leakage during positive-pressure ventilation that makes NIV more efficient, reduction of the aerosol spreading of the SARS-CoV-2 virus, and its high tolerability by the patients. Its main disadvantages include the need for high flow of gases (>100 L per minute and high consumption of oxygen supplies), with the difficulty of humidification, the potential rebreathing causing carbon dioxide retention especially at low inspiratory flow and the patient's movement of head and body. Although Helmet seemed to be widely used in COVID-19 patients in Italy, its effectiveness and side effects to treat pulmonary failure and reduce mortality are not clear at this time. No international expert recommendation could be provided (Alhazzani et al., 2020).

Conventional mechanical ventilation

A conventional mechanical ventilator is needed in 10% −17% of COVID-19 patients (Huang et al., 2020). High mortality in COVID-19 patients was reported after tracheal intubation and CMV (Bhatraju et al., 2020). That may be related to late intubation after severe hypoxia and its subsequent multiple organ damage (Sorbello et al., 2020) or complications associated with high-pressure ventilation such as pneumothorax during CMV, which increase the lung damage (Yao et al., 2020).

However, the consistent high mortality after tracheal intubation during this pandemic around the world has inspired alternative techniques, such as HFNO, for oxygen/ventilation treatment in COVID-19 patients and avoiding CMV, although there is no clear conclusion at this moment (Jiang and Wei, 2020).

High-frequency jet ventilation

High-frequency jet ventilation (HFJV) is characterized by its open system, high frequency (respiratory rate >60/minutes), small tidal volume, and low airway pressures (Ihra et al., 2000). High frequency can minimize the diaphragm movement and therefore benefit the AF ablation. Low airway pressure and low tidal volume may benefit the oxygen/ventilation in ARDS treatment or during hypovolemic shock. HFJV at a frequency close to heart rate or synchronized with heart rate also assists cardiovascular function (Angus et al., 1997).

Previous studies suggested that HFJV may provide better oxygenation than CMV in the treatment of respiratory failure or ARDS caused by various reasons. Although there was no improvement of mortality in HFJV over CMV, it acts as an alternative mechanical ventilation with similar efficacy (Bingold et al., 2012).

HFJV is not widely used now for the treatment of pulmonary failure or ARDS as it is difficult to monitor FiO_2, airway pressure, $PetCO_2$ due to its open system, and the difficulty of humidification of the inhaled gases. With its characteristics of better oxygenation under the condition of small tidal volume and low airway pressure, HFJV is expected to treat hypoxia in COVID-19 patients efficiently, especially for those with severe pulmonary failure or ARDS (Jiang and Wei, 2020).

High-frequency two-way jet ventilation

High-frequency two-way jet ventilation (HFTJV) is composed of both active inspiratory and expiratory phases. During the inspiratory phase, a jet pulse is injected into the lung, while a jet pulse is injected out of the lung during the expiratory phase (Wei et al., 1992).

Compared with regular HFJV, the active exhalation by the reverse jet pulse during the expiratory phase not only further decreases mean airway pressures but also enhances oxygenation/ventilation and improvement of circulatory function (Wei et al., 1992). HFTJV theoretically provides the greater capability of improving cardiopulmonary functions than CMV or regular HFJV. Additionally, the reverse jet pulse inside the trachea generates active expiration and may eliminate the SARS-CoV-2 virus out of the lungs due to the Venturi effects generated by the reverse jet pulses. HFTJV is expected to lower mortality in COVID-19 patients, compared with traditional CMV. Therefore, it is important and urgent to investigate the effectiveness and side effects of HFTJV in the treatment of COVID-19 patients with ARDS.

Supraglottic jet oxygenation and ventilation

HFJV is an infraglottic jet ventilation, with jet pulses originated below the vocal cord, which necessitates tracheal intubation and deep patients' sedation. Many studies (Liang et al., 2019a, 2019b; Li et al., 2017; Wu et al., 2017) suggested that supraglottic jet oxygenation and ventilation (SJOV) with jet pulses originated above the vocal cords can also maintain similar efficacy of oxygenation/ventilation as HFJV, as long as the jet pulses are directed toward vocal cord. Compared with the regular infraglottic HFJV, SJOV has the following advantages: (1) easy, quick, and convenient to set up and use; (2) easy to learn and train, even patients can do it themselves through synchronizing their inhalation with the inspiratory jet pulses; (3) monitoring of breathing function with the ability to monitor $PetCO_2$; and (4) minimizing the barotrauma complications frequency seen in the transtracheal jet ventilation (up to 30% in emergent airway management) (Craft et al., 1990), due to its guarantee of opening systems by opened mouth and nose during SJOV.

SJOV is expected to treat hypoxia in COVID-19 patients, especially during the early phase of the disease. Beside the aforementioned advantages, SJOV may have other advantages in the treatment of hypoxia in COVID-19 patients. These include the following: it can be easily adjusted from treating mild hypoxia to moderate or severe hypoxia by increasing driving pressures and change of position of Wei Nasal Jet (WNJ) from mouth to nose under mild sedation (Qin et al., 2017); it requires less sedation than NIV but provides efficient oxygenation/ventilation and may be used to avoid tracheal intubation; it may provide similar efficacy on oxygenation and ventilation, but reduced use of sedation requirement compared with the CMV.

Similar to HFNO, SJOV is a ventilation technique that has the potential to generate aerosol transmission of the SARS-CoV-2 virus. If SJOV is used to treat hypoxia/hypercapnia in COVID-19 patients, it should be performed in a negative pressure room with an anteroom between patients' rooms and clean area. Adequate PPE should be worn to protect healthcare workers from cross-infection (Jiang and Wei, 2020).

EXTRACORPOREAL MEMBRANE OXYGENATION

In ECMO, blood is pumped outside the body to a heart—lung machine that removes carbon dioxide from the blood, supplies it with oxygen, and pumps it again to the tissues. Blood flows from the right side of the heart to the membrane oxygenator in the heart—lung machine and then is rewarmed and sent back to the body. This method allows the blood to "bypass" the heart and lungs, allowing these organs to rest and heal from any injury. ECMO is used in critical care situations, when the heart and/or the lungs are injured and need time to repair so it may be used in care for COVID-19, ARDS, and other infections (MayoClinic, 2020).

The WHO recommended that well-equipped expert health places with sufficient ECMO volume to maintain proficiency consider ECMO support in COVID-19-related ARDS with refractory hypoxemia if lung protective mechanical ventilation (Brower et al., 2000) was insufficient to support the patient (World Health Organization, 2020a). Despite such optimism for a possible role for ECMO in both acute respiratory and cardiac failure, early reports of patients with COVID-19 requiring ECMO suggested that mortality could be greater than 90% (Henry and Lippi, 2020).

High mortality in the initial published experience led some clinicians and investigators to recommend withholding ECMO support in patients with COVID-19 (Ñamendys-Silva, 2020).

RECOMMENDATION

There are insufficient data to recommend either for or against the use of ECMO in patients with COVID-19 and refractory hypoxemia.

RATIONALE

ECMO has been used as a short-term rescue therapy in patients with ARDS caused by COVID-19 and refractory hypoxemia. However, there is no conclusive evidence that ECMO is responsible for better clinical outcomes regardless of the cause of hypoxemic respiratory failure (Combes et al., 2018; Munshi et al., 2019).

The clinical outcomes for patients with ARDS who are treated with ECMO are variable and depend on multiple factors including the cause of respiratory failure, the severity of lung injury, the presence of comorbidities, and the ECMO experience of the healthcare providers (Henry and Lippi, 2020; Bullen et al., 2020; Mustafa et al., 2020). A recent case series of 83 COVID-19 patients in Paris reported a 60-day mortality of 31% for patients on ECMO (Schmidt et al., 2020). This mortality was similar to the mortality observed in a 2018 study of non-COVID-19 patients with ARDS who were treated with ECMO during the ECMO to Rescue Lung Injury in Severe ARDS (EOLIA) trial; that study reported a mortality of 35% at day 60 (Combes et al., 2018).

The Extracorporeal Life Support Organization (ELSO) Registry provides the largest multicenter outcome data set of patients with confirmed COVID-19 who received ECMO support. A recent cohort study evaluated ELSO Registry data for 1035 COVID-19 patients who initiated EMCO between January 16 and May 1, 2020, at 213 hospitals in 36 countries. This study reported an estimated cumulative in-hospital mortality of 37.4% in these patients 90 days after they initiated ECMO (95% CI: 34.4%—40.4%) (Barbaro et al., 2020). Without a controlled trial that evaluates the use of ECMO in patients with COVID-19 and hypoxemic respiratory failure (e.g., ARDS), the benefits of ECMO cannot be clearly defined for this patient population (National Institutes of Health, 2020a).

Rating of recommendations: A = strong; B = moderate; C = optional rating of evidence: I = one or more randomized trials with clinical outcomes and/or validated laboratory endpoints; II = one or more well-designed, nonrandomized trials or observational cohort studies; III = expert opinion (National Institutes of Health, 2020a).

REFERENCES

ACOG practice bulletin no. 196 summary: thromboembolism in pregnancy. Obstet. Gynecol. 132 (1), 2018, 243—248. https://doi.org/10.1097/AOG.0000000000002707.

ACOG practice bulletin no. 207: thrombocytopenia in pregnancy. Obstet. Gynecol. 133 (3), 2019, e181—e193. https://doi.org/10.1097/AOG.0000000000003100.

Adam, C., Tseng, E.K, Nieuwlaat, R., Angchaisuksiri, P., Blair, C., Dane, K.E., Davila, J., DeSancho, M., Diuguid, D.L., Griffin, D., Kahn, S.R., Klok, F.A., Lee, A.I., Neumann, I., Pai, A., Pai, M., Schünemann, H.J., 2020. ASH 2020 guidelines on the use of anticoagulation in patients with COVID-19: draft recommendations. Am. Soc. Hematol. http://www.hematology.org/COVIDguidelines.

Agarwal, A., Mukherjee, A., Kumar, G., Chatterjee, P., Bhatnagar, T., Malhotra, P., Latha, B., et al., 2020. Convalescent plasma in the management of moderate COVID-19 in India: an open-label parallel-arm phase II multicentre randomized controlled trial (PLACID trial). medRxiv. https://doi.org/10.1101/2020.09.03.20187252.

Al-Tawfiq, Jaffar, A., Momattin, H., Jean, D., Memish, Z.A., 2014. Ribavirin and interferon therapy in patients infected with the Middle East respiratory syndrome coronavirus: an observational study. Int. J. Infect. Dis. 20 (March), 42–46. https://doi.org/10.1016/j.ijid.2013.12.003.

Aldeyab, M.A., Kearney, M.P., McElnay, J.C., Magee, F.A., Conlon, G., MacIntyre, J., McCullagh, B., et al., 2012. A point prevalence survey of antibiotic use in four acute-care teaching hospitals utilizing the European surveillance of antimicrobial consumption (ESAC) audit tool. Epidemiol. Infect. 140 (9), 1714–1720. https://doi.org/10.1017/S095026881100241X.

Alhazzani, W., Møller, M.H., Arabi, Y.M., Loeb, M., Gong, M.N., Fan, E., Simon, O., et al., 2020. Surviving sepsis campaign: guidelines on the management of critically ill adults with coronavirus disease 2019 (COVID-19). Crit. Care Med. 48 (6). https://journals.lww.com/ccmjournal/Fulltext/2020/06000/Surviving_Sepsis_Campaign__Guidelines_on_the.29.aspx.

American College of Cardiology, 2020. Ventricular Arrhythmia Risk Due to Hydroxychloroquine-Azithromycin Treatment for COVID-19. https://www.acc.org/latest-in-cardiology/articles/2020/03/27/14/00/ventricular-arrhythmia-risk-due-to-hydroxychloroquine-azithromycin-treatment-for-covid-19.

American College of Obstetricians and Gynecologists, 2020. COVID-19 FAQs for Obstetricians-Gynecologists.

American Society of Hematology, 2020. Should DOACs, LMWH, UFH, Fondaparinux, Argatroban, or Bivalirudin at Intermediate-Intensity or Therapeutic-Intensity vs. Prophylactic Intensity be Used for Patients with COVID-19 Related Critical Illness Who Do Not Have Suspected or Confirmed VTE? https://guidelines.ash.gradepro.org/profile/3CQ7J0SWt58.

Amin, S., Lux, A., O'Callaghan, F., 2019. The journey of metformin from glycaemic control to MTOR inhibition and the suppression of tumour growth. Br. J. Clin. Pharmacol. 85 (1), 37–46. https://doi.org/10.1111/bcp.13780.

Ang, A., Pullar, J.M., Currie, M.J., Vissers, M.C.M., 2018. Vitamin C and immune cell function in inflammation and cancer. Biochem. Soc. Trans. 46 (5), 1147–1159. https://doi.org/10.1042/BST20180169.

Angus, D.C., Lidsky, N.M., Dotterweich, L.M., Pinsky, M.R., 1997. The influence of high-frequency jet ventilation with varying cardiac-cycle specific synchronization on cardiac output in ARDS. Chest 112 (6), 1600–1606. https://doi.org/10.1378/chest.112.6.1600.

Aoki, Y., Maeno, T., Aoyagi, K., Ueno, M., Aoki, F., Aoki, N., Nakagawa, J., et al., 2009. Pioglitazone, a peroxisome proliferator-activated receptor gamma ligand, suppresses bleomycin-induced acute lung injury and fibrosis. Respiration 77 (3), 311–319. https://doi.org/10.1159/000168676.

Arabi, Y.M., Mandourah, Y., Al-Hameed, F., Anees, A.S., Almekhlafi, G.A., Hussein, M.A., Jose, J., et al., 2018. Corticosteroid therapy for critically ill patients with Middle East respiratory syndrome. Am. J. Respir. Crit. Care Med. 197 (6), 757–767. https://doi.org/10.1164/rccm.201706-1172OC.

Arthur, Y., Kim, R.T.G., 2020. Coronavirus Disease 2019 (COVID-19): Management in Hospitalized Adults. UpToDate. https://www.uptodate.com/contents/coronavirus-disease-2019-covid-19-management-in-hospitalized-adults?search= coronavirus-disease-2019-covid-19-management-inhospitalized- adults&source=search_result&selectedTitle=1~150&usage_type=default&display_rank=1.

Arshad, S., Kilgore, P., Chaudhry, Z.S., Jacobsen, G., Wang, D.D., Huitsing, K., Brar, I., Alangaden, G.J., Ramesh, M.S., McKinnon, J.E., O'Neill, W., Zervos, M., Henry Ford COVID-19 Task Force, 2020a. Treatment with hydroxychloroquine, azithromycin, and combination in patients hospitalized with COVID-19. Int. J. Infect. Dis. 97, 396–403.

Arshad, U., Pertinez, H., Box, H., Tatham, L., Rajoli, R.K.R., Curley, P., Neary, M., Sharp, J., Liptrott, N.J., Valentijn, A., David, C., Rannard, S.P., O'Neill, P.M., Aljayyoussi, G., Pennington, S.H., Ward, S.A., Hill, A., Back, D.J., Khoo, S.H., Bray, P.G., Biagini, G.A., Owen, A., 2020b. Prioritization of anti-SARS-CoV-2 drug repurposing opportunities based on plasma and target site concentrations derived from their established human pharmacokinetics. Clin. Pharmacol. Therapeut. 108 (4), 775–790.

Atherton, J.G., Kratzing, C.C., Fisher, A., 1978. The effect of ascorbic acid on infection chick-embryo ciliated tracheal organ cultures by coronavirus. Arch. Virol. 56 (3), 195–199. https://doi.org/10.1007/BF01317848.

Auriant, I., Jallot, A., Hervé, P., Cerrina, J., Le Roy Ladurie, F., Fournier, J.L., Lescot, B., Parquin, F., 2001. Noninvasive ventilation reduces mortality in acute respiratory failure following lung resection. Am. J. Respir. Crit. Care Med. 164 (7), 1231–1235. https://doi.org/10.1164/ajrccm.164.7.2101089.

Bahrampour Juybari, K., Pourhanifeh, M.H., Hosseinzadeh, A., Hemati, K., Mehrzadi, S., 2020. Melatonin potentials against viral infections including COVID-19: current evidence and new findings. Virus Res. 287 (October), 198108. https://doi.org/10.1016/j.virusres.2020.198108.

Bamford, A.P., Bentley, A., Dean, J., 2020. ICS guidance for prone positioning of the conscious COVID patient 2020. Intensive Care Soc. https://emcrit.org/wp-content/uploads/2020/04/2020-04-12-Guidance-for-conscious-proning.pdf.

Barbarin, V., Nihoul, A., Misson, P., Arras, M., Delos, M., Leclercq, I., Lison, D., Huaux, F., 2005. The role of pro- and anti-inflammatory responses in silica-induced lung fibrosis. Respir. Res. 6 (1), 112. https://doi.org/10.1186/1465-9921-6-112.

Barbaro, R.P., MacLaren, G., Boonstra, P.S., Iwashyna, T.J., Slutsky, A.S., Fan, E., Bartlett, R.H., et al., 2020. Extracorporeal membrane oxygenation support in COVID-19: an international cohort study of the extracorporeal life support organization registry. Lancet 396 (10257), 1071–1078. https://doi.org/10.1016/S0140-6736(20)32008-0.

Barnes, G.D., Burnett, A., Allen, A., Blumenstein, M., Clark, N.P., Cuker, A., Dager, W.E., et al., 2020. Thromboembolism and anticoagulant therapy during the COVID-19 pandemic: interim clinical guidance from the anticoagulation forum. J. Thromb. Thrombolysis 50 (1), 72–81. https://doi.org/10.1007/s11239-020-02138-z.

Barrot, L., Asfar, P., Mauny, F., Winiszewski, H., Montini, F., Badie, J., Quenot, J.-P., et al., 2020. Liberal or conservative oxygen therapy for acute respiratory distress syndrome. N. Engl. J. Med. 382 (11), 999–1008. https://doi.org/10.1056/NEJMoa1916431.

Bates, S.M., Rajasekhar, A., Middeldorp, S., McLintock, C., Rodger, M.A., James, A.H., Sara, R.V., et al., 2018. American Society of Hematology 2018 guidelines for management of venous thromboembolism: venous thromboembolism in the context of pregnancy. Blood Adv. 2 (22), 3317–3359. https://doi.org/10.1182/bloodadvances.2018024802.

Baveye, S., Elass, E., Mazurier, J., Spik, G., Legrand, D., 1999. Lactoferrin: a multifunctional glycoprotein involved in the modulation of the inflammatory process. Clin. Chem. Lab. Med. 37 (3), 281–286. https://doi.org/10.1515/CCLM.1999.049.

Beigel, J.H., Nam, H.H., Adams, P.L., Amy, K., Ince, W.L., El-Kamary, S.S., Sims, A.C., 2019. Advances in respiratory virus therapeutics - a meeting report from the 6th isirv antiviral group conference. Antivir. Res. 167 (July), 45–67. https://doi.org/10.1016/j.antiviral.2019.04.006.

Beigel, J.H., Tomashek, K.M., Dodd, L.E., Mehta, A.K., Zingman, B.S., Kalil, A.C., Hohmann, E., et al., 2020. "Remdesivir for the treatment of covid-19 — final report. N. Engl. J. Med. 383 (19), 1813–1826. https://doi.org/10.1056/nejmoa2007764.

Bell, B.G., Schellevis, F., Stobberingh, E., Goossens, H., Pringle, M., 2014. A systematic review and meta-analysis of the effects of antibiotic consumption on antibiotic resistance. BMC Infect. Dis. 14 (1), 1–25. https://doi.org/10.1186/1471-2334-14-13.

Bessière, F., Roccia, H., Delinière, A., Charrière, R., Chevalier, P., Argaud, L., Cour, M., 2020. Assessment of QT intervals in a case series of patients with coronavirus disease 2019 (COVID-19) infection treated with hydroxychloroquine alone or in combination with azithromycin in an intensive care unit. JAMA Cardiol. 5 (9), 1067–1069.

Bhala, N., Emberson, J., Merhi, A., Abramson, S., Arber, N., Baron, J.A., Bombardier, C., et al., 2013. Vascular and upper gastrointestinal effects of non-steroidal anti-inflammatory drugs: meta-analyses of individual participant data from randomised trials. Lancet 382 (9894), 769–779. https://doi.org/10.1016/S0140-6736(13)60900-9.

Bhatraju, P.K., Ghassemieh, B.J., Nichols, M., Kim, R., Jerome, K.R., Nalla, A.K., Greninger, A.L., et al., 2020. Covid-19 in critically ill patients in the seattle region - case series. N. Engl. J. Med. 382 (21), 2012–2022. https://doi.org/10.1056/NEJMoa2004500.

Bikdeli, B., Madhavan, M.V., Jimenez, D., Taylor, C., Dreyfus, I., Driggin, E., Nigoghossian, C.Der, et al., 2020. COVID-19 and thrombotic or thromboembolic disease: implications for prevention, antithrombotic therapy, and follow-up: JACC state-of-the-art review. J. Am. Coll. Cardiol. 75 (23), 2950–2973. https://doi.org/10.1016/j.jacc.2020.04.031.

Bingold, T.M., Scheller, B., Wolf, T., Meier, J., Koch, A., Zacharowski, K., Rosenberger, P., Iber, T., 2012. Superimposed high-frequency jet ventilation combined with continuous positive airway pressure/assisted spontaneous breathing improves oxygenation in patients with H1N1-associated ARDS. Ann. Intensive Care 2 (March), 7. https://doi.org/10.1186/2110-5820-2-7.

Blüher, M., Fasshauer, M., Tönjes, A., Kratzsch, J., Schön, M.R., Paschke, R., 2005. Association of interleukin-6, C-reactive protein, interleukin-10 and adiponectin plasma concentrations with measures of obesity, insulin sensitivity and glucose metabolism. Exp. Clin. Endocrinol. Diabetes 113 (9), 534–537. https://doi.org/10.1055/s-2005-872851.

Borba, M.G.S., Val, F.F.A., Sampaio, V.S., Alexandre, M.A.A., Melo, G.C., Brito, M., Mourão, M.P.G., Brito-Sousa, J.D., Baía-da-Silva, D., Guerra, M.V.F., Hajjar, L.A., Pinto, R.C., Balieiro, A.A.S., Pacheco, A.G.F., Santos Jr., J.D.O., Naveca, F.G., Xavier, M.S., Siqueira, A.M., Schwarzbold, A., Croda, J., Nogueira, M.L., Lacerda, M.V.G., CloroCovid-19 Team, 2020. Effect of high vs low doses of chloroquine diphosphate as adjunctive therapy for patients hospitalized with severe acute respiratory syndrome coronavirus 2 (SARS-CoV-2) infection: a randomized clinical trial. JAMA 3 (4), e208857.

Bozzette, S.A., Sattler, F.R., Chiu, J., Wu, A.W., Gluckstein, D., Kemper, C., Bartok, A., Niosi, J., Abramson, I., Coffman, J., 1990. A controlled trial of early adjunctive treatment with corticosteroids for *Pneumocystis carinii* pneumonia in the acquired immunodeficiency syndrome. California Collaborative Treatment Group. N. Engl. J. Med. 323 (21), 1451–1457. https://doi.org/10.1056/NEJM199011223232104.

Bramante, C., Nicholas Ingraham, Murray, T., Marmor, S., Hoversten, S., Gronski, J., McNeil, C., et al., 2020. Observational study of metformin and risk of mortality in patients hospitalized with covid-19. medRxiv. https://doi.org/10.1101/2020.06.19.20135095.

Bray, M., Rayner, C., Noël, F., Jans, D., Wagstaff, K., 2020. "Ivermectin and COVID-19: a report in antiviral research, widespread interest, an FDA warning, two letters to the editor and the authors' responses. Antivir. Res. 178, 104805.

Breslin, N., Baptiste, C., Gyamfi-Bannerman, C., Miller, R., Martinez, R., Bernstein, K., Ring, L., et al., 2020. Coronavirus disease 2019 infection among asymptomatic and symptomatic pregnant women: two weeks of confirmed presentations to an affiliated pair of New York city hospitals. Am. J. Obstet. Gynecol. MFM 2 (2), 100118. https://doi.org/10.1016/j.ajogmf.2020.100118.

Brewster, D.J., Chrimes, N., Do, T.B.T., Fraser, K., Groombridge, C.J., Higgs, A., Humar, M.J., et al., 2020. Consensus statement: safe airway society principles of airway management and tracheal intubation specific to the COVID-19 adult patient group. Med. J. Aust. 212 (10), 472–481. https://doi.org/10.5694/mja2.50598.

Briel, M., Meade, M., Mercat, A., Brower, R.G., Talmor, D., Walter, S.D., Arthur, S.S., et al., 2010. Higher vs lower positive end-expiratory pressure in patients with acute lung injury and acute respiratory distress syndrome: systematic review and meta-analysis. JAMA 303 (9), 865–873. https://doi.org/10.1001/jama.2010.218.

Brotman, D.J., Girod, J.P., Posch, A., Jani, J.T., Patel, J.V., Gupta, M., Lip, G.Y.H., Reddy, S., Kickler, T.S., 2006. Effects of short-term glucocorticoids on hemostatic factors in healthy volunteers. Thromb. Res. 118 (2), 247–252. https://doi.org/10.1016/j.thromres.2005.06.006.

Brower, R.G., Matthay, M.A., Morris, A., David Schoenfeld, Thompson, B.T., Wheeler, A., 2000. Ventilation with lower tidal volumes as compared with traditional tidal volumes for acute lung injury and the acute respiratory distress syndrome. N. Engl. J. Med. 342 (18), 1301–1308. https://doi.org/10.1056/NEJM200005043421801.

Bullen, E.C., Teijeiro-Paradis, R., Fan, E., 2020. How I select which patients with ARDS should be treated with venovenous extracorporeal membrane oxygenation. Chest 158 (3), 1036–1045. https://doi.org/10.1016/j.chest.2020.04.016.

Burkill, S., Vattulainen, P., Geissbuehler, Y., Espin, M.S., Popescu, C., Suzart-Woischnik, K., Hillert, J., et al., 2019. The association between exposure to interferon-beta during pregnancy and birth measurements in offspring of women with multiple sclerosis. PloS One 14 (12), e0227120. https://doi.org/10.1371/journal.pone.0227120.

Calder, P.C., Carr, A.C., Gombart, A.F., Eggersdorfer, M., 2020. Optimal nutritional status for a well-functioning immune system is an important factor to protect against viral infections. Nutrients 12 (4). https://doi.org/10.3390/nu12041181.

Campione, E., Lanna, C., Cosio, T., Rosa, L., Conte, M.P., Iacovelli, F., Romeo, A., et al., 2020. Lactoferrin as potential supplementary nutraceutical agent in COVID-19 patients: in vitro and in vivo preliminary evidences. bioRxiv. https://doi.org/10.1101/2020.08.11.244996.

Canga, A.G., Prieto, A.M.S., Liébana, M.J.D., Martínez, N.F., Vega, M.S., Vieitez, J.J.G., 2008. "The pharmacokinetics and interactions of ivermectin in humans—a mini-review. Am. Assoc. Pharmaceut. Sci. J. 10 (1), 42–46.

Cantorna, M.T., Snyder, L., Lin, Y.-D., Yang, L., 2015. Vitamin D and 1,25(OH)2D regulation of T cells. Nutrients 7 (4), 3011–3021. https://doi.org/10.3390/nu7043011.

Cao, B., Wang, Y., Wen, D., Liu, W., Wang, J., Fan, G., Ruan, L., et al., 2020a. A trial of lopinavir–ritonavir in adults hospitalized with severe covid-19. N. Engl. J. Med. 382 (19), 1787–1799. https://doi.org/10.1056/nejmoa2001282.

Cao, Y., Jia, W., Zou, L., Jiang, T., Wang, G., Chen, L., Huang, L., et al., 2020b. Ruxolitinib in treatment of severe coronavirus disease 2019 (COVID-19): a multicenter, single-blind, randomized controlled trial. J. Allergy Clin. Immunol. 146 (1), 137–146.e3. https://doi.org/10.1016/j.jaci.2020.05.019.

Caputo, N.D., Strayer, R.J., Levitan, R., 2020. Early self-proning in awake, non-intubated patients in the emergency department: a single ED's experience during the COVID-19 pandemic. Acad. Emerg. Med. 27 (5), 375–378. https://doi.org/10.1111/acem.13994.

Carlucci, P.M., Ahuja, T., Petrilli, C., Rajagopalan, H., Jones, S., Joseph, R., 2020. Hydroxychloroquine and azithromycin plus zinc vs hydroxychloroquine and azithromycin alone: outcomes in hospitalized COVID-19 patients. medRxiv. https://doi.org/10.1101/2020.05.02.20080036.

Carr, A.C., Maggini, S., 2017. Vitamin C and immune function. Nutrients 9 (11). https://doi.org/10.3390/nu9111211.

Casadevall, A., Pirofski, L.-a, 2020. The convalescent sera option for containing COVID-19. J. Clin. Invest. 130 (4), 1545–1548. https://doi.org/10.1172/JCI138003.

Cavalcanti, A.B., Zampieri, F.G., Rosa, R.G., Azevedo, L.C.P., Veiga, V.C., Avezum, A., Damiani, L.P., Marcadenti, A., Kawano-Dourado, L., Lisboa, T., Junqueira, D.L.M., de Barros E Silva, P.G.M., Tramujas, L., Abreu-Silva, E.O., Laranjeira, L.N., Soares, A.T., Echenique, L.S., Pereira, A.J., Freitas, F.G.R., Berwanger, O., Coalition Covid-19 Brazil I Investigators, 2020. Hydroxychloroquine with or without azithromycin in mild-to-moderate covid-19. N. Engl. J. Med. 383 (21), 2041–2052.

Center for Disease Control and Prevention, 2019. Stem Cell and Exosome Products. https://www.cdc.gov/hai/outbreaks/stem-cell-products.html.

Chaccour, C., Hammann, F., Ramón-García, S., Rabinovich, N.R., 2020. Ivermectin and COVID-19: keeping rigor in times of urgency. Am. J. Trop. Med. Hyg. 102 (6), 1156–1157.

Chang, P., Liao, Y., Guan, J., Guo, Y., Zhao, M., Hu, J., Zhou, J., et al., 2020. Combined treatment with hydrocortisone, vitamin C, and thiamine for sepsis and septic shock: a randomized controlled trial. Chest 158 (1), 174–182. https://doi.org/10.1016/j.chest.2020.02.065.

Chen, F., Chan, K.H., Jiang, Y., Kao, R.Y.T., Lu, H.T., Fan, K.W., Cheng, V.C.C., et al., 2004. In vitro susceptibility of 10 clinical isolates of SARS coronavirus to selected antiviral compounds. J. Clin. Virol. 31 (1), 69–75. https://doi.org/10.1016/j.jcv.2004.03.003.

Chen, J.-M., Fan, Y.-C., Lin, J.-W., Chen, Y.-Y., Hsu, W.-L., Chiou, S.-S., 2017. Bovine lactoferrin inhibits dengue virus infectivity by interacting with heparan sulfate, low-density lipoprotein receptor, and DC-SIGN. Int. J. Mol. Sci. 18 (9) https://doi.org/10.3390/ijms18091957.

Chen, G., Wu, Di, Guo, W., Cao, Y., Huang, D., Wang, H., Wang, T., et al., 2020a. Clinical and immunological features of severe and moderate coronavirus disease 2019. J. Clin. Invest. 130 (5), 2620–2629. https://doi.org/10.1172/JCI137244.

Chen, J., Hu, C., Chen, L., Tang, L., Zhu, Y., Xu, X., Chen, L., et al., 2020b. Clinical study of mesenchymal stem cell treatment for acute respiratory distress syndrome induced by epidemic influenza A (H7N9) infection: a hint for COVID-19 treatment. Engineering 6 (10), 1153–1161. https://doi.org/10.1016/j.eng.2020.02.006.

Chen, J., Xia, L., Liu, L., Xu, Q., Ling, Y., Huang, D., Huang, W., et al., 2020c. Antiviral activity and safety of darunavir/cobicistat for the treatment of COVID-19. Open Forum Infect. Dis. 7 (7), 1–5. https://doi.org/10.1093/ofid/ofaa241.

Cheng, M.H., Zhang, S., Porritt, R.A., Rivas, M.N., Paschold, L., Willscher, E., Binder, M., Arditi, M., Bahar, I., 2020. Superantigenic character of an insert unique to SARS-CoV-2 spike supported by skewed TCR repertoire in patients with hyperinflammation. Proc. Natl. Acad. Sci. U. S. A. 117 (41), 25254–25262. https://doi.org/10.1073/pnas.2010722117.

Chorin, E., Dai, M., Shulman, E., Wadhwani, L., Bar-Cohen, R., Barbhaiya, C., Aizer, A., Holmes, D., Bernstein, S., Spinelli, M., Park, D.S., Chinitz, L.A., Jankelson, L., 2020. The QT interval in patients with COVID-19 treated with hydroxychloroquine and azithromycin. Nat. Med. 26, 808–809.

Chu, C.M., Cheng, V.C.C.C., Hung, I.F.N.N., Wong, M.M.L.L., Chan, K.S.H.S., Chan, K.S.H.S., T Kao, R.Y.T., et al., 2004. Role of lopinavir/ritonavir in the treatment of SARS: initial virological and clinical findings. Thorax 59 (3), 252–256. https://doi.org/10.1136/thorax.2003.012658.

Chu, D.K., Kim, L.H.-Y., Young, P.J., Zamiri, N., Almenawer, S.A., Jaeschke, R., Szczeklik, W., Schünemann, H.J., Neary, J.D., Alhazzani, W., 2018. Mortality and morbidity in acutely ill adults treated with liberal versus conservative oxygen therapy (IOTA): a systematic review and meta-analysis. Lancet 391 (10131), 1693–1705. https://doi.org/10.1016/S0140-6736(18)30479-3.

Cleveland Clinic, 2020. Non-Steroidal Anti-inflammatory Drugs. https://my.clevelandclinic.org/health/drugs/11086-non-steroidal-anti-inflammatory-medicines.nsaids.

Cohen, A.T., Harrington, R.A., Goldhaber, S.Z., Hull, R.D., Wiens, B.L., Gold, A., Hernandez, A.F., Gibson, C.M., 2016. Extended thromboprophylaxis with Betrixaban in acutely ill medical patients. N. Engl. J. Med. 375 (6), 534–544. https://doi.org/10.1056/NEJMoa1601747.

Colson, P., Rolain, J.M., Lagier, J.C., Brouqui, P., Raoult, D., 2020. Chloroquine and hydroxychloroquine as available weapons to fight COVID-19. Int. J. Antimicrob. Agents 55 (4), 105932.

Colunga Biancatelli, R.M.L., Berrill, M., Marik, P.E., 2020. The antiviral properties of vitamin C. Expert Rev. Anti-Infect. Ther. https://doi.org/10.1080/14787210.2020.1706483.

Combes, A., Hajage, D., Capellier, G., Demoule, A., Lavoué, S., Guervilly, C., Da Silva, D., et al., 2018. Extracorporeal membrane oxygenation for severe acute respiratory distress syndrome. N. Engl. J. Med. 378 (21), 1965–1975. https://doi.org/10.1056/NEJMoa1800385.

Craft, T.M., Chambers, P.H., Ward, M.E., Goat, V.A., 1990. Two cases of barotrauma associated with transtracheal jet ventilation. Br. J. Anaesth. 64 (4), 524–527. https://doi.org/10.1093/bja/64.4.524.

CredibleMeds, 2020. Combined List of Drugs That Prolong QT and/or Cause Torsades de Pointes (TDP). https://crediblemeds.org/pdftemp/pdf/CombinedList.pdf.

Cummings, M.J., Baldwin, M.R., Abrams, D., Jacobson, S.D., Meyer, B.J., Balough, E.M., Aaron, J.G., et al., 2020. Epidemiology, clinical course, and outcomes of critically ill adults with COVID-19 in New York city: a prospective cohort study. Lancet 395 (10239), 1763–1770. https://doi.org/10.1016/S0140-6736(20)31189-2.

Czock, D., Keller, F., Rasche, F.M., Häussler, U., 2005. Pharmacokinetics and pharmacodynamics of systemically administered glucocorticoids. Clin. Pharmacokinet. 44 (1), 61–98. https://doi.org/10.2165/00003088-200544010-00003.

Dashraath, P., Wong, J.L.J., Lim, M.X.K., Lim, L.M., Li, S., Biswas, A., Choolani, M., Mattar, C., Su, L.L., 2020. Coronavirus disease 2019 (COVID-19) pandemic and pregnancy. Am. J. Obstet. Gynecol. 222 (6), 521–531. https://doi.org/10.1016/j.ajog.2020.03.021.

David, A.J., Martin, A.J., Wagstaff, K.M., 2019. Inhibitors of nuclear transport. Curr. Opin. Cell Biol. 58, 50–60.

Davoudi-Monfared, E., Rahmani, H., Khalili, H., Hajiabdolbaghi, M., Salehi, M., Abbasian, L., Kazemzadeh, H., Yekaninejad, M.S., 2020. A randomized clinical trial of the efficacy and safety of interferon β-1a in treatment of severe COVID-19. Antimicrob. Agents Chemother. 64 (9) https://doi.org/10.1128/AAC.01061-20.

De Meyera, S., Bojkovab, D., Cinatlb, J., Van Dammea, E., Buycka, S.C.C., Van Loocka, M., Woodfallc, B., 2020. Lack of antiviral activity of darunavir against SARS-CoV-2. Int. J. Infect. Dis. 97, 7–10.

Delahoy, M.J., Whitaker, M., O'Halloran, A., Chai, S.J., Kirley, P.D., Alden, N., Kawasaki, B., et al., 2020. Characteristics and maternal and birth outcomes of hospitalized pregnant women with laboratory-confirmed COVID-19 - COVID-NET, 13 states, March 1–August 22, 2020. Morb. Mortal. Wkly. Rep. 69 (38), 1347–1354. https://doi.org/10.15585/mmwr.mm6938e1.

Díez, J.-M., Romero, C., Gajardo, R., 2020. Currently available intravenous immunoglobulin contains antibodies reacting against severe acute respiratory syndrome coronavirus 2 antigens. Immunotherapy 12 (8), 571–576. https://doi.org/10.2217/imt-2020-0095.

Ding, Q., Lu, P., Fan, Y., Xia, Y., Liu, M., 2020. The clinical characteristics of pneumonia patients coinfected with 2019 novel coronavirus and influenza virus in Wuhan, China. J. Med. Virol. 92 (9), 1549–1555. https://doi.org/10.1002/jmv.25781.

Driggin, E., Madhavan, M.V., Bikdeli, B., Chuich, T., Laracy, J., Biondi-Zoccai, G., Brown, T.S., Der Nigoghossian, C., Zidar, D.A., Haythe, J., Brodie, D., Beckman, J.A., Kirtane, A.J., Stone, G.W., Krumholz, H.M., Parikh, S.A., 2020. Cardiovascular considerations for patients, health care workers, and health systems during the COVID-19 pandemic. J. Am. Coll. Cardiol. 75 (18), 2352–2371.

Drugscom, 2020. Drugs A to Z. Melatonin. https://www.drugs.com/sfx/melatonin-side-effects.html.

Du, Y., Tu, L., Zhu, P., Mu, M., Wang, R., Yang, P., Wang, X., et al., 2020. Clinical features of 85 fatal cases of COVID-19 from wuhan: a retrospective observational study. Am. J. Respir. Crit. Care Med. 201 (11), 1372–1379. https://doi.org/10.1164/rccm.202003-0543OC.

Elass-Rochard, E., Legrand, D., Salmon, V., Roseanu, A., Trif, M., Tobias, P.S., Mazurier, J., Spik, G., 1998. Lactoferrin inhibits the endotoxin interaction with CD14 by competition with the lipopolysaccharide-binding protein. Infect. Immun. 66 (2), 486–491. https://doi.org/10.1128/IAI.66.2.486-491.1998.

Elharrar, X., Trigui, Y., Dols, A.-M., Touchon, F., Martinez, S., Prud'homme, E., Papazian, L., 2020. Use of prone positioning in nonintubated patients with COVID-19 and hypoxemic acute respiratory failure. JAMA 323 (22), 2336–2338. https://doi.org/10.1001/jama.2020.8255.

Estcourt, L.J., Roberts, D.J., 2020. Convalescent plasma for covid-19. BMJ. https://doi.org/10.1136/bmj.m3516.

Fan, E., Del Sorbo, L., Goligher, E.C., Hodgson, C.L., Munshi, L., Walkey, A.J., Adhikari, N.K.J., et al., 2017. An Official American Thoracic Society/European Society of Intensive Care Medicine/Society of Critical Care Medicine Clinical Practice Guideline: mechanical ventilation in adult patients with acute respiratory distress syndrome. Am. J. Respir. Crit. Care Med. 195 (9), 1253–1263. https://doi.org/10.1164/rccm.201703-0548ST.

Fang, C.-H., Li, J.-J., Hui, R.-T., 2005. Statin, like aspirin, should be given as early as possible in patients with acute coronary syndrome. Med. Hypotheses 64 (1), 192–196. https://doi.org/10.1016/j.mehy.2004.06.018.

Fang, L., Karakiulakis, G., Roth, M., 2020. Are patients with hypertension and diabetes mellitus at increased risk for COVID-19 infection? Lancet Respir. Med. 8 (4), e21. https://doi.org/10.1016/S2213-2600(20)30116-8.

Favilli, A., Mattei Gentili, M., Raspa, F., Giardina, I., Parazzini, F., Vitagliano, A., Borisova, A.V., Gerli, S., 2020. Effectiveness and safety of available treatments for COVID-19 during pregnancy: a critical review. J. Matern. Fetal Neonatal Med. 1–14. https://doi.org/10.1080/14767058.2020.1774875.

FDA, 2020a. FDA Issues Emergency Use Authorization for Potential COVID-19 Treatment. Coronavirus (COVID-19) Update. https://www.fda.gov/news-events/press-announcements/coronavirus-covid-19-update-fda-issues-emergency-use-authorization-potential-covid-19-treatment.

FDA, 2020b. Remdesivir (VEKLURY).

Fei, Z., Yu, T., Du, R., Fan, G., Liu, Y., Liu, Z., Xiang, J., Wang, Y., Song, B., Gu, X., Guan, L., Wei, Y., Li, B.C.H., Wu, X., Xu, J., Tu, S., Zhang, Y., Chen, H., 2020. Clinical course and risk factors for mortality of adult inpatients with COVID-19 in Wuhan, China: a retrospective cohort study. Lancet 395, 1054–1062.

Ferrando, C., Mellado-Artigas, R., Gea, A., Arruti, E., Aldecoa, C., Adalia, R., Ramasco, F., et al., 2020. Awake prone positioning does not reduce the risk of intubation in COVID-19 treated with high-flow nasal oxygen therapy: a multicenter, adjusted cohort study. Crit. Care 24 (1), 597. https://doi.org/10.1186/s13054-020-03314-6.

Fisher, B.J., Seropian, I.M., Kraskauskas, D., Thakkar, J.N., Voelkel, N.F., Fowler 3rd, A.A., Natarajan, R., 2011. Ascorbic acid attenuates lipopolysaccharide-induced acute lung injury. Crit. Care Med. 39 (6), 1454–1460. https://doi.org/10.1097/CCM.0b013e3182120cb8.

Food and Drug Administration, 2018. Interferon Alfa-2b, Recombinant, pp. 1–40. https://www.accessdata.fda.gov/drugsatfda_docs/label/2018/103132Orig1s5199lbl.pdf.

Food and Drug Administration, 2019a. Approved Cellular and Gene Therapy Products. https://www.fda.gov/vaccines-blood-biologics/cellular-gene-therapy-products/approved-cellular-and-gene-therapy-products.

Food and Drug Administration, 2019b. FDA Warns about Stem Cell Therapies. https://www.fda.gov/consumers/consumer-updates/fda-warns-about-stem-cell-therapies.

Food and Drug Administration, 2019c. Interferon Beta-1a (Rebif). https://www.accessdata.fda.gov/drugsatfda_docs/label/2019/103780s5204lbl.pdf.

Food and Drug Administration, 2020a. Emergency Use Authorization (EUA) of COVID-19 Convalescent Plasma for Treatment of COVID-19 in Hospitalized Patients, pp. 1–36.

Food and Drug Administration, 2020b. FDA Advises Patients on Use of Non-steroidal Anti-inflammatory Drugs (NSAIDs) for COVID-19. https://www.fda.gov/drugs/drug-safety-and-availability/fda-advises-patients-use-non-steroidal-anti-inflammatory-drugs-nsaids-covid-19.

Food and Drug Administration, 2020c. FDA Emergency Use Authorization (EUA) Request for COVID-19 Convalescent Plasma (CCP), pp. 1–17. https://www.fda.gov/media/141480/download.

Food and Drug Administration, 2020d. Letter of Authorization, Reissuance of Convalescent Plasma EUA November 30, 2020, pp. 1–8, 18250.

Food and Drug Administration, 2020e. Remdesivir by Gilead Sciences: FDA Warns of Newly Discovered Potential Drug Interaction That May Reduce Effectiveness of Treatment.

Fowler 3rd, A.A., Syed, A.A., Knowlson, S., Sculthorpe, R., Don, F., DeWilde, C., Farthing, C.A., et al., 2014. Phase I safety trial of intravenous ascorbic acid in patients with severe sepsis. J. Transl. Med. 12 (January), 32. https://doi.org/10.1186/1479-5876-12-32.

Fowler 3rd, A.A., Truwit, J.D., Hite, R.D., Morris, P.E., DeWilde, C., Priday, A., Fisher, B., et al., 2019. Effect of vitamin C infusion on organ failure and biomarkers of inflammation and vascular injury in patients with sepsis and severe acute respiratory failure: the CITRIS-ALI randomized clinical trial. JAMA 322 (13), 1261–1270. https://doi.org/10.1001/jama.2019.11825.

Fowler 3rd, A.A., Fisher, B.J., Kashiouris, M.G., 2020. Vitamin C for sepsis and acute respiratory failure-reply. JAMA. https://doi.org/10.1001/jama.2019.21987.

Fraisse, F., Holzapfel, L., Couland, J.M., Simonneau, G., Bedock, B., Feissel, M., Herbecq, P., Pordes, R., Poussel, J.F., Roux, L., 2000. Nadroparin in the prevention of deep vein thrombosis in acute decompensated COPD. The association of non-university affiliated intensive care specialist physicians of France. Am. J. Respir. Crit. Care Med. 161 (4 Pt 1), 1109–1114.

Frat, J.-P., Thille, A.W., Mercat, A., Girault, C., Ragot, S., Perbet, S., Prat, G., et al., 2015. High-flow oxygen through nasal cannula in acute hypoxemic respiratory failure. N. Engl. J. Med. 372 (23), 2185–2196. https://doi.org/10.1056/NEJMoa1503326.

Fu, Y., Cheng, Y., Wu, Y., 2020. Understanding SARS-CoV-2-mediated inflammatory responses: from mechanisms to potential therapeutic tools. Virol. Sin. 35 (3), 266–271. https://doi.org/10.1007/s12250-020-00207-4.

Fujii, T., Luethi, N., Young, P.J., Frei, D.R., Eastwood, G.M., French, C.J., Adam, M.D., et al., 2020. Effect of vitamin C, hydrocortisone, and thiamine vs hydrocortisone alone on time alive and free of vasopressor support among patients with septic shock: the vitamins randomized clinical trial. JAMA 323 (5), 423–431. https://doi.org/10.1001/jama.2019.22176.

Furtado, R.H.M., Berwanger, O., Fonseca, H.A., Corrêa, T.D., Ferraz, L.R., Lapa, M.G., Zampieri, F.G., Veiga, V.C., Azevedo, L.C.P., Rosa, R.G., Lopes, R.D., Avezum, A., Manoel, A.L.O., Piza, F.M.T., Martins, P.A., Lisboa, T.C., Pereira, A.J., Olivato, G.B., Dantas, V.C.S., Milan, E.P., Gebara, O.C.E., Amazonas, R.B., Olive, Cavalcanti, A.B., Coalition COVID-19 Brazil II Investigators, 2020. Azithromycin in addition to standard of care versus standard of care alone in the treatment of patients admitted to the hospital with severe COVID-19 in Brazil (Coalition II): a randomised clinical trial. Lancet 396 (10256), 959–967.

Galeotti, C., Kaveri, S.V., Bayry, J., 2017. IVIG-mediated effector functions in autoimmune and inflammatory diseases. Int. Immunol. 29 (11), 491–498. https://doi.org/10.1093/intimm/dxx039.

Galeotti, C., Kaveri, S., Bayry, J., 2020. Intravenous immunoglobulin immunotherapy for coronavirus disease-19 (COVID-19). Clinical & translational immunology 9 (10), e1198. https://doi.org/10.1002/cti2.1198.

García-de-Acilu, M., Patel, B.K., Roca, O., 2019. Noninvasive approach for de Novo acute hypoxemic respiratory failure: noninvasive ventilation, high-flow nasal cannula, both or none? Curr. Opin. Crit. Care 25 (1), 54–62. https://doi.org/10.1097/MCC.0000000000000570.

Gebistorf, F., Karam, O., Wetterslev, J., Afshari, A., 2016. Inhaled nitric oxide for acute respiratory distress syndrome (ARDS) in children and adults. Cochrane Database Syst. Rev. 2016 (6), CD002787. https://doi.org/10.1002/14651858.CD002787.pub3.

Geleris, J., Sun, Y., Platt, J., Zucker, J., Baldwin, M., Hripcsak, G., Labella, A., Manson, D.K., Kubin, C., Barr, R.G., Sobieszczyk, M.E., Schluger, N.W., 2020. Observational study of hydroxychloroquine in hospitalized patients with COVID-19. N. Engl. J. Med. 382 (25), 2411–2418.

George, L.A., 2020. Coronavirus Disease 2019 (COVID-19): Critical Care and Airway Management Issues. UpToDate. https://www.uptodate.com/contents/coronavirus-disease-2019-covid-19-critical-care-and-airway-management-issues.

Gharbharan, A., Jordans, C.C.E., GeurtsvanKessel, C., Hollander, J.G den, Karim, F., Mollema, F.P.N., Stalenhoef, J.E., et al., 2020. Convalescent plasma for COVID-19. A randomized clinical trial. medRxiv. https://doi.org/10.1101/2020.07.01.20139857.

Gilbert, C., Valois, M., Koren, G., 2006. Pregnancy outcome after first-trimester exposure to metformin: a meta-analysis. Fertil. Steril. 86 (3), 658–663. https://doi.org/10.1016/j.fertnstert.2006.02.098.

Ginde, A.A., Brower, R.G., Caterino, J.M., Finck, L., Banner-Goodspeed, V.M., Grissom, C.K., Hayden, D., et al., 2019. Early high-dose vitamin D(3) for critically ill, vitamin D-deficient patients. N. Engl. J. Med. 381 (26), 2529–2540. https://doi.org/10.1056/NEJMoa1911124.

Goldman, J.D., Lye, D.C.B., Hui, D.S., Marks, K.M., Bruno, R., Montejano, R., Spinner, C.D., et al., 2020. Remdesivir for 5 or 10 days in patients with severe COVID-19. N. Engl. J. Med. 383 (19), 1827–1837. https://doi.org/10.1056/nejmoa2015301.

Goldstein, M.R., Poland, G.A., Graeber, C.W., 2020. Are certain drugs associated with enhanced mortality in COVID-19? QJM 113 (7), 509–510. https://doi.org/10.1093/qjmed/hcaa103.

Goligher, E.C., Hodgson, C.L., Adhikari, N.K.J., Meade, M.O., Wunsch, H., Uleryk, E., Gajic, O., et al., 2017. Lung recruitment maneuvers for adult patients with acute respiratory distress syndrome. A systematic review and meta-analysis. Ann. Am. Thoracic Soc. 14 (Suppl. 4), S304–S311. https://doi.org/10.1513/AnnalsATS.201704-340OT.

Gombart, A.F., Pierre, A., Maggini, S., 2020. A review of micronutrients and the immune system-working in harmony to reduce the risk of infection. Nutrients 12 (1). https://doi.org/10.3390/nu12010236.

González-Chávez, S.A., Arévalo-Gallegos, S., Rascón-Cruz, Q., 2009. Lactoferrin: structure, function and applications. Int. J. Antimicrob. Agents 33 (4). https://doi.org/10.1016/j.ijantimicag.2008.07.020, 301.e1-8.

Greiller, C.L., Martineau, A.R., 2015. Modulation of the immune response to respiratory viruses by vitamin D. Nutrients 7 (6), 4240–4270. https://doi.org/10.3390/nu7064240.

Gritti, G., Raimondi, F., Ripamonti, D., Riva, I., Landi, F., Alborghetti, L., Frigeni, M., et al., 2020. IL-6 signalling pathway inactivation with siltuximab in patients with COVID-19 respiratory failure: an observational cohort study. medRxiv. https://doi.org/10.1101/2020.04.01.20048561.

Gröber, U., Kisters, K., 2012. Influence of drugs on vitamin D and calcium metabolism. Derm. Endocrinol. 4 (2), 158–166. https://doi.org/10.4161/derm.20731.

Grosser, T., Ricciotti, E., FitzGerald, G.A., 2017. The cardiovascular pharmacology of nonsteroidal anti-inflammatory drugs. Trends Pharmacol. Sci. 38 (8), 733–748. https://doi.org/10.1016/j.tips.2017.05.008.

Guan, W.-j, Ni, Z.-y, Hu, Y.-h Y., Liang, W.-h, Ou, C.-q, He, J.-x, Liu, L., et al., 2020. Clinical characteristics of coronavirus disease 2019 in China. N. Engl. J. Med. 382 (18), 1708–1720. https://doi.org/10.1056/nejmoa2002032.

Guzzo, C.A., Furtek, C.I., Porras, A.G., Chen, C., Tipping, R., Clineschmidt, C.M., Sciberras, D.G., Hsieh, J.Y., Lasseter, K.C., 2002. Safety, tolerability, and pharmacokinetics of escalating high doses of ivermectin in healthy adult subjects. J. Clin. Pharmacol. 42 (10), 1122–1133.

Hambridge, K., 2007. The management of lipohypertrophy in diabetes care. Br. J. Nurs. 16 (9), 520–524. https://doi.org/10.12968/bjon.2007.16.9.23428.

Han, J., Liu, Y., Liu, H., Li, Y., 2019. Genetically modified mesenchymal stem cell therapy for acute respiratory distress syndrome. Stem Cell Res. Ther. 10 (1), 386. https://doi.org/10.1186/s13287-019-1518-0.

Han, H., Yang, L., Liu, R., Liu, F., Wu, K.L., Li, J., Liu, X.H., Zhu, C.L., 2020. Prominent changes in blood coagulation of patients with SARS-CoV-2 infection. Clin. Chem. Lab. Med. 58 (7), 1116–1120.

Hao, L., Shan, Q., Wei, J., Ma, F., Sun, P., 2019. Lactoferrin: major physiological functions and applications. Curr. Protein Pept. Sci. 20 (2), 139–144. https://doi.org/10.2174/1389203719666180514150921.

Hasani, A., Chapman, T.H., McCool, D., Smith, R.E., Dilworth, J.P., Agnew, J.E., 2008. Domiciliary humidification improves lung mucociliary clearance in patients with bronchiectasis. Chron. Respir. Dis. 5 (2), 81–86. https://doi.org/10.1177/1479972307087190.

Heit, J.A., Kobbervig, C.E., James, A.H., Petterson, T.M., Bailey, K.R., Melton 3rd, L.J., 2005. Trends in the incidence of venous thromboembolism during pregnancy or postpartum: a 30-year population-based study. Ann. Intern. Med. 143 (10), 697–706. https://doi.org/10.7326/0003-4819-143-10-200511150-00006.

Hellwig, K., Caron, F.D., Wicklein, E.-M., Bhatti, A., Adamo, A., 2020. Pregnancy outcomes from the global pharmacovigilance database on interferon beta-1b exposure. Ther. Adv. Neurol. Disord. 13 https://doi.org/10.1177/1756286420910310.

Hemilä, H., 1997. Vitamin C intake and susceptibility to pneumonia. Pediatr. Infect. Dis. J. 16 (9), 836–837. https://doi.org/10.1097/00006454-199709000-00003.

Hemilä, H., 2003. Vitamin C and SARS coronavirus. J. Antimicrob. Chemother. https://doi.org/10.1093/jac/dkh002.

Hemilä, H., 2017. Vitamin C and infections. Nutrients 9 (4). https://doi.org/10.3390/nu9040339.

Henry, B.M., Lippi, G., 2020. Poor survival with extracorporeal membrane oxygenation in acute respiratory distress syndrome (ARDS) due to coronavirus disease 2019 (COVID-19): pooled analysis of early reports. J. Crit. Care. https://doi.org/10.1016/j.jcrc.2020.03.011.

Herr, C., Shaykhiev, R., Bals, R., 2007. The role of cathelicidin and defensins in pulmonary inflammatory diseases. Expet Opin. Biol. Ther. 7 (9), 1449–1461. https://doi.org/10.1517/14712598.7.9.1449.

Hoffman 2nd, H.N., Phyliky, R.L., Fleming, C.R., 1988. Zinc-induced copper deficiency. Gastroenterology 94 (2), 508–512. https://doi.org/10.1016/0016-5085(88)90445-3.

Horby, P., Lim, W.S., Emberson, J., Mafham, M., Bell, J., Linsell, L., Staplin, N., et al., 2020. Effect of dexamethasone in hospitalized patients with COVID-19: preliminary report. medRxiv. https://doi.org/10.1101/2020.06.22.20137273.

Hu, W., Wang, Y., Li, J., Huang, J., Pu, Y., Jiang, Y., Dong, X., Ding, Z., Zhao, B., Luo, Q., 2020. The predictive value of D-dimer test for venous thromboembolism during puerperium: a prospective cohort study. Clin. Appl. Thromb. Hemostasis 26. https://doi.org/10.1177/1076029620901786.

Huang, H.-B., Peng, J.-M., Weng, L., Liu, G.-Y., Du, B., 2018. High-flow oxygen therapy in immunocompromised patients with acute respiratory failure: a review and meta-analysis. J. Crit. Care 43 (February), 300–305. https://doi.org/10.1016/j.jcrc.2017.09.176.

Huang, C., Wang, Y., Li, X., Ren, L., Zhao, J., Hu, Y., Zhang, L., et al., 2020. Clinical features of patients infected with 2019 novel coronavirus in Wuhan, China. Lancet 395 (10223), 497–506. https://doi.org/10.1016/S0140-6736(20)30183-5.

Hui, D.S., Chow, B.K., Lo, T., Tsang, O.T.Y., Ko, F.W., Ng, S.S., Gin, T., Chan, M.T.V., 2019. Exhaled air dispersion during high-flow nasal cannula therapy versus CPAP via different masks. Eur. Respir. J. 53 (4) https://doi.org/10.1183/13993003.02339-2018.

Hwang, S.Y., Ryoo, S.M., Park, J.E., Jo, Y.H., Jang, D.-H., Suh, G.J., Kim, T., et al., 2020. Combination therapy of vitamin C and thiamine for septic shock: a multi-centre, double-blinded randomized, controlled study. Intensive Care Med. 46 (11), 2015–2025. https://doi.org/10.1007/s00134-020-06191-3.

Iba, T., Di Nisio, M., Levy, J.H., Kitamura, N., Thachil, J., 2017. New criteria for sepsis-induced coagulopathy (SIC) following the revised sepsis definition: a retrospective analysis of a nationwide survey. BMJ Open 7 (9), e017046. https://doi.org/10.1136/bmjopen-2017-017046.

Iglesias, J., Vassallo, A.V., Patel, V.V., Sullivan, J.B., Cavanaugh, J., Elbaga, Y., 2020. Outcomes of metabolic resuscitation using ascorbic acid, thiamine, and glucocorticoids in the early treatment of sepsis: the ORANGES trial. Chest 158 (1), 164–173. https://doi.org/10.1016/j.chest.2020.02.049.

Ihra, G., Gockner, G., Kashanipour, A., Aloy, A., 2000. High-frequency jet ventilation in European and North American institutions: developments and clinical practice. Eur. J. Anaesthesiol. 17 (7), 418–430. https://doi.org/10.1046/j.1365-2346.2000.00692.x.

Ishikawa, H., Awano, N., Fukui, T., Sasaki, H., Kyuwa, S., 2013. The protective effects of lactoferrin against murine norovirus infection through inhibition of both viral attachment and replication. Biochem. Biophys. Res. Commun. 434 (4), 791–796. https://doi.org/10.1016/j.bbrc.2013.04.013.

Jiang, B., Wei, H., 2020. Oxygen therapy strategies and techniques to treat hypoxia in COVID-19 patients. Eur. Rev. Med. Pharmacol. Sci. 24 (19), 10239–10246. https://doi.org/10.26355/eurrev_202010_23248.

Joyner, M.J., Senefeld, J.W., Klassen, S.A., Mills, J.R., Johnson, P.W., Theel, E.S., Wiggins, C.C., et al., 2020b. Effect of convalescent plasma on mortality among hospitalized patients with COVID-19: initial three-month experience. medRxiv. https://doi.org/10.1101/2020.08.12.20169359.

Joyner, M.J., Bruno, K.A., Klassen, S.A., Kunze, K.L., Wiggins, C.C., Johnson, P.W., Lesser, E.R., Hodge, D.O., Senefeld, J.W., Klompas, A.M., Shepherd, J.R.A., Rea, R.F.,

Whelan, E.R., et al., 2020a. Safety update: COVID-19 convalescent plasma in 20,000 hospitalized patients. J. Plant Physiol. 95 (x), 153237. https://doi.org/10.1016/j.jplph.2020.153237.

Juliana, C.R., Sherman, M.S., Fatteh, N., Vogel, F., Sacks, J., Rajter, J.-J., 2020. ICON (ivermectin in COvid nineteen) study: use of ivermectin is associated with lower mortality in hospitalized patients with COVID19. medRxiv, 20124461.

Juybari, K.B., Hosseinzadeh, A., Ghaznavi, H., Kamali, M., Sedaghat, A., Mehrzadi, S., Naseripour, M., 2019. Melatonin as a modulator of degenerative and regenerative signaling pathways in injured retinal ganglion cells. Curr. Pharmaceut. Des. 25 (28), 3057–3073. https://doi.org/10.2174/1381612825666190829151314.

Kahn, S.R., Lim, W., Dunn, A.S., Cushman, M., Dentali, F., Akl, E.A., Cook, D.J., Balekian, A.A., Klein, R.C., Le, H., Schulman, S., Murad, M.H., 2012. Prevention of VTE in nonsurgical patients: antithrombotic therapy and prevention of thrombosis, 9th ed: American College of Chest Physicians evidence-based clinical practice guidelines. Chest 141 (2 Suppl. l), e195S–e226S.

Kakoulidis, I., Ilias, I., Koukkou, E., 2020. SARS-CoV-2 infection and glucose homeostasis in pregnancy. What about antenatal corticosteroids? Diabetes Metab. Syndr. 14 (4), 519–520. https://doi.org/10.1016/j.dsx.2020.04.045.

Kalil, A.C., Metersky, M.L., Klompas, M., Muscedere, J., Sweeney, D.A., Palmer, L.B., Napolitano, L.M., et al., 2016. Management of adults with hospital-acquired and ventilator-associated pneumonia: 2016 clinical practice guidelines by the infectious diseases Society of America and the American Thoracic Society. Clin. Infect. Dis. 63 (5), e61–111. https://doi.org/10.1093/cid/ciw353.

Karalis, D.G., Hill, A.N., Clifton, S., Wild, R.A., 2016. The risks of statin use in pregnancy: a systematic review. J. Clin. Lipidol. 10 (5), 1081–1090. https://doi.org/10.1016/j.jacl.2016.07.002.

Keyaerts, E., Li, S., Vijgen, L., Rysman, E., Verbeeck, J., Van Ranst, M., Maes, P., 2009. Antiviral activity of chloroquine against human coronavirus OC43 infection in newborn mice. Antimicrob. Agents Chemother. 53 (8), 3416–3421. https://doi.org/10.1128/AAC.01509-08.

Kim, W.-Y., Jo, E.-J., Eom, J.S., Mok, J., Kim, M.-H., Kim, K.U., Park, H.-K., Lee, M.K., Lee, K., 2018. Combined vitamin C, hydrocortisone, and thiamine therapy for patients with severe pneumonia who were admitted to the intensive care unit: propensity score-based analysis of a before-after cohort study. J. Crit. Care 47 (October), 211–218. https://doi.org/10.1016/j.jcrc.2018.07.004.

King, G.L., Park, K., Li, Q., 2016. Selective insulin resistance and the development of cardiovascular diseases in diabetes: the 2015 Edwin Bierman award lecture. Diabetes 65 (6), 1462–1471. https://doi.org/10.2337/db16-0152.

Kirkby, N.S., Chan, M.V., Zaiss, A.K., Garcia-Vaz, E., Jiao, J., Berglund, L.M., Verdu, E.F., et al., 2016. Systematic study of constitutive cyclooxygenase-2 expression: role of NF-KB and NFAT transcriptional pathways. Proc. Natl. Acad. Sci. U. S. A. 113 (2), 434–439. https://doi.org/10.1073/pnas.1517642113.

Knight, M., Bunch, K., Vousden, N., Morris, E., Simpson, N., Gale, C., O'Brien, P., Quigley, M., Brocklehurst, P., Kurinczuk, J.J., 2020. Characteristics and outcomes of pregnant women admitted to hospital with confirmed SARS-CoV-2 infection in UK: national population based cohort study. Br. Med. J. 369 (June), m2107. https://doi.org/10.1136/bmj.m2107.

Koga, Y., Kaneda, K., Fujii, N., Tanaka, R., Miyauchi, T., Fujita, M., Hidaka, K., Oda, Y., Tsuruta, R., 2020. Comparison of high-flow nasal cannula oxygen therapy and non-invasive ventilation as first-line therapy in respiratory failure: a multicenter retrospective study. Acute Med. Surg. 7 (1), e461. https://doi.org/10.1002/ams2.461.

Koss, C.A., Natureeba, P., Plenty, A., Luwedde, F., Mwesigwa, J., Ades, V., Charlebois, E.D., et al., 2014. Risk factors for preterm birth among HIV-infected pregnant Ugandan women randomized to lopinavir/ritonavir-or efavirenz-based antiretroviral therapy. J. Acquir. Immune Defic. Syndr. 67 (2), 128–135. https://doi.org/10.1097/QAI.0000000000000281.

Kow, C.S., Hasan, S.S., 2020. Meta-analysis of effect of statins in patients with COVID-19. Am. J. Cardiol. 134 (November), 153–155. https://doi.org/10.1016/j.amjcard.2020.08.004.

Kutsukake, M., Matsutani, T., Tamura, K., Matsuda, A., Kobayashi, M., Tachikawa, E., Uchida, E., 2014. Pioglitazone attenuates lung injury by modulating adipose inflammation. J. Surg. Res. 189 (2), 295–303. https://doi.org/10.1016/j.jss.2014.03.007.

Kylie, M.W., Sivakumaran, H., Heaton, S.M., Harrich, D., Jans, D.A., 2012. Ivermectin is a specific inhibitor of importin α/β-mediated nuclear import able to inhibit replication of HIV-1 and dengue virus. Biochem. J. 443 (3), 851–856.

Laaksi, I., 2012. Vitamin D and respiratory infection in adults. Proc. Nutr. Soc. 71 (1), 90–97. https://doi.org/10.1017/S0029665111003351.

Le, R.Q., Liang, L., Yuan, W., Shord, S.S., Nie, L., Habtemariam, B.A., Przepiorka, D., Farrell, A.T., Pazdur, R., 2018. FDA approval summary: tocilizumab for treatment of chimeric antigen receptor T cell-induced severe or life-threatening cytokine release syndrome. Oncologist 23 (8), 943–947. https://doi.org/10.1634/theoncologist.2018-0028.

Lebovitz, H.E., 2019. Thiazolidinediones: the forgotten diabetes medications. Curr. Diabetes Rep. 19 (12), 151. https://doi.org/10.1007/s11892-019-1270-y.

Lee, M., Sankatsing, R., Schippers, E., Vogel, M., Fätkenheuer, G., van der Ven, A., Kroon, F., Rockstroh, J., Wyen, C., Bäumer, A., de Groot, E., Koopmans, P., Stroes, E., Reiss, P., van der Burger, D., 2007. Pharmacokinetics and pharmacodynamics of combined use of lopinavir/ritonavir and rosuvastatin in HIV-infected patients. Antivir. Ther. 12 (7), 1127–1132.

Lee, J.W., Gupta, N., Serikov, V., Matthay, M.A., 2009. Potential application of mesenchymal stem cells in acute lung injury. Expet Opin. Biol. Ther. 9 (10), 1259–1270. https://doi.org/10.1517/14712590903213651.

Lee, K.-H., Tseng, W.-C., Yang, C.-Y., Tarng, D.-C., 2019. The anti-inflammatory, anti-oxidative, and anti-apoptotic benefits of stem cells in acute ischemic kidney injury. Int. J. Mol. Sci. 20 (14) https://doi.org/10.3390/ijms20143529.

Lei, G.-S., Zhang, C., Cheng, B.-H., Lee, C.-H., 2017. Mechanisms of action of vitamin D as supplemental therapy for pneumocystis pneumonia. Antimicrob. Agents Chemother. 61 (10) https://doi.org/10.1128/AAC.01226-17.

Leizorovicz, A., Cohen, A.T., Turpie, A.G., Olsson, C.G., Vaitkus, P.T., Goldhaber, S.Z., PREVENT Medical Thromboprophylaxis Study Group, 2004. Randomized, placebo-controlled trial of dalteparin for the prevention of venous thromboembolism in acutely ill medical patients. Circulation 110 (7), 874–879.

Lejal, N., Tarus, B., Bouguyon, E., Chenavas, S., Bertho, N., Delmas, B., Ruigrok, R.W.H., Di Primo, C., Slama-Schwok, A., 2013. Structure-based discovery of the novel antiviral properties of naproxen against the nucleoprotein of influenza A virus. Antimicrob. Agents Chemother. 57 (5), 2231–2242. https://doi.org/10.1128/AAC.02335-12.

Lemos, A.C.B., do Espírito Santo, D.A., Salvetti, M.C., Gilio, R.N., Agra, L.B., Pazin-Filho, A., Miranda, C.H., 2020. Therapeutic versus prophylactic anticoagulation for severe COVID-19: a randomized phase II clinical trial (HESACOVID). Thromb. Res. 196, 359–366.

Leng, Z., Zhu, R., Hou, W., Feng, Y., Yang, Y., Han, Q., Shan, G., et al., 2020. Transplantation of ACE2(-) mesenchymal stem cells improves the outcome of patients with COVID-19 pneumonia. Aging Dis. 11 (2), 216–228. https://doi.org/10.14336/AD.2020.0228.

Leon, C., Druce, J.D., Catton, M.G., Jans, D.A., Wagstaff, K.M., 2020. The FDA-approved drug ivermectin inhibits the replication of SARS-CoV-2 in vitro. Antivir. Res. 178, 104787.

Leone, M., Einav, S., Chiumello, D., Constantin, J.-M., De Robertis, E., De Abreu, M.G., Gregoretti, C., et al., 2020. Noninvasive respiratory support in the hypoxaemic perioperative/periprocedural patient: a joint ESA/ESICM guideline. Intensive Care Med. 46 (4), 697–713. https://doi.org/10.1007/s00134-020-05948-0.

Levy, S.D., Alladina, J.W., Hibbert, K.A., Harris, R.S., Bajwa, E.K., Hess, D.R., 2016. High-flow oxygen therapy and other inhaled therapies in intensive care units. Lancet 387 (10030), 1867–1878. https://doi.org/10.1016/S0140-6736(16)30245-8.

Li, T., 2020. Diagnosis and clinical management of severe acute respiratory syndrome coronavirus 2 (SARS-CoV-2) infection: an operational recommendation of peking union medical college hospital (V2.0). Emerg. Microb. Infect. 9 (1), 582–585. https://doi.org/10.1080/22221751.2020.1735265.

Li, G., Zhao, S., Cui, S., Li, L., Xu, Y., Li, Y., 2015. Effect comparison of metformin with insulin treatment for gestational diabetes: a meta-analysis based on RCTs. Arch. Gynecol. Obstet. 292 (1), 111–120. https://doi.org/10.1007/s00404-014-3566-0.

Li, Q., Xie, P., Zha, B., Wu, Z.Y., Wei, H., 2017. Supraglottic jet oxygenation and ventilation saved a patient with 'cannot intubate and cannot ventilate' emergency difficult airway. J. Anesth. 31 (1), 144–147. https://doi.org/10.1007/s00540-016-2279-x.

Li, L., Li, R., Wu, Z., Yang, X., Zhao, M., Liu, J., Chen, D., 2020a. Therapeutic strategies for critically ill patients with COVID-19. Ann. Intensive Care 10 (1), 45. https://doi.org/10.1186/s13613-020-00661-z.

Li, L., Zhang, W., Hu, Y., Tong, X., Zheng, S., Yang, J., Kong, Y., et al., 2020b. Effect of convalescent plasma therapy on time to clinical improvement in patients with severe and life-threatening COVID-19: a randomized clinical trial. JAMA 324 (5), 460–470. https://doi.org/10.1001/jama.2020.10044.

Liang, H., Acharya, G., 2020. Novel corona virus disease (COVID-19) in pregnancy: what clinical recommendations to follow? Acta Obstet. Gynecol. Scand. https://doi.org/10.1111/aogs.13836.

Liang, H., Hou, Y., Sun, L., Li, Q., Wei, H., Yi, F., 2019a. Supraglottic jet oxygenation and ventilation for obese patients under intravenous anesthesia during hysteroscopy: a randomized controlled clinical trial. BMC Anesthesiol. 19 (1), 151. https://doi.org/10.1186/s12871-019-0821-8.

Liang, H., Hou, Y., Wei, H., Feng, Y., 2019b. Supraglottic jet oxygenation and ventilation assisted fiberoptic intubation in a paralyzed patient with morbid obesity and obstructive sleep apnea: a case report. BMC Anesthesiol. 19 (1), 40. https://doi.org/10.1186/s12871-019-0709-7.

Liang, B., Chen, J., Li, T., Wu, H., Yang, W., Li, Y., Li, J., et al., 2020. Clinical remission of a critically ill COVID-19 patient treated by human umbilical cord mesenchymal stem cells: a case report. Medicine 99 (31), e21429. https://doi.org/10.1097/MD.0000000000021429.

Lier, A.J., Tuan, J.J., Davis, M.W., Paulson, N., McManus, D., Campbell, S., Peaper, D.R., Topal, J.E., 2020. Case report: disseminated strongyloidiasis in a patient with COVID-19. Am. J. Trop. Med. Hyg. 103 (4), 1590–1592. https://doi.org/10.4269/ajtmh.20-0699.

Little, P., 2020. Non-steroidal anti-inflammatory drugs and covid-19. BMJ. https://doi.org/10.1136/bmj.m1185.

Liu, X., Wang, X.-J., 2020. Since January 2020 Elsevier has created a COVID-19 resource centre with free information in English and Mandarin on the novel coronavirus COVID-19. The COVID-19 resource centre is hosted on Elsevier connect, the company' s public news and information. J. Genet. Genom. J. 47 (2), 119–121.

Liu, C., Feng, X., Qi, L., Wang, Y., Qian, L., Hua, M., 2016a. Adiponectin, TNF-α and inflammatory cytokines and risk of type 2 diabetes: a systematic review and meta-analysis. Cytokine 86, 100–109. https://doi.org/10.1016/j.cyto.2016.06.028.

Liu, Q., Zhou, Y.-h, Yang, Z.-qiu, 2016b. The cytokine storm of severe influenza and development of immunomodulatory therapy. Cell. Mol. Immunol. 13 (1), 3–10. https://doi.org/10.1038/cmi.2015.74.

Liu, J., Li, X., Lu, Q., Ren, Di, Sun, X., Rousselle, T., Li, J., Leng, J., 2019. AMPK: a balancer of the renin-angiotensin system. Biosci. Rep. 39 (9) https://doi.org/10.1042/BSR20181994.

Liu, J., Cao, R., Xu, M., Wang, X., Zhang, H., Hu, H., Li, Y., Hu, Z., Zhong, W., Wang, M., 2020a. Hydroxychloroquine, a less toxic derivative of chloroquine, is effective in inhibiting SARS-CoV-2 infection in vitro. Cell Discov. 6 (1), 6–9. https://doi.org/10.1038/s41421-020-0156-0.

Liu, J., Wang, T., Cai, Q., Sun, L., Huang, D., Zhou, G., He, Q., Wang, F.-S., Liu, L., Chen, J., 2020b. Longitudinal changes of liver function and hepatitis B reactivation in COVID-19 patients with pre-existing chronic hepatitis B virus infection. Hepatol. Res. 50 (11), 1211–1221. https://doi.org/10.1111/hepr.13553.

Lo, K.B., Bhargav, R., Salacup, G., Pelayo, J., Albano, J., McCullough, P.A., Rangaswami, J., 2020. Angiotensin converting enzyme inhibitors and angiotensin II receptor blockers and outcomes in patients with COVID-19: a systematic review and meta-analysis. Expet Rev. Cardiovasc. Ther. 18 (12), 919–930. https://doi.org/10.1080/14779072.2020.1826308.

Lönnerdal, B., Iyer, S., 1995. Lactoferrin: molecular structure and biological function. Annu. Rev. Nutr. 15, 93–110. https://doi.org/10.1146/annurev.nu.15.070195.000521.

Lu, L., Hangoc, G., Oliff, A., Chen, L.T., Shen, R.N., Broxmeyer, H.E., 1987. Protective influence of lactoferrin on mice infected with the polycythemia-inducing strain of friend virus complex. Cancer Res. 47 (15), 4184–4188.

Lu, D., Zhang, J., Ma, C., Yue, Y., Zou, Z., Yu, C., Yin, F., 2018. Link between community-acquired pneumonia and vitamin D levels in older patients. Z. Gerontol. Geriatr. 51 (4), 435–439. https://doi.org/10.1007/s00391-017-1237-z.

Lucchini, A., Giani, M., Isgrò, S., Rona, R., Foti, G., 2020. "The 'helmet bundle' in COVID-19 patients undergoing non invasive ventilation. Intensive Crit. Care Nurs. https://doi.org/10.1016/j.iccn.2020.102859.

Lukomska, B., Stanaszek, L., Zuba-Surma, E., Legosz, P., Sarzynska, S., Drela, K., 2019. Challenges and controversies in human mesenchymal stem cell therapy. Stem Cell. Int. 2019, 9628536. https://doi.org/10.1155/2019/9628536.

Lundberg, L., Pinkham, C., Baer, A., Amaya, M., Narayanan, A., Wagstaff, K.M., Jans, D.A., Kehn-Hall, K., 2013. Nuclear import and export inhibitors alter capsid protein distribution in mammalian cells and reduce venezuelan equine encephalitis virus replication. Antivir. Res. 100 (3), 662–672.

Maiese, A., Manetti, A.C., La Russa, R., Di Paolo, M., Turillazzi, E., Frati, P., Fineschi, V., 2020. Autopsy findings in COVID-19-related deaths: a literature review. Forensic Sci. Med. Pathol. https://doi.org/10.1007/s12024-020-00310-8.

Mair-Jenkins, J., Saavedra-Campos, M., Baillie, J.K., Cleary, P., Khaw, F.-M., Lim, W.S., Makki, S., Rooney, K.D., Nguyen-Van-Tam, J.S., Beck, C.R., 2015. The effectiveness of convalescent plasma and hyperimmune immunoglobulin for the treatment of severe acute respiratory infections of viral etiology: a systematic review and exploratory meta-analysis. J. Infect. Dis. 211 (1), 80–90. https://doi.org/10.1093/infdis/jiu396.

Maisonnasse, P., Guedj, J., Contreras, V., Behillil, S., Solas, C., Marlin, R., Naninck, T., Pizzorno, A., Lemaitre, J., Gonçalves, A., Kahlaoui, N., Terrier, O., Fang, R.H.T., Enouf, V., Dereuddre-Bosquet, N., Brisebarre, A.,

Touret, F., Chapon, C., Hoen, B., Lina, B., Calatrava, M.R., van der Werf, Le Grand, R., 2020. Hydroxychloroquine use against SARS-CoV-2 infection in non-human primates. Nature 585 (7826), 584–587.

Mammen, M.J., Aryal, K., Alhazzani, W., Alexander, P.E., 2020. Corticosteroids for patients with acute respiratory distress syndrome: a systematic review and meta-analysis of randomized trials. Pol. Arch. Intern. Med. 130 (4), 276–286. https://doi.org/10.20452/pamw.15239.

Marietta, M., Ageno, W., Artoni, A., De Candia, E., Gresele, P., Marchetti, M., Marcucci, R., Tripodi, A., 2020. COVID-19 and haemostasis: a position paper from Italian society on thrombosis and haemostasis (SISET). Blood Transfus. 18 (3), 167–169. https://doi.org/10.2450/2020.0083-20.

Marik, P.E., 2020. Vitamin C: an essential 'stress hormone' during sepsis. J. Thorac. Dis. 12 (Suppl. 1), S84–S88. https://doi.org/10.21037/jtd.2019.12.64.

Marik, P.E., Khangoora, V., Rivera, R., Hooper, M.H., Catravas, J., 2017. Hydrocortisone, vitamin C, and thiamine for the treatment of severe sepsis and septic shock: a retrospective before-after study. Chest 151 (6), 1229–1238. https://doi.org/10.1016/j.chest.2016.11.036.

Marini, J.J., Gattinoni, L., 2020. Management of COVID-19 respiratory distress. JAMA 323 (22), 2329–2330. https://doi.org/10.1001/jama.2020.6825.

Marovich, M., Mascola, J.R., Cohen, M.S., 2020. Monoclonal antibodies for prevention and treatment of COVID-19. JAMA 324 (2), 131–132. https://doi.org/10.1001/jama.2020.10245.

Martineau, A.R., Jolliffe, D.A., Hooper, R.L., Greenberg, L., Aloia, J.F., Bergman, P., Dubnov-Raz, G., et al., 2017. Vitamin D supplementation to prevent acute respiratory tract infections: systematic review and meta-analysis of individual participant data. Br. Med. J. 356 (February), i6583. https://doi.org/10.1136/bmj.i6583.

Matsiukevich, D., Piraino, G., Lahni, P., Hake, P.W., Wolfe, V., O'Connor, M., James, J., Zingarelli, B., 2017. Metformin ameliorates gender-and age-dependent hemodynamic instability and myocardial injury in murine hemorrhagic shock. Biochim. Biophys. Acta 1863 (10 Pt B), 2680–2691. https://doi.org/10.1016/j.bbadis.2017.05.027.

Mauri, T., Turrini, C., Eronia, N., Grasselli, G., Alberto Volta, C., Bellani, G., Pesenti, A., 2017. Physiologic effects of high-flow nasal cannula in acute hypoxemic respiratory failure. Am. J. Respir. Crit. Care Med. 195 (9), 1207–1215. https://doi.org/10.1164/rccm.201605-0916OC.

MayoClinic, 2020. Patient Care and Health Information. Extracorporeal Membrane Oxygenation. In: https://www.mayoclinic.org/tests-procedures/ecmo/about/pac-20484615.

McBane 2nd, R.D., 2020. Anticoagulation in COVID-19: A Systematic Review, Meta-analysis, and Rapid Guidance From Mayo Clinic. Mayo Clinic proceedings 95 (11), 2467–2486. https://doi.org/10.1016/j.mayocp.2020.08.030.

Meffre, E., Iwasaki, A., 2020. Interferon deficiency can lead to severe COVID. Nature 587 (7834), 374–376. https://doi.org/10.1038/d41586-020-03070-1.

Mehrzadi, S., Hemati, K., Reiter, R.J., Hosseinzadeh, A., 2020. Mitochondrial dysfunction in age-related macular degeneration: melatonin as a potential treatment. Expert Opin. Ther. Targets 24 (4), 359–378. https://doi.org/10.1080/14728222.2020.1737015.

Mehta, N., Kalra, A., Nowacki, A.S., Anjewierden, S., Han, Z., Bhat, P., Carmona-Rubio, A.E., et al., 2020. Association of use of angiotensin-converting enzyme inhibitors and angiotensin II receptor blockers with testing positive for coronavirus disease 2019 (COVID-19). JAMA Cardiol. 5 (9), 1020–1026. https://doi.org/10.1001/jamacardio.2020.1855.

Meng, L., Qiu, H., Wan, L., Ai, Y., Xue, Z., Guo, Q., Deshpande, R., et al., 2020. Intubation and ventilation amid the COVID-19 outbreak: Wuhan's experience. Anesthesiology 132 (6), 1317–1332. https://doi.org/10.1097/ALN.0000000000003296.

Mercuro, N.J., Yen, C.F., Shim, D.J., Maher, T.R., McCoy, C.M., Zimetbaum, P.J., Gold, H.S., 2020. Risk of QT interval prolongation associated with use of hydroxychloroquine with or without concomitant azithromycin among hospitalized patients testing positive for coronavirus disease 2019 (COVID-19). JAMA Cardiol. 5 (9), 1036–1041.

Mirabelli, C., Wotring, J., Zhang, C., McCarty, S., Fursmidt, R., Frum, T., Kadambi, N., et al., 2020. Morphological cell profiling of SARS-CoV-2 infection identifies drug repurposing candidates for COVID-19. bioRxiv. https://doi.org/10.1101/2020.05.27.117184.

Mishra, G.P., Mulani, J., 2020. Corticosteroids for COVID-19: the search for an optimum duration of therapy. Lancet Respir. Med. 2600 (20), 30530. https://doi.org/10.1016/s2213-2600(20)30530-0.

Mitjà, O., Corbacho-Monné, M., Ubals, M., Tebe, C., Peñafiel, J., Tobias, A., Ballana, E., Alemany, A., Riera-Martí, N., Pérez, C.A., Suñer, C., Laporte, P., Admella, P., Mitjà, J., Clua, M., Bertran, L., Sarquella, M., Gavilán, S., Ara, J., Argimon, J.M., Casabona, J., Cuatrecasas, G., Cañadas, P., El, Vall-Mayans, M., BCN PEP-CoV-2 Research Group, 2020. Hydroxychloroquine for early treatment of adults with mild covid-19: a randomized-controlled trial. Clin. Infect. Dis. https://doi.org/10.1093/cid/ciaa1009 (online ahead of print).

Moores, L.K., Tritschler, T., Brosnahan, S., Carrier, M., Collen, J.F., Doerschug, K., Holley, A.B., et al., 2020. Prevention, diagnosis, and treatment of VTE in patients with coronavirus disease 2019: CHEST guideline and expert panel report. Chest. https://doi.org/10.1016/j.chest.2020.05.559.

Mortensen, E.M., Restrepo, M.I., Anzueto, A., Pugh, J., 2005. The effect of prior statin use on 30-day mortality for patients hospitalized with community-acquired pneumonia. Respir. Res. 6 (1), 82. https://doi.org/10.1186/1465-9921-6-82.

Moskowitz, A., Huang, D.T., Hou, P.C., Gong, J., B Doshi, P., Grossestreuer, A.V., Andersen, L.W., et al., 2020. Effect of ascorbic acid, corticosteroids, and thiamine on organ injury in septic shock: the ACTS randomized clinical trial. JAMA 324 (7), 642–650. https://doi.org/10.1001/jama.2020.11946.

Mousavi, S., Bereswill, S., Heimesaat, M.M., 2019. Immunomodulatory and antimicrobial effects of vitamin C. Eur. J. Microbiol. Immunol. 9 (3), 73–79. https://doi.org/10.1556/1886.2019.00016.

Mulangu, S., Dodd, L.E., Davey, R.T., Tshiani Mbaya, O., Proschan, M., Mukadi, D., Lusakibanza Manzo, M., et al., 2019. A randomized, controlled trial of Ebola virus disease therapeutics. N. Engl. J. Med. 381 (24), 2293–2303. https://doi.org/10.1056/nejmoa1910993.

Munshi, L., Walkey, A., Goligher, E., Pham, T., Uleryk, E.M., Fan, E., 2019. Venovenous extracorporeal membrane oxygenation for acute respiratory distress syndrome: a systematic review and meta-analysis. Lancet Respir. Med. 7 (2), 163–172. https://doi.org/10.1016/S2213-2600(18)30452-1.

Murdaca, G., Tonacci, A., Negrini, S., Greco, M., Borro, M., Puppo, F., Gangemi, S., 2019. Emerging role of vitamin D in autoimmune diseases: an update on evidence and therapeutic implications. Autoimmun. Rev. 18 (9), 102350. https://doi.org/10.1016/j.autrev.2019.102350.

Murphy, M.B., Moncivais, K., Caplan, A.I., 2013. Mesenchymal stem cells: environmentally responsive therapeutics for regenerative medicine. Exp. Mol. Med. 45 (11), e54. https://doi.org/10.1038/emm.2013.94.

Mustafa, A.K., Alexander, P.J., Joshi, D.J., Tabachnick, D.R., Cross, C.A., Pappas, P.S., Tatooles, A.J., 2020. Extracorporeal membrane oxygenation for patients with COVID-19 in severe respiratory failure. JAMA Surg. 155 (10), 990–992. https://doi.org/10.1001/jamasurg.2020.3950.

Myint, Z.W., Oo, T.H., Thein, K.Z., Tun, A.M., Saeed, H., 2018. Copper deficiency anemia: review article. Ann. Hematol. 97 (9), 1527–1534. https://doi.org/10.1007/s00277-018-3407-5.

Ñamendys-Silva, S.A., 2020. ECMO for ARDS due to COVID-19. Heart Lung J. Crit. Care. https://doi.org/10.1016/j.hrtlng.2020.03.012.

National Institutes of Health, 2020a. COVID-19 Treatment Guidelines Panel. https://www.covid19treatmentguidelines.nih.gov/.

National Institutes of Health, 2020b. Office of Dietary Supplements. Zinc Fact Sheet for Health Professionals. https://ods.od.nih.gov/factsheets/Zinc-HealthProfessional/.

Neunert, C., Lim, W., Crowther, M., Cohen, A., Solberg, L.J., Crowther, M.A., 2011. The American Society of Hematology 2011 evidence-based practice guideline for immune thrombocytopenia. Blood 117 (16), 4190–4207. https://doi.org/10.1182/blood-2010-08-302984.

Nguyen, L.S., Dolladille, C., Drici, M.D., Fenioux, C., Alexandre, J., Mira, J.P., Moslehi, J.J., Roden, D.M., Funck-Brentano, C., Salem, J.E., 2020. Cardiovascular toxicities associated with hydroxychloroquine and azithromycin: an analysis of the world health organization pharmacovigilance database. Circulation 142 (3), 303–305.

Ni, Y.-N., Luo, J., Yu, H., Liu, D., Liang, B.-M., Liang, Z.-A., 2018. The effect of high-flow nasal cannula in reducing the mortality and the rate of endotracheal intubation when used before mechanical ventilation compared with conventional oxygen therapy and noninvasive positive pressure ventilation. A systematic review and meta-analysis. Am. J. Emerg. Med. 36 (2), 226–233. https://doi.org/10.1016/j.ajem.2017.07.083.

Nopp, S., Moik, F., Jilma, B., Pabinger, I., Ay, C., 2020. Risk of venous thromboembolism in patients with COVID-19: a systematic review and meta-analysis. Res. Pract. Thromb. Hemostasis 4 (7), 1178–1191.

Novack, V., Eisinger, M., Amit, F., Terblanche, M., Adhikari, N.K.J., Douvdevani, A., Amichay, D., Almog, Y., 2009. The effects of statin therapy on inflammatory cytokines in patients with bacterial infections: a randomized double-blind placebo controlled clinical trial. Intensive Care Med. 35 (7), 1255–1260. https://doi.org/10.1007/s00134-009-1429-0.

Novick, S.G., Godfrey, J.C., Godfrey, N.J., Wilder, H.R., 1996. How does zinc modify the common cold? Clinical observations and implications regarding mechanisms of action. Med. Hypotheses 46 (3), 295–302. https://doi.org/10.1016/s0306-9877(96)90259-5.

Oda, H., Wakabayashi, H., Tanaka, M., Yamauchi, K., Sugita, C., Yoshida, H., Abe, F., Sonoda, T., Kurokawa, M., 2020. Effects of lactoferrin on infectious diseases in Japanese summer: a randomized, double-blinded, placebo-controlled trial. J. Microbiol. Immunol. Infect. https://doi.org/10.1016/j.jmii.2020.02.010.

Omrani, A.S., Saad, M.M., Baig, K., Bahloul, A., Abdul-Matin, M., Alaidaroos, A.Y., Almakhlafi, G.A., Albarrak, M.M., Memish, Z.A., Albarrak, A.M., 2014. Ribavirin and interferon alfa-2a for severe Middle East respiratory syndrome coronavirus infection: a retrospective cohort study. Lancet Infect. Dis. 14 (11), 1090–1095. https://doi.org/10.1016/S1473-3099(14)70920-X.

Ota, H., Goto, T., Yoshioka, T., Ohyama, N., 2008. Successful pregnancies treated with pioglitazone in infertile patients with polycystic ovary syndrome. Fertil. Steril. 90 (3), 709–713. https://doi.org/10.1016/j.fertnstert.2007.01.117.

Padhy, B.M., Mohanty, R.R., Das, S., Meher, B.R., 2020. Therapeutic potential of ivermectin as add on treatment in COVID 19: a systematic review and meta-analysis. J. Pharmacol. Pharma. Sci. 23, 462–469.

Paranjpe, I., Fuster, V., Lala, A., Russak, A.J., Glicksberg, B.S., Levin, M.A., Charney, A.W., Narula, J., Fayad, Z.A., Bagiella, E., Zhao, S., Nadkarni, G.N., 2020. Association of treatment dose anticoagulation with in-hospital survival among hospitalized patients with COVID-19. J. Am. Coll. Cardiol. 76 (1), 122–124.

Park, J.W., Lee, J.H., Park, Y.H., Park, S.J., Cheon, J.H., Kim, W.H., Kim, T.I., 2017. Sex-dependent difference in the effect of metformin on colorectal cancer-specific mortality of diabetic colorectal cancer patients. World J. Gastroenterol. 23 (28), 5196–5205. https://doi.org/10.3748/wjg.v23.i28.5196.

Patell, R., Bogue, T., Koshy, A., Bindal, P., Merrill, M., Aird, W.C., Bauer, K.A., Zwicker, J.I., 2020. Postdischarge thrombosis and hemorrhage in patients with COVID-19. Blood 136 (11), 1342–1346.

Pereda, R., González, D., Rivero, H.B., Rivero, J.C., Pérez, A., Del Rosario López, L., Mezquia, N., Venegas, R.,

Betancourt, J.R., Domínguez, R.E., 2020. Therapeutic effectiveness of interferon-α2b against COVID-19: the Cuban experience. J. Interferon Cytokine Res. 40 (9), 438–442. https://doi.org/10.1089/jir.2020.0124.

Perez, E.E., Orange, J.S., Bonilla, F., Chinen, J., Chinn, I.K., Dorsey, M., El-Gamal, Y., et al., 2017. Update on the use of immunoglobulin in human disease: a review of evidence. J. Allergy Clin. Immunol. 139 (3S), S1–S46. https://doi.org/10.1016/j.jaci.2016.09.023.

Peter, H., Mafham, M., Linsell, L., Bell, J.L., Staplin, N., Emberson, J.R., Wiselka, M., Ustianowski, A., Elmahi, E., Prudon, B., Whitehouse, A., Felton, T., Williams, J., Faccenda, J., Jonathan Underwood, J., Landray, M.J., 2020. Effect of hydroxychloroquine in hospitalized patients with COVID-19: preliminary results from a multicentre, randomized, controlled trial. medRxiv. https://doi.org/10.1101/2020.07.15.20151852.

Pfützner, A., Schöndorf, T., Hanefeld, M., Forst, T., 2010. High-sensitivity C-reactive protein predicts cardiovascular risk in diabetic and nondiabetic patients: effects of insulin-sensitizing treatment with pioglitazone. J. Diabetes Sci. Technol. 4 (3), 706–716. https://doi.org/10.1177/193229681000400326.

Piazza, G., Morrow, D.A., 2020. Diagnosis, management, and pathophysiology of arterial and venous thrombosis in COVID-19. JAMA 324 (24), 2548–2549. https://doi.org/10.1001/jama.2020.23422.

Picheta, R., 2020. France says ibuprofen may aggravate coronavirus. Experts say more evidence is needed. CNN Health. https://cnn.com/2020/03/16/health/coronavirus-ibuprofen-french-health-minister-scn-intl-scli/index.html.

Pietrantoni, A., Fortuna, C., Elena Remoli, M., Ciufolini, M.G., Superti, F., 2015. Bovine lactoferrin inhibits toscana virus infection by binding to heparan sulphate. Viruses 7 (2), 480–495. https://doi.org/10.3390/v7020480.

Pittenger, M.F., Discher, D.E., Péault, B.M., Phinney, D.G., Hare, J.M., Caplan, A.I., 2019. Mesenchymal stem cell perspective: cell biology to clinical progress. NPJ Regen. Med. 4 (1), 22. https://doi.org/10.1038/s41536-019-0083-6.

Plattner, F., Bibb, J.A., 2012. Serine and threonine phosphorylation. In: Basic Neurochemistry, 467–92. Elsevier.

Pollack, P.S., Shields, K.E., Burnett, D.M., Osborne, M.J., Cunningham, M.L., Stepanavage, M.E., 2005. Pregnancy outcomes after maternal exposure to simvastatin and lovastatin. Birth Defects Res. A Clin. Mol. Teratol. 73 (11), 888–896. https://doi.org/10.1002/bdra.20181.

Poon, L.C., Yang, H., Kapur, A., Melamed, N., Dao, B., Divakar, H., David McIntyre, H., et al., 2020. Global interim guidance on coronavirus disease 2019 (COVID-19) during pregnancy and puerperium from FIGO and allied partners: information for healthcare professionals. Int. J. Gynecol. Obstet. 149 (3), 273–286. https://doi.org/10.1002/ijgo.13156.

Protect Investigators for the Canadian Critical Care Trials Group and the Australian and New Zealand Intensive Care Society Clinical Trials Group, Cook, D., Meade, M., Guyatt, G., Walter, S., Heels-Ansdell, D., Warkentin, T.E.,

Zytaruk, N., Crowther, M., Geerts, W., Cooper, Vlahakis, N.E., 2011. Dalteparin versus unfractionated heparin in critically ill patients. N. Engl. J. Med. 364 (14), 1305–1314.

Qin, Y., Li, L.Z., Zhang, X.Q., Wei, Y., Wang, Y.L., Wei, H.F., Wang, X.R., Yu, W.F., Su, D.S., 2017. Supraglottic jet oxygenation and ventilation enhances oxygenation during upper gastrointestinal endoscopy in patients sedated with propofol: a randomized multicentre clinical trial. Br. J. Anaesth. 119 (1), 158–166. https://doi.org/10.1093/bja/aex091.

Qiu, D., Li, X.-N., 2015. Pioglitazone inhibits the secretion of proinflammatory cytokines and chemokines in astrocytes stimulated with lipopolysaccharide. Int. J. Clin. Pharm. Ther. 53 (9), 746–752. https://doi.org/10.5414/CP202339.

Quan, H., Zhang, H., Wei, W., Fang, T., 2016. Gender-related different effects of a combined therapy of exenatide and metformin on overweight or obesity patients with type 2 diabetes mellitus. J. Diabetes Complicat. 30 (4), 686–692. https://doi.org/10.1016/j.jdiacomp.2016.01.013.

RCOG guideline, 2020. Coronavirus (COVID-19) Infection in Pregnancy.

Recovery Collaborative Group, Horby, P., Lim, W.S., Emberson, J.R., Mafham, M., Bell, J.L., Linsell, L., Staplin, N., Brightling, C., Ustianowski, A., Elmahi, E., Prudon, B., Green, C., Felton, T., Chadwick, D., Rege, K., Fegan, C., Chappell, L.C., Faust, S.N., Jaki, T., Jeffery, K., Montgomery A, R., Landray, M.J., 2020. Dexamethasone in hospitalized patients with covid-19 - preliminary report. N. Engl. J. Med. (Online ahead of print).

Réger, B., Péterfalvi, A., Litter, I., Pótó, L., Mózes, R., Tóth, O., Kovács, G.L., Losonczy, H., 2013. Challenges in the evaluation of D-dimer and fibrinogen levels in pregnant women. Thromb. Res. 131 (4), e183–e187. https://doi.org/10.1016/j.thromres.2013.02.005.

Rinott, E., Kozer, E., Shapira, Y., Bar-Haim, A., Youngster, I., 2020. Ibuprofen use and clinical outcomes in COVID-19 patients. Clin. Microbiol. Infect. 26 (9), 1259.e5–1259.e7. https://doi.org/10.1016/j.cmi.2020.06.003.

Roberts, S.S., Martinez, M., Covington, D.L., Rode, R.A., Mary, V., 2009. Lopinavir/ritonavir in pregnancy. J. Acquir. Immune Defic. Syndr. 51 (4), 456–461.

Roberts, L.N., Whyte, M.B., Georgiou, L., Giron, G., Czuprynska, J., Rea, C., Vadher, B., Patel, R.K., Gee, E., Arya, R., 2020. Postdischarge venous thromboembolism following hospital admission with COVID-19. Blood 136 (11), 1347–1350.

Roche, 2020. Roche Provides an Update on the Phase III COVA-CTA Trial of Actemra/RoActemra in Hospitalised Patients with Severe COVID-19 Associated Pneumonia. https://www.roche.com/investors/updates/inv-update-2020-07-29.htm.

Rodrigo, C., Leonardi-Bee, J., Nguyen-Van-Tam, J., Lim, W.S., 2016. Corticosteroids as adjunctive therapy in the treatment of influenza. Cochrane Database Syst. Rev. 3 (March), CD010406. https://doi.org/10.1002/14651858.CD010406.pub2.

Romero, R., Erez, O., Hüttemann, M., Maymon, E., Panaitescu, B., Conde-Agudelo, A., Pacora, P., Yoon, B.H., Grossman, L.I., 2017. Metformin, the aspirin of the 21st century: its role in gestational diabetes mellitus, prevention of preeclampsia and cancer, and the promotion of longevity. Am. J. Obstet. Gynecol. 217 (3), 282–302. https://doi.org/10.1016/j.ajog.2017.06.003.

Rondanelli, M., Miccono, A., Lamburghini, S., Avanzato, I., Riva, A., Allegrini, P., Anna Faliva, M., Peroni, G., Nichetti, M., Perna, S., 2018. Self-care for common colds: the pivotal role of vitamin D, vitamin C, zinc, and echinacea in three main immune interactive clusters (physical barriers, innate and adaptive immunity) involved during an episode of common colds-practical advice on dosages A. Evid. Based Compl. Alternative Med. 2018, 5813095. https://doi.org/10.1155/2018/5813095.

Rosenberg, E.S., Dufort, E.M., Udo, T., Wilberschied, L.A., Kumar, J., Tesoriero, J., Weinberg, P., Kirkwood, J., Muse, A., DeHovitz, J., Blog, D.S., Hutton, B., Holtgrave, D.R., Zucker, H.A., 2020. Association of treatment with hydroxychloroquine or azithromycin with in-hospital mortality in patients with COVID-19 in New York state. JAMA 323 (24), 2493–2502.

Ross, A.C., Taylor, C.L., Yaktine, A.L., Del Valle, H.B., 2011. In: Ross, A.C., Taylor, C.L., Yaktine, A.L., et al. (Eds.), Institute of Medicine (US) Committee to Review Dietary Reference Intakes for Vitamin D and Calcium. National Academies Press, Washington (DC). https://www.ncbi.nlm.nih.gov/books/NBK56070/.

Russell, B., Moss, C., Rigg, A., Van Hemelrijck, M., 2020. COVID-19 and treatment with NSAIDs and corticosteroids: should we be limiting their use in the clinical setting? Ecancermedicalscience 14, 1023. https://doi.org/10.3332/ecancer.2020.1023.

Sadeghi, S., Mosaffa, N., Mahmoud Hashemi, S., Mehdi Naghizadeh, M., Ghazanfari, T., 2020a. The immunomodulatory effects of mesenchymal stem cells on long term pulmonary complications in an animal model exposed to a sulfur mustard analog. Int. Immunopharm. 80 (March), 105879. https://doi.org/10.1016/j.intimp.2019.105879.

Sadeghi, S., Soudi, S., Shafiee, A., Mahmoud Hashemi, S., 2020b. Mesenchymal stem cell therapies for COVID-19: current status and mechanism of action. Life Sci. 262 (December), 118493. https://doi.org/10.1016/j.lfs.2020.118493.

Sakamoto, A., Hongo, M., Saito, Kan, Nagai, R., Ishizaka, N., 2012. Reduction of renal lipid content and proteinuria by a PPAR-γ agonist in a rat model of angiotensin II-induced hypertension. Eur. J. Pharmacol. 682 (1–3), 131–136. https://doi.org/10.1016/j.ejphar.2012.02.027.

Samama, M.M., Cohen, A.T., Darmon, J.Y., Desjardins, L., Eldor, A., Janbon, C., Leizorovicz, A., Nguyen, H., Olsson, C.G., Turpie, A.G., Weisslinger, N., 1999. A comparison of enoxaparin with placebo for the prevention of venous thromboembolism in acutely ill medical patients. Prophylaxis in medical patients with enoxaparin study group. N. Engl. J. Med. 341 (11), 793–800.

Sánchez-Barceló, E.J., Mediavilla, M.D., Tan, D.X., Reiter, R.J., 2010. Clinical uses of melatonin: evaluation of human trials. Curr. Med. Chem. 17 (19), 2070–2095. https://doi.org/10.2174/092986710791233689.

Sandberg-Wollheim, M., Alteri, E., Stam Moraga, M., Kornmann, G., 2011. Pregnancy outcomes in multiple sclerosis following subcutaneous interferon beta-1a therapy. Mult. Scler. 17 (4), 423–430. https://doi.org/10.1177/1352458510394610.

Sanders, J.M., Monogue, M.L., Jodlowski, T.Z., Cutrell, J.B., 2020. Pharmacologic treatments for coronavirus disease 2019 (COVID-19): a review. JAMA 323 (18), 1824–1836. https://doi.org/10.1001/jama.2020.6019.

Sanofi and Regeneron Provide Update on Kevzara® (Sarilumab) Phase 3 U.S. Trial in COVID-19 Patients, 2020. https://www.sanofi.com/en/media-room/press-releases/2020/2020-07-02-22-30-00.

Sartini, C., Tresoldi, M., Scarpellini, P., Tettamanti, A., Carcò, F., Landoni, G., Zangrillo, A., 2020. Respiratory parameters in patients with COVID-19 after using noninvasive ventilation in the prone position outside the intensive care unit. JAMA 323 (22), 2338–2340. https://doi.org/10.1001/jama.2020.7861.

Sarzani, R., Giulietti, F., Di Pentima, C., Giordano, P., Spannella, F., 2020. Disequilibrium between the classic renin-angiotensin system and its opposing arm in SARS-CoV-2-related lung injury. Am. J. Physiol. Lung Cell Mol. Physiol. 319 (2), L325–L336. https://doi.org/10.1152/ajplung.00189.2020.

Schenck, E.J., Hoffman, K., Goyal, P., Choi, J., Torres, L., Rajwani, K., Tam, C.W., Ivascu, N., Martinez, F.J., Berlin, D.A., 2020. Respiratory mechanics and gas exchange in COVID-19-associated respiratory failure. Ann. Am. Thoracic Soc. https://doi.org/10.1513/AnnalsATS.202005-427RL.

Schmidt, M., Hajage, D., Lebreton, G., Monsel, A., Voiriot, G., Levy, D., Baron, E., et al., 2020. Extracorporeal membrane oxygenation for severe acute respiratory distress syndrome associated with COVID-19: a retrospective cohort study. Lancet Respir. Med. 8 (11), 1121–1131. https://doi.org/10.1016/S2213-2600(20)30328-3.

Schoergenhofer, C., Jilma, B., Stimpfl, T., Karolyi, M., Zoufaly, A., 2020. Pharmacokinetics of lopinavir and ritonavir in patients hospitalized with coronavirus disease 2019 (COVID-19). Ann. Intern. Med. 173 (8), 670–672. https://doi.org/10.7326/M20-1550.

Schoggins, J.W., Wilson, S.J., Panis, M., Murphy, M.Y., Jones, C.T., Bieniasz, P., Rice, C.M., 2011. A diverse range of gene products are effectors of the type I interferon antiviral response. Nature 472 (7344), 481–485. https://doi.org/10.1038/nature09907.

Schwaiger, J., Karbiener, M., Aberham, C., Farcet, M.R., Kreil, T.R., 2020. No SARS-CoV-2 neutralization by intravenous immunoglobulins produced from plasma collected before the 2020 pandemic. J. Infect. Dis. 222 (12), 1960–1964. https://doi.org/10.1093/infdis/jiaa593.

Schwalfenberg, G.K., 2011. A review of the critical role of vitamin D in the functioning of the immune system and the clinical implications of vitamin D deficiency. Mol. Nutr. Food Res. 55 (1), 96–108. https://doi.org/10.1002/mnfr.201000174.

Science, M., Maguire, J.L., Russell, M.L., Smieja, M., Walter, S.D., Loeb, M., 2013. Low serum 25-hydroxyvitamin D level and risk of upper respiratory tract infection in children and adolescents. Clin. Infect. Dis. 57 (3), 392–397. https://doi.org/10.1093/cid/cit289.

Shakoory, B., Carcillo, J.A., Chatham, W.W., Amdur, R.L., Zhao, H., Dinarello, C.A., Cron, R.Q., Opal, S.M., 2016. Interleukin-1 receptor blockade is associated with reduced mortality in sepsis patients with features of macrophage activation syndrome: reanalysis of a prior phase III trial. Crit. Care Med. 44 (2), 275–281. https://doi.org/10.1097/CCM.0000000000001402.

Shalhoub, S., Farahat, F., Al-Jiffri, A., Simhairi, R., Shamma, O., Siddiqi, N., Mushtaq, A., 2015. IFN-A2a or IFN-B1a in combination with ribavirin to treat Middle East respiratory syndrome coronavirus pneumonia: a retrospective study. J. Antimicrob. Chemother. 70 (7), 2129–2132. https://doi.org/10.1093/jac/dkv085.

Shao, Z., Feng, Y., Zhong, L., Xie, Q., Lei, M., Liu, Z., Wang, C., et al., 2020. Clinical efficacy of intravenous immunoglobulin therapy in critical patients with COVID-19: a multicenter retrospective cohort study. medRxiv. https://doi.org/10.1101/2020.04.11.20061739.

Sharma, S., Ray, A., Sadasivam, B., 2020. Metformin in COVID-19: a possible role beyond diabetes. Diabetes Res. Clin. Pract. 164 (June), 108183. https://doi.org/10.1016/j.diabres.2020.108183.

Shetty, A.K., 2020. Mesenchymal stem cell infusion shows promise for combating coronavirus (COVID-19)- induced pneumonia. Aging Dis. https://doi.org/10.14336/AD.2020.0301.

Shorr, A.F., Williams, M.D., 2009. Venous thromboembolism in critically ill patients. Observations from a randomized trial in sepsis. Thromb. Haemostasis 101 (1), 139–144.

Shrestha, S.K., 2020. Statin drug therapy may increase COVID-19 infection. Nepalese Med. J. 3 (1), 326–327. https://doi.org/10.3126/nmj.v3i1.28256.

Shu, L., Niu, C., Li, R., Huang, T., Wang, Y., Huang, M., Ji, N., et al., 2020. Treatment of severe COVID-19 with human umbilical cord mesenchymal stem cells. Stem Cell Res. Ther. 11 (1), 361. https://doi.org/10.1186/s13287-020-01875-5.

Siddiqi, H.K., Mehra, M.R., 2020. "COVID-19 illness in native and immunosuppressed states: a clinical–therapeutic staging proposal. J. Heart Lung Transpl. 39 (5), 405–407.

Sieswerda, E., de Boer, M.G.J., Bonten, M.M.J., Boersma, W.G., Jonkers, R.E., Aleva, R.M., Jan Kullberg, B., et al., 2020. "Recommendations for antibacterial therapy in adults with COVID-19 — an evidence based guideline. Clin. Microbiol. Infect. 27, 9–14. https://doi.org/10.1016/j.cmi.2020.09.041.

Simonovich, V.A., Burgos Pratx, L.D., Scibona, P., Beruto, M.V., Vallone, M.G., Vázquez, C., Savoy, N., et al., 2020. A randomized trial of convalescent plasma in covid-19 severe pneumonia. N. Engl. J. Med. https://doi.org/10.1056/NEJMoa2031304.

Skalny, A.V., Rink, L., Ajsuvakova, O.P., Aschner, M., Gritsenko, V.A., Alekseenko, S.I., Svistunov, A.A., et al., 2020. "Zinc and respiratory tract infections: perspectives for COVID-19 (review). Int. J. Mol. Med. 46 (1), 17–26. https://doi.org/10.3892/ijmm.2020.4575.

Skipper, C.P., Pastick, K.A., Engen, N.W., Bangdiwala, A.S., Abassi, M., Lofgren, S.M., Williams, D.A., Okafor, E.C., Pullen, M.F., Nicol, M.R., Nascene, A.A., Hullsiek, K.H., Cheng, M.P., Luke, D., Lother, S.A., MacKenzie, L.J., Drobot, G., Kelly, L.E., Schwartz, I.S., Zarychanski, R., McDonald, E.G., Lee, T.C., Boulware, D.R., 2020. Hydroxychloroquine in nonhospitalized adults with early COVID-19: a randomized trial. Annu. Intern. Med. 173 (8), 623–631.

Society for Maternal Fetal Medicine, 2020. Management Considerations for Pregnant Patients with COVID-19. https://s3.amazonaws.com/cdn.smfm.org/media/2336/SMFM_COVID_Management_of_COVID_pos_preg_patients_4-30-20_final.pdf.

Sorbello, M., El-Boghdadly, K., Di Giacinto, I., Cataldo, R., Esposito, C., Falcetta, S., Merli, G., et al., 2020. The Italian coronavirus disease 2019 outbreak: recommendations from clinical practice. Anaesthesia 75 (6), 724–732. https://doi.org/10.1111/anae.15049.

Spinner, C.D., Gottlieb, R.L., Criner, G.J., López, J.R.A., Cattelan, A.M., Viladomiu, A.S., Ogbuagu, O., et al., 2020. Effect of remdesivir vs standard care on clinical status at 11 days in patients with moderate COVID-19: a randomized clinical trial. JAMA 324 (11), 1048–1057. https://doi.org/10.1001/jama.2020.16349.

Spyropoulos, A.C., Levy, J.H., Ageno, W., Connors, J.M., Hunt, B.J., Iba, T., Levi, M., et al., 2020a. Scientific and standardization committee communication: clinical guidance on the diagnosis, prevention, and treatment of venous thromboembolism in hospitalized patients with COVID-19. J. Thromb. Haemostasis 18 (8), 1859–1865. https://doi.org/10.1111/jth.14929.

Spyropoulos, A.C., Lipardi, C., Xu, J., Peluso, C., Spiro, T.E., De Sanctis, Y., Barnathan, E.S., Raskob, G.E., 2020b. Modified IMPROVE VTE risk score and elevated D-dimer identify a high venous thromboembolism risk in acutely ill medical population for extended thromboprophylaxis. TH Open 4 (1), e59–65. https://doi.org/10.1055/s-0040-1705137.

Squillaro, T., Peluso, G., Galderisi, U., 2016. Clinical trials with mesenchymal stem cells: an update. Cell Transplant. 25 (5), 829–848. https://doi.org/10.3727/096368915X689622.

Stauffer, W.M., Alpern, J.D., Walker, P.F., 2020. COVID-19 and dexamethasone: a potential strategy to avoid steroid-related strongyloides hyperinfection. JAMA. https://doi.org/10.1001/jama.2020.13170.

Steinberg, K.P., Hudson, L.D., Goodman, R.B., Lee Hough, C., Lanken, P.N., Hyzy, R., Thompson, B.T., Ancukiewicz, M., 2006. Efficacy and safety of corticosteroids for persistent acute respiratory distress syndrome. N. Engl. J. Med. 354 (16), 1671–1684. https://doi.org/10.1056/NEJMoa051693.

Stéphan, F., Barrucand, B., Petit, P., Rézaiguia-Delclaux, S., Médard, A., Delannoy, B., Cosserant, B., et al., 2015. High-flow nasal oxygen vs noninvasive positive airway pressure in hypoxemic patients after cardiothoracic surgery:

a randomized clinical trial. JAMA 313 (23), 2331–2339. https://doi.org/10.1001/jama.2015.5213.

Stevens, V., Dumyati, G., Fine, L.S., Fisher, S.G., Van Wijngaarden, E., 2011. Cumulative antibiotic exposures over time and the risk of clostridium difficile infection. Clin. Infect. Dis. 53 (1), 42–48. https://doi.org/10.1093/cid/cir301.

Stockman, L.J., Bellamy, R., Garner, P., 2006. SARS: systematic review of treatment effects. PLoS Med. 3 (9), e343. https://doi.org/10.1371/journal.pmed.0030343.

Sun, Q., Qiu, H., Huang, M., Yang, Y., 2020. Lower mortality of COVID-19 by early recognition and intervention: experience from Jiangsu province. Ann. Intensive Care. https://doi.org/10.1186/s13613-020-00650-2.

Synairgen plc, 2020. Press Release: Synairgen Announces Positive Results from Trial of SNG001 in Hospitalised COVID-19 Patients. https://www.lsegissuerservices.com/spark/Synairgen/events/97cda0b9-0529-4be1-.

Tahir ul Qamar, M., Alqahtani, S.M., Alamri, M.A., Chen, L.L., 2020. Structural basis of SARS-CoV-2 3CLpro and anti-COVID-19 drug discovery from medicinal plants. J. Pharmaceut. Anal. 10 (4), 313–319. https://doi.org/10.1016/j.jpha.2020.03.009.

Tai, F.W.D., McAlindon, M.E., 2018. NSAIDs and the small bowel. Curr. Opin. Gastroenterol. 34 (3), 175–182. https://doi.org/10.1097/MOG.0000000000000427.

Tan, W.Y.T., Young, B.E., Lye, D.C., Chew, D.E.K., Dalan, R., 2020. Statin use is associated with lower disease severity in COVID-19 infection. Sci. Rep. 10 (1), 17458. https://doi.org/10.1038/s41598-020-74492-0.

Tang, N., Bai, H., Chen, X., Gong, J., Li, D., Sun, Z., Tang, N., Bai, H., Chen, X., Gong, J., Li, D., Sun, Z., 2020. Anticoagulant treatment is associated with decreased mortality in severe coronavirus disease 2019 patients with coagulopathy. J. Thromb. Hemostasis 18 (5), 1094–1099. https://doi.org/10.1111/jth.14817.

Tay, M.Y.F., Fraser, J.E., Chan, W.K.K., Moreland, N.J., Rathore, A.P., Wang, C., Vasudevan, S.G., Jans, D.A., 2013. Nuclear localization of dengue virus (DENV) 1-4 non-structural protein 5; protection against all 4 DENV serotypes by the inhibitor ivermectin. Antivir. Res. 99 (3), 301–306.

Thachil, J., Tang, N., Gando, S., Falanga, A., Cattaneo, M., Levi, M., Clark, C., Iba, T., 2020. ISTH interim guidance on recognition and management of coagulopathy in COVID-19. J. Thromb. Haemostasis 18 (5), 1023–1026. https://doi.org/10.1111/jth.14810.

The Queensland Health, 2020. Interim Infection Prevention and Control Guidelines for the Management of COVID-19 in Healthcare Settings, pp. 1–54.

Theodore, G.K., 1965. Safety of chloroquine in pregnancy. JAMA 191 (9), 765.

Toner, P., McAuley, D.F., Shyamsundar, M., 2015. Aspirin as a potential treatment in sepsis or acute respiratory distress syndrome. Crit. Care 19 (October), 374. https://doi.org/10.1186/s13054-015-1091-6.

Totura, A.L., Whitmore, A., Agnihothram, S., Schäfer, A., Katze, M.G., Heise, M.T., Baric, R.S., 2015. Toll-like receptor

3 signaling via TRIF contributes to a protective innate immune response to severe acute respiratory syndrome coronavirus infection. mBio 6 (3). https://doi.org/10.1128/mBio.00638-15 e00638−15.

Tran, K., Cimon, K., Severn, M., Pessoa-Silva, C.L., Conly, J., 2012. Aerosol generating procedures and risk of transmission of acute respiratory infections to healthcare workers: a systematic review. PloS One 7 (4), e35797. https://doi.org/10.1371/journal.pone.0035797.

Tresguerres, M., Parks, S.K., Goss, G.G., 2006. V-H(+)-ATPase, Na(+)/K(+)-ATPase and NHE2 immunoreactivity in the gill epithelium of the pacific hagfish (*Epatretus stoutii*). Comp. Biochem. Physiol. Mol. Integr. Physiol. 145 (3), 312−321. https://doi.org/10.1016/j.cbpa.2006.06.045.

Trounson, A., McDonald, C., 2015. Stem cell therapies in clinical trials: progress and challenges. Cell Stem Cell 17 (1), 11−22. https://doi.org/10.1016/j.stem.2015.06.007.

Tsolaki, V., Siempos, I., Magira, E., Kokkoris, S., Zakynthinos, G.E., Zakynthinos, S., 2020. PEEP levels in COVID-19 pneumonia. Crit. Care. https://doi.org/10.1186/s13054-020-03049-4.

University of Liverpool, 2020. COVID-19 Drug Interactions. https://www.covid19-druginteractions.org/.

Vaduganathan, M., Vardeny, O., Michel, T., McMurray, J.J.V., Pfeffer, M.A., Solomon, S.D., 2020. Renin-angiotensin-aldosterone system inhibitors in patients with covid-19. N. Engl. J. Med. 382 (17), 1653−1659. https://doi.org/10.1056/NEJMsr2005760.

Vandermeer, M.L., Thomas, A.R., Kamimoto, L., Reingold, A., Gershman, K., Meek, J., Farley, M.M., et al., 2012. Association between use of statins and mortality among patients hospitalized with laboratory-confirmed influenza virus infections: a multistate study. J. Infect. Dis. 205 (1), 13−19. https://doi.org/10.1093/infdis/jir695.

Vásárhelyi, B., Sátori, A., Olajos, F., Szabó, A., Beko, G., 2011. Low vitamin D levels among patients at Semmelweis University: retrospective analysis during a one-year period. Orv. Hetil. 152 (32), 1272−1277. https://doi.org/10.1556/OH.2011.29187.

Velthuis, A.J.W.T., van den Worm, S.H.E., Sims, A.C., Baric, R.S., Snijder, E.J., van Hemert, M.J., 2010. Zn(2+) inhibits coronavirus and arterivirus RNA polymerase activity in vitro and zinc ionophores block the replication of these viruses in cell culture. PLoS Pathog. 6 (11), e1001176. https://doi.org/10.1371/journal.ppat.1001176.

Veronika, G., Magar, L., Dornfeld, D., Giese, S., Pohlmann, A., Höper, D., Kong, B.-W., Jans, D.A., Beer, M., Haller, O., Schwemmle, M., 2016. Influenza A viruses escape from MxA restriction at the expense of efficient nuclear VRNP import. Sci. Rep. 6, 23138.

Villar, J., Confalonieri, M., Pastores, S.M., Umberto Meduri, G., 2020a. Rationale for prolonged corticosteroid treatment in the acute respiratory distress syndrome caused by coronavirus disease 2019. Crit. Care Explor. 2 (4). https://journals.lww.com/ccejournal/Fulltext/2020/04000/Rationale_for_Prolonged_Corticosteroid_Treatment.18.aspx.

Villar, J., Ferrando, C., Martínez, D., Ambrós, A., Muñoz, T., Soler, J.A., Aguilar, G., et al., 2020b. Dexamethasone treatment for the acute respiratory distress syndrome: a multicentre, randomised controlled trial. Lancet Respir. Med. 8 (3), 267−276. https://doi.org/10.1016/S2213-2600(19)30417-5.

Vincenzo, B., Hughes, B., 2020. Coronavirus Disease 2019 (COVID-19): Pregnancy Issues and Antenatal Care. UpToDate. https://www.uptodate.com/contents/coronavirus-disease-2019-covid-19-pregnancy-issues-and-antenatal-care. (Accessed 14 November 2020).

Vincent, M.J., Bergeron, E., Benjannet, S., Erickson, B.R., Rollin, P.E., Ksiazek, T.G., Seidah, N.G., Nichol, S.T., 2005. Chloroquine is a potent inhibitor of SARS coronavirus infection and spread. Virol. J. 2, 1−10. https://doi.org/10.1186/1743-422X-2-69.

Wang, H., Ma, S., 2008. The cytokine storm and factors determining the sequence and severity of organ dysfunction in multiple organ dysfunction syndrome. Am. J. Emerg. Med. 26 (6), 711−715. https://doi.org/10.1016/j.ajem.2007.10.031.

Wang, T.-T., Nestel, F.P., Bourdeau, V., Nagai, Y., Wang, Q., Liao, J., Tavera-Mendoza, L., et al., 2004. Cutting edge: 1,25-dihydroxyvitamin D3 is a direct inducer of antimicrobial peptide gene expression. J. Immunol. 173 (5), 2909−2912. https://doi.org/10.4049/jimmunol.173.5.2909.

Wang, S., Qu, X., Zhao, R.C., 2012. Clinical applications of mesenchymal stem cells. J. Hematol. Oncol. 5 (April), 19. https://doi.org/10.1186/1756-8722-5-19.

Wang, M., Lu, S., Li, S., Shen, F., 2013. Reference intervals of D-dimer during the pregnancy and puerperium period on the STA-R evolution coagulation analyzer. Clin. Chim. Acta 425 (October), 176−180. https://doi.org/10.1016/j.cca.2013.08.006.

Wang, B., Timilsena, Y.P., Blanch, E., Adhikari, B., 2019. Lactoferrin: structure, function, denaturation and digestion. Crit. Rev. Food Sci. Nutr. 59 (4), 580−596. https://doi.org/10.1080/10408398.2017.1381583.

Wang, D., Hu, B., Hu, C., Zhu, F., Liu, X., Zhang, J., Wang, B., et al., 2020a. Clinical characteristics of 138 hospitalized patients with 2019 novel coronavirus-infected pneumonia in Wuhan, China. JAMA 323 (11), 1061−1069. https://doi.org/10.1001/jama.2020.1585.

Wang, K., Gheblawi, M., Oudit, G.Y., 2020b. Angiotensin converting enzyme 2: a double-edged sword. Circulation 142 (5), 426−428. https://doi.org/10.1161/CIRCULATIONAHA.120.047049.

Wang, M., Cao, R., Zhang, L., Yang, X., Liu, J., Xu, M., Shi, Z., Hu, Z., Zhong, W., Xiao, G., 2020c. Remdesivir and chloroquine effectively inhibit the recently emerged novel coronavirus (2019-NCoV) in vitro. Cell Res. 30 (3), 269−271. https://doi.org/10.1038/s41422-020-0282-0.

Wang, X., Guo, X., Xin, Q., Pan, Y., Li, J., Chu, Y., Feng, Y., Wang, Q., 2020d. Neutralizing antibodies responses to SARS-CoV-2 in COVID-19 inpatients and convalescent patients. medRxiv. https://doi.org/10.1101/2020.04.15.20065623.

Wang, Y., Zhang, D., Du, G., Du, R., Zhao, J., Jin, Y., Fu, S., Gao, L., Cheng, Z., Lu, Q., Hu, Y., Luo, G., Wang, K., Lu, Y., Li, H., Wang, S., Ruan, S., Yang, C., Mei, C.,

Wang, Y., Ding, D., Wu, F., Tang, X., Ye, X., Ye, Y., Liu, B., Yang, J., Yin, W., Wang, A., Fan, G., Zhou, F., Liu, Z., Gu, X., Xu, J., Shang, L., Zhan, W.C., 2020e. Remdesivir in adults with severe COVID-19: a randomised, double-blind, placebo-controlled, multicentre trial. Lancet 395 (10236), 1569–1578.

Wang, Z., Yang, B., Li, Q., Wen, L., Zhang, R., 2020f. Clinical features of 69 cases with coronavirus disease 2019 in Wuhan, China. Clin. Infect. Dis. 71 (15), 769–777. https://doi.org/10.1093/cid/ciaa272.

Warrington, T.P., Bostwick, J.M., 2006. Psychiatric adverse effects of corticosteroids. Mayo Clin. Proc. 81 (10), 1361–1367. https://doi.org/10.4065/81.10.1361.

Webmed, 2020a. "Vitamin D." Vitamins & Supplements. https://www.webmd.com/vitamins/vitamin.

Webmed, 2020b. Vitamins & Supplements. Zinc. https://www.webmd.com/vitamins/zin.

Webmed, 2020c. Vitamins and Supplements. Lactoferrin. https://www.webmd.com/vitamins/ai/ingredientmono-49/lactoferrin.

Wei, P.-F., 2020. Diagnosis and treatment protocol for novel coronavirus pneumonia. Chin. Med J 133 (9), 1087–1095.

Wei, H.F., Jin, S.A., Ma, Z.C., Bi, H.S., Ba, X.Y., 1992. Clinical studies on high frequency two-way jet ventilation. J. Tongji Med. Univ. 12 (3), 183–188. https://doi.org/10.1007/BF02887822.

Wei, X.-B., Wang, Z.-H., Liao, X.-L., Guo, W.-X., Wen, J.-Y., Qin, T.-H., Wang, S.-H., 2020. Efficacy of vitamin C in patients with sepsis: an updated meta-analysis. Eur. J. Pharmacol. 868 (February), 172889. https://doi.org/10.1016/j.ejphar.2019.172889.

Weingart, S.D., Levitan, R.M., 2012. Preoxygenation and prevention of desaturation during emergency airway management. Ann. Emerg. Med. 59 (3), 165–175.e1. https://doi.org/10.1016/j.annemergmed.2011.10.002.

Williamson, B.N., Feldmann, F., Schwarz, B., Meade-white, K., Porter, D.P., Schulz, J., Van Doremalen, N., et al., 2020. Clinical benefit of remdesivir in rhesus macaques infected with SARS-CoV-2. Nature 585 (7824), 273–276. https://doi.org/10.1038/s41586-020-2423-5.Clinical.

World Health Organization, 2019. AWARE Classification of Antibiotics: World Health Organization Model List of Essential Medicines 21st List. Geneva. https://apps.who.int/iris/bitstream/handle/10665/325771/WHO-MVP-EMP-IAU-2019.06-eng.pdf?ua=1.

World Health Organization, 2020a. Clinical Management of Severe Acute Respiratory Infection (SARI) When COVID-19 Disease Is Suspected- Interim Guidance.

World Health Organization, 2020b. Corticosteroids for COVID-19. https://www.who.int/publications/i/item/WHO-2019-nCoV-Corticosteroids-2020.1.

Wu, C., Wei, J., Cen, Q., Sha, X., Cai, Q., Ma, W., Cao, Y., 2017. Supraglottic jet oxygenation and ventilation-assisted fibre-optic bronchoscope intubation in patients with difficult airways. Intern. Emerg. Med. 12 (5), 667–673. https://doi.org/10.1007/s11739-016-1531-6.

Wu, C., Chen, X., Cai, Y., Xia, Jia'An, Zhou, X., Xu, S., Huang, H., et al., 2020. Risk factors associated with acute respiratory distress syndrome and death in patients with coronavirus disease 2019 pneumonia in Wuhan, China. JAMA Intern. Med. 180 (7), 934–943. https://doi.org/10.1001/jamainternmed.2020.0994.

Xie, X., Sinha, S., Yi, Z., Langlais, P.R., Madan, M., Bowen, B.P., Willis, W., Meyer, C., 2017. Role of adipocyte mitochondria in inflammation, lipemia and insulin sensitivity in humans: effects of pioglitazone treatment. Int. J. Obes. https://doi.org/10.1038/ijo.2017.192.

Xie, J., Tong, Z., Guan, X., Du, B., Qiu, H., 2020. Clinical characteristics of patients who died of coronavirus disease 2019 in China. JAMA Netw. Open 3 (4), e205619. https://doi.org/10.1001/jamanetworkopen.2020.5619.

Xu, X., Han, M., Li, T., Sun, W., Wang, D., Fu, B., Zhou, Y., et al., 2020. Effective treatment of severe COVID-19 patients with tocilizumab. Proc. Natl. Acad. Sci. U. S. A. 117 (20), 10970–10975. https://doi.org/10.1073/pnas.2005615117.

Xue, J., Moyer, A., Peng, B., Wu, J., Hannafon, B.N., Ding, W.-Q., 2014. Chloroquine is a zinc ionophore. PloS One 9 (10), e109180. https://doi.org/10.1371/journal.pone.0109180.

Yao, W., Wang, T., Jiang, B., Gao, F., Wang, L., Zheng, H., Xiao, W., et al., 2020. Emergency tracheal intubation in 202 patients with COVID-19 in Wuhan, China: lessons learnt and international expert recommendations. Br. J. Anaesth. 125 (1), e28–37. https://doi.org/10.1016/j.bja.2020.03.026.

Yaris, F., Yaris, E., Kadioglu, M., Ulku, C., Kesim, M., Kalyoncu, N.I., 2004. Normal pregnancy outcome following inadvertent exposure to rosiglitazone, gliclazide, and atorvastatin in a diabetic and hypertensive woman. Reprod. Toxicol. 18 (4), 619–621. https://doi.org/10.1016/j.reprotox.2004.02.014.

Yoshikawa, T., Hill, T., Li, K., Peters, C.J., Tseng, C.-T.K., 2009. Severe acute respiratory syndrome (SARS) coronavirus-induced lung epithelial cytokines exacerbate SARS pathogenesis by modulating intrinsic functions of monocyte-derived macrophages and dendritic cells. J. Virol. 83 (7), 3039–3048. https://doi.org/10.1128/JVI.01792-08.

Yu, I.T., Xie, Z.H., Tsoi, K.K., Chiu, Y.L., Lok, S.W., Tang, X.P., Hui, D.S., et al., 2007. Why did outbreaks of severe acute respiratory syndrome occur in some hospital wards but not in others? Clin. Infect. Dis. 44 (8), 1017–1025. https://doi.org/10.1086/512819.

Yuan, S., 2015. Statins may decrease the fatality rate of Middle East respiratory syndrome infection. mBio. https://doi.org/10.1128/mBio.01120-15.

Zhang, W.-Y., Schwartz, E.A., Permana, P.A., Reaven, P.D., 2008. Pioglitazone inhibits the expression of inflammatory cytokines from both monocytes and lymphocytes in patients with impaired glucose tolerance. Arterioscler. Thromb. Vasc. Biol. 28 (12), 2312–2318. https://doi.org/10.1161/ATVBAHA.108.175687.

Zhang, P., Zhu, L., Cai, J., Lei, F., Qin, J.-J., Xie, J., Liu, Y.-M., et al., 2020a. Association of inpatient use of angiotensin-converting enzyme inhibitors and angiotensin II receptor blockers with mortality among patients with hypertension hospitalized with COVID-19. Circ. Res. 126 (12), 1671–1681. https://doi.org/10.1161/CIRCRESAHA. 120.317134.

Zhang, R., Wang, X., Ni, L., Xiao, D., Ma, B., Niu, S., Liu, C., Reiter, R.J., 2020b. COVID-19: melatonin as a potential adjuvant treatment. Life Sci. 250, 117583. https://doi.org/10.1016/j.lfs.2020.117583.

Zhang, W., Zhao, Y., Zhang, F., Wang, Q., Li, T., Liu, Z., Wang, J., et al., 2020c. The use of anti-inflammatory drugs in the treatment of people with severe coronavirus disease 2019 (COVID-19): the perspectives of clinical immunologists from China. Clin. Immunol. 214 (May), 108393. https://doi.org/10.1016/j.clim.2020.108393.

Zhao, X., Jiang, Y., Zhao, Y., Xi, H., Chang, L., Fan, Q., Feng, X., 2020. Analysis of the susceptibility to COVID-19 in pregnancy and recommendations on potential drug screening. Eur. J. Clin. Microbiol. Infect. Dis. 39 (7), 1209–1220. https://doi.org/10.1007/s10096-020-03897-6.

Zhong, J., Tang, J., Ye, C., Dong, L., 2020. The immunology of COVID-19: is immune modulation an option for treatment? Lancet Rheumatol. 2 (7), e428–e436. https://doi.org/10.1016/S2665-9913(20)30120-X.

Zhou, G., Myers, R., Li, Y., Chen, Y., Shen, X., Fenyk-Melody, J., Wu, M., et al., 2001. Role of AMP-activated protein kinase in mechanism of metformin action. J. Clin. Invest. 108 (8), 1167–1174. https://doi.org/10.1172/JCI13505.

Zhou, F., Ting, Y., Du, R., Fan, G., Liu, Y., Liu, Z., Xiang, J., et al., 2020a. Clinical course and risk factors for mortality of adult inpatients with COVID-19 in Wuhan, China: a retrospective cohort study. Lancet 395 (10229), 1054–1062. https://doi.org/10.1016/S0140-6736(20)30566-3.

Zhou, Q., Chen, V., Shannon, C.P., Wei, X.-S., Xiang, X., Wang, X., Wang, Z.-H., Tebbutt, S.J., Kollmann, T.R., Fish, E.N., 2020b. Interferon-A2b treatment for COVID-19. medRxiv. https://doi.org/10.1101/2020.04.06.20042580.

Zhu, Y., Yin, H., Zhang, R., Ye, X., Wei, J., 2019. High-flow nasal cannula oxygen therapy versus conventional oxygen therapy in patients after planned extubation: a systematic review and meta-analysis. Crit. Care 23 (1), 180. https://doi.org/10.1186/s13054-019-2465-y.

Ziehr, D.R., Alladina, J., Petri, C.R., Maley, J.H., Moskowitz, A., Medoff, B.D., Hibbert, K.A., Thompson, B.T., Hardin, C.C., 2020. Respiratory pathophysiology of mechanically ventilated patients with COVID-19: a cohort study. Am. J. Respir. Crit. Care Med. https://doi.org/10.1164/rccm.202004-1163LE.

Zmijewski, J.W., Lorne, E., Zhao, X., Tsuruta, Y., Sha, Y., Liu, G., Siegal, G.P., Abraham, E., 2008. Mitochondrial respiratory complex I regulates neutrophil activation and severity of lung injury. Am. J. Respir. Crit. Care Med. 178 (2), 168–179. https://doi.org/10.1164/rccm.200710-1602OC.

Zumla, A., Azhar, E.I., Arabi, Y., Alotaibi, B., Rao, M., McCloskey, B., Petersen, E., Maeurer, M., 2015. Host-directed therapies for improving poor treatment outcomes associated with the Middle East respiratory syndrome coronavirus infections. Int. J. Infect. Dis. https://doi.org/10.1016/j.ijid.2015.09.005.

Zumla, A., Chan, J.F.W.W., Azhar, E.I., Hui, D.S.C.C., Yuen, K.-Y.Y., 2016. Coronaviruses-drug discovery and therapeutic options. Nat. Rev. Drug Discov. 15 (5), 327–347. https://doi.org/10.1038/nrd.2015.37.

CHAPTER 6

Prognosis and Outcomes of COVID-19 infection During Pregnancy

AHMED A. WALI, MD • SHIMAA M. ABD-EL-FATAH, MD

(Dr. Prof. M Fadel Shaltout . Prof. of Obstetrics and Gynecology, Cairo University, Faculty of Medicine)

Covid-19 Infection and Pregnancy. https://doi.org/10.1016/B978-0-323-90595-4.00003-0

Susceptibility to infection: At the beginning of the pandemic, there was a controversy whether the immunological changes occurring in pregnancy would make the pregnant woman more vulnerable to contract SARS-CoV-2 (severe acute respiratory syndrome coronavirus 2) or not, and whether or not would the maternal COVID-19 (Coronavirus Disease, 2019) be more severe. Some authors initially argued that pregnancy could be protective against COVID-19 cytokine storm by lowering TNF-α (tumor necrosis factor-alpha) through downregulation of the proinflammatory activity of Th1 (T-helper cell type 1). It is now established that the pregnant population are, in general, not more vulnerable to acquire SARS-CoV-2; however, the clinical course of the disease may be more severe during pregnancy, and rapid deterioration may occur, especially after 28 weeks. Pregnancy increases a woman's chance of severe disease, hospitalization, ICU (intensive care unit) admission, and mechanical ventilation. Having severe COVID-19 increases the odds of preterm and cesarean delivery (CD) (Allotey et al., 2020; Badr et al., 2020; Delahoy et al., 2020; Ellington et al., 2020; Elshafeey et al., 2020; Hanna et al., 2020; Hessami et al., 2020; Khoury et al., 2020; Mor and Cardenas, 2010; Panagiotakopoulos et al., 2020; Poon et al., 2020; Zambrano et al., 2020a).

Clinical course of the disease and disease severity during pregnancy: The overall characteristics of the disease in pregnant women are similar to nonpregnant women, but with differences in the course of the disease and its severity. The commonest reported symptoms are fever and cough with rates 32.8% and 41.3% respectively, compared with 77% and 68% in the general population. Average duration of symptoms is approximately 1 month, although a quarter of patients can have some persistent symptoms for more than 2 months. Asymptomatic infection is hard to calculate: the Centers for Disease Control and Prevention (2020) gives an estimate of 40% in the general population, and the calculated range of asymptomatic infection for pregnant women in a very recent metaanalysis is 9%−30.9%. Initially asymptomatic pregnant women can present with subsequent severe COVID-19 and deteriorate rapidly. Pregnant women are less likely to report symptoms, which mimic physiological changes of pregnancy such as dyspnea, fatigue, myalgia, gastrointestinal symptoms, and even fever, which contributes in the delayed diagnosis amongst the pregnant population and first presentation in a severe form of the disease (Afshar et al., 2020; Allotey et al., 2020; Breslin et al., 2020; Centers for Disease Control and Prevention, 2020; Matar et al., 2021; Papapanou et al., 2021; World

Health Organization, 2020a; Xie et al., 2020; Zambrano et al., 2020a).

Commonest laboratory findings among COVID-19 pregnant women are lymphopenia (35%−69.6%), elevated C-reactive protein levels (48%−69%), and elevated D-dimers (82%−84.6%). However, interpretation of D-dimer levels differs during pregnancy due to the normally elevated levels during pregnancy. Regarding radiological findings, abnormal chest CT (computed tomography) scans are observed in 68.6% of pregnant women with COVID-19, with the most common patterns being ground glass opacities, almost always bilaterally (Allotey et al., 2020; Papapanou et al., 2021; Righini et al., 2008; Szecsi et al., 2010).

Severe disease and hospitalization: Definitions of severe and critical COVID-19 in pregnant women follow those of the general population (Table 6.1). Pneumonia occurs in 40%−49% of pregnant women with COVID-19. Around 31.9% of pregnant women are hospitalized, which is a high fraction knowing that hospitalization rate in nonpregnant women is 5.8%. However, it should be taken in consideration while interpreting these figures that part of the reported hospital admissions could be for the mere reason of delivery and not the disease severity itself. Severe illness necessitating hospital admission usually starts around 7 days after the onset of symptoms and requires an average hospital stay of 6 and 12 days for severe and critical disease, respectively (Allotey et al., 2020; Berghella and Hughes, 2021; Ellington et al., 2020; Narang et al., 2020; National Institutes of Health, 2020; Pierce-Williams et al., 2020; World Health Organization, 2020a).

It is not fully comprehended why pregnant women show a higher trend of having severe form of the disease. Angiotensin-converting enzyme 2 (ACE2) receptors have been verified to be the way through which SARS-CoV-2 gains access to the human cell. Perhaps the upregulation of ACE2 receptors, which occurs during pregnancy, could be an explanation of the different course of the illness in the unique pregnant population. Other contributing factors are increased oxygen consumption and reduced functional residual capacity normally occurring, which could aggravate the severity in patients with pneumonia (Ahlberg et al., 2020; Narang et al., 2020; Pringle et al., 2011; Stephens et al., 2020).

Black women seem more prone to SARS-CoV-2, along with Asian and other minority races. Risk factors for severe COVID-19 in the pregnant population include maternal age above 35 years, comorbidities such as hypertension and diabetes, obesity, and gestational diabetes. It is possible that comorbidities and COVID-19 share synergistic immunological and

TABLE 6.1
Definitions of Severe and Critical COVID-19.

	WHO: COVID-19 Disease Severity	NIH: Clinical Classification of COVID-19

Severe	• Clinical signs of pneumonia (i.e., fever, cough, dyspnea, fast breathing)
	+
	• One of the following:
	  • Respiratory rate >30 breaths/ minute
	  • Severe respiratory distress
	  • SpO₂ <90% on room air
		One of the following:
		• Respiratory frequency >30 breaths per minute
		• SpO₂ <94% on room air at sea level
		• PaO₂/FiO₂ <300 mmHg
		• Lung infiltrates >50%

Critical	• Presence of ARDS, sepsis, or septic shock • Other complications include acute pulmonary embolism, acute coronary syndrome, acute stroke, and delirium	• People who have respiratory failure, septic shock, and/or multiple organ dysfunction

ARDS, acute respiratory distress syndrome; *NIH*, National Institutes of Health; *PaO₂/FiO₂*, ratio of arterial partial pressure of oxygen to fraction of inspired oxygen; *SpO₂*, saturation of peripheral oxygen; *WHO*, World Health Organization.

inflammatory pathways (Allotey et al., 2020; Ellington et al., 2020; Knight et al., 2020; Narang et al., 2020).

ICU admission and mechanical ventilation: ICU admission rates in maternal COVID-19 range between 3% and 13%, and rates of mechanical ventilation are between 1.5% and 4.3%. Risk factors for ICU admission are similar to the ordinary population, including obesity and preexisting morbidities such as diabetes and hypertension. Tables 6.2 and 6.3 show rates of ICU admission and mechanical ventilation in pregnant women with COVID-19, respectively (Allotey et al., 2020; Capobianco et al., 2020; Juan et al., 2020; Khalil

et al., 2020a; Matar et al., 2021; Segars et al., 2020; Smith et al., 2020; Trippella et al., 2020; Zaigham and Andersson, 2020). A systematic review by Turan et al. (2020) specifically reported the percent of mechanically ventilated patients among SARS-CoV-2-positive pregnant women admitted in the ICU and was as high as 83.6% (51 ventilated out of 61 admitted). The COVID-19 Response Pregnancy and Infant Linked Outcomes Team of the Centers for Disease Control and Prevention (CDC) reports a large study comparing 23,434 SARS-CoV-2 positive pregnant women to 386,028 positive nonpregnant women in their

TABLE 6.2
Frequency and Percent of ICU Admission in Pregnant Women With COVID-19.

Study (First Author, Year)	n/N	Percent (%)
Allotey (2020)	232/10901	3.0
Capobianco (2020)	–	13.0
Juan (2020)	12/253	4.7
Khalil et al. (2020a)	–	7.0
Smith (2020)	1/23	4.3
Trippella (2020)	10/275	3.6
Zaigham (2020)	3/108	2.8

ICU, intensive care unit; *n*, the number of patients admitted in ICU; *N*, the total number of pregnant women with COVID-19 in the study.

TABLE 6.3

Frequency and Percent of Pregnant Women With COVID-19 Who Needed Mechanical Ventilation.

Study (First Author, Year)	n/N	Percent (%)
Allotey (2020)		
Invasive ventilation	155/10713	1.5
ECMO	16/1935	0.8
Juan (2020)	3/170	1.8
Khalil et al. (2020a)		
Invasive ventilation	–	3.4
ECMO	–	0.7
Matar et al. (2021)	2/136	1.5
Segars (2020)	4/162	2.5
Smith (2020)	1/23	4.3
Trippella (2020)	5/275	1.8

ECMO, extracorporeal membrane oxygenation; *n*, the number of patients who needed mechanical ventilation; *N*, the total number of pregnant women with COVID-19 in the study.

reproductive age: the pregnant women showed a higher risk of ICU admission (10.5 vs. 3.9 per 1000 cases), invasive ventilation (2.9 vs. 1.1 per 1000 cases), ECMO (extracorporeal membrane oxygenation) (0.7 vs. 0.3 per 1000 cases), and death (1.5 vs. 1.2 per 1000 cases) (Zambrano et al., 2020a).

Complications of COVID-19 in pregnancy: All complications of COVID-19 that occur in the nonpregnant population can occur in pregnant women, including respiratory, cardiovascular, thromboembolic, renal, neurological, gastrointestinal, hepatic, hematologic and cutaneous complications, psychiatric illness, secondary infections, and post-COVID syndrome. In this section, emphasis is on complications with specific maternal impact or specific presentation.

Acute respiratory failure occurs in up to 8% of COVID-19 patients and is considered the principal cause of death in COVID-19. The vulnerability of a pregnant woman lies in their increased oxygen consumption, decreased functional residual capacity, and physiological hyperventilation with compensated respiratory alkalosis, which can make the progress rapidly if pneumonia occurs. Moreover, due to different anatomy and physiology of the pregnant woman, and also the presence of a fetus, there are some special considerations in management of ARDS in pregnant women. Unlike nonpregnant patients in who SpO_2 (saturation of peripheral oxygen) should be above 90%, it should be above 95% and maintained between 92% and 95% once stabilized in pregnant women, which is to ensure PaO_2 (partial pressure of oxygen) above

70 mmHg with the aim of preserving adequate oxygen diffusion across the placenta to the fetus. The prone position, which is encouraged in COVID-19 patients, might be challenging in pregnant ladies especially in the late trimester and can be substituted by the semi-prone or left lateral positions. Also, the uterus can be displaced to minimize aortocaval compression by applying pads above and below the level of the uterus. Lines such as permissive hypercapnia, high PEEP (positive end-expiratory pressure), and ECMO should be used with close fetal surveillance especially that studies on fetal effect are not abundant (Campbell and Klocke, 2001; Caputo et al., 2020; Chen et al., 2020c; Liu et al., 2020a; Pacheco and Saad, 2018; Ruan et al., 2020; Soma-Pillay et al., 2016; Stephens et al., 2020; Syeda et al., 2020; Tolcher et al., 2020; Webster et al., 2020; World Health Organization, 2020b).

Thromboembolic complications: COVID-19 has been described to cause microthromboembolism and DIC and also to increase the risk of venous thromboembolism (VTE) due to the classic components of the Virchow's triad: (1) endothelial injury (2) imbalance between procoagulants and anticoagulants caused by the inflammatory and immunological course of the disease, and (3) immobilization in hospitalized patients. VTE can occur in one-third of COVID-19 ICU patients, mainly deep venous thrombosis and pulmonary embolism. Pregnancy is typically known to induce a hypercoagulable state, which increases chances of VTE in COVID-19 patients especially if hospitalized. Therefore, hospitalized pregnant women with COVID-19 should be prescribed prophylactic anticoagulation—unless contraindicated—during their hospital stay. Whether to utilize unfractionated or low-molecular-weight heparin should be guided by the time frame to delivery. Therapeutic anticoagulation should only be administered according to VTE risk scores and presence of other indications (Begbie et al., 2000; James, 2010; Jose and Manuel, 2020; Libby and Lüscher, 2020; Lowenstein and Solomon, 2020; Maier et al., 2020; National Institutes of Health, 2020; Panigada et al., 2020; Ranucci et al., 2020; Tang et al., 2020b; Teuwen et al., 2020; Turan et al., 2020; Zhou et al., 2020).

Cardiac complications that can occur in COVID-19 patients include acute myocardial injury, cardiomyopathy, acute coronary syndrome, arrhythmias, and acute heart failure. Causes may be the direct viral myocarditis, the indirect cytokine-mediated cardiac injury, microvascular thrombosis, the cardiac overload during infection, or as a part of multiple organ system failure. Acute myocardial injury—defined as elevation of troponin level above the 99th centile—can occur in up to 16% of hospitalized patients with COVID-19. Myocardial injury has been described in pregnant women with COVID-19, with a high mortality rate of 13%, mainly due to associated fatal arrhythmias. The vulnerability of the pregnant women lies in their hyperdynamic circulation, increased blood volume, and cardiac output. Aggravating cofactors include advanced maternal age, preexisting comorbidities, and obesity. Cardiomyopathy, which may occur in nonpregnant COVID-19 patients, has also been described in pregnant women with guarded prognosis (Juusela et al., 2020; de Lorenzo et al., 2020; Guo et al., 2020; Hendren et al., 2020; Long et al., 2020; Mercedes et al., 2021; Prasitlumkum et al., 2020; Verity et al., 2020; Williams, 2003).

Acute kidney injury (AKI) has been described in COVID-19 patients with an incidence of 10.6%, which is higher than that in non-COVID-19 hospitalized patients. The pathophysiology of AKI in COVID-19 patients is not fully understood. Suggested mechanisms include (1) hypovolemia and hemodynamic changes—secondary to shock or multiple organ system failure—leading to acute tubal necrosis (ATN), (2) renal microvascular thromboembolism, (3) rhabdomyolysis, (4) direct viral effect on the kidney, as viral evidence in renal tissue has been confirmed in autopsy, and/or (5) side effect to the antiviral remdesivir is under study. So hypothetically, the pregnant women due to the normal physiological changes during pregnancy, which include extravascular fluid shift and procoagulopathic changes, are more vulnerable to AKI. A 33-year-old woman from Iran has been described who was 34 weeks pregnant and was diagnosed to be SARS-Co-V-2 positive by nasopharyngeal reverse-transcription polymerase chain reaction (RT-PCR) and developed severe form of the disease necessitating mechanical ventilation; during her hospital stay, she developed ATN, which—with no other possible etiology—resolved after improvement of the disease (European Medicines Agency, 2020; Farkash et al., 2020; Lin et al., 2020; Mitchell and Chiwele, 2021; Naicker et al., 2020; National Institute for Health and Care Excellence, 2020; Taghizadieh et al., 2020; Zaim et al., 2020).

Hematological conditions, which can complicate COVID-19, include SARS-Co-V-2-induced immune thrombocytopenia or immune thrombocytopenic purpura (ITP). These cases need meticulous diagnostic approach to exclude other possible causes. Contributing factors are advanced age and disease severity. ITP has been diagnosed in asymptomatic patients and during recovery. Proposed theories include immunological changes and genetic mutations. Some reports have described ITP newly diagnosed in pregnancy with COVID-19. Flare of previously known ITP occurring in pregnancy with COVID-19 has been also described (Bhattacharjee and Banerjee, 2020; Nesr et al., 2020; Tang et al., 2020a).

Maternal death: Although pregnancy is associated with higher odds of hospitalization, ICU admission, and mechanical ventilation compared with the nonpregnancy state, fortunately, maternal mortality due to COVID-19 is below 2%. In the large systematic review conducted by Allotey et al. (2020), the maternal mortality was 0.6% (73 out of 11,580 mothers). The initial report issued by the CDC in June 2020 stated that COVID-19 pregnant women do not have a higher mortality compared with their nonpregnant peers; however, this was later followed by an update in November 2020 calculating that the pregnant population have a slightly higher risk of death compared with the nonpregnant (1.5 vs. 1.2 per 1000 cases) population with an adjusted risk ratio of 1.7. Mortalities are mostly due to ARDS (acute respiratory distress syndrome) with its cardiopulmonary consequences and venous thromboembolism. Unfortunately, mortalities happen to be more in middle- and low-income countries where healthcare facilities are more limited (Allotey et al., 2020; Ellington et al., 2020; Hantoushzadeh et al., 2020; Hessami et al., 2020; Nakamura-Pereira et al., 2020; Takemoto et al., 2020; Zambrano et al., 2020a).

Maternal deaths amid different studies are summarized in Table 6.4 (Allotey et al., 2020; Capobianco et al., 2020; Juan et al., 2020; Khalil et al., 2020a; Matar et al., 2021; Segars et al., 2020; Smith et al., 2020; Trippella et al., 2020; Zaigham and Andersson, 2020).

TABLE 6.4
Frequency and Percent of Maternal Mortality in Pregnant Women With COVID-19.

Study (First Author, Year)	n/N	Percent (%)
Allotey (2020)	73/11580	0.6
Capobianco (2020)	–	0.0
Juan (2020)		
Case series and reports	9/324	2.8
Case series only	7/304	2.3
Khalil et al. (2020a)	–	0.0
Matar et al. (2021)	1/136	0.7
Segars (2020)	2/162	1.2
Smith (2020)	1/92	0.0
Trippella (2020)	10/275	0.4
Zaigham (2020)	0/108	0.0

n, the number of patients admitted in ICU; *N*, the total number of pregnant women with COVID-19 in the study.

In summary, pregnant women with COVID-19 are more probable to need hospitalization, ICU admission, and mechanical ventilation, compared with their nonpregnant counterparts, but their risk of death is similar or slightly higher (Ellington et al., 2020; Papapanou et al., 2021; Zambrano et al., 2020a).

OBSTERTIC OUTCOMES AND MATERNAL, FETAL, AND NEONATAL PROGNOSIS IN COVID-19

CD rates: According to the National Center for Health Statistics, the overall CD rate in the United States was 31.7% in 2019 (Hamilton et al., 2020).

CD rates among pregnant women diagnosed with COVID-19 are very high with rates ranging between 48.3% and 94%. In several instances, being diagnosed with COVID-19 is the only indication for CD. The mere maternal COVID-19 diagnosis accounts for 7.7% and 60.4% of CDs, and being symptomatic for COVID-19 or not does not change the odds of having a CD. Other indications include severe/critical maternal disease and fetal compromise. Dubey et al. (2020) analyzed CD rates geographically and found rates to be considerably higher in China (91%) compared with the United States (40%) and Europe (38%). A Swedish study observed that SARS-CoV-2-positive women are less likely to undergo induction of labor with a percentage of 18.7% of deliveries compared with 29.6% in their matched SARS-CoV-2 negative controls. Table 6.5 summarizes CD rates along several different studies (Ahlberg et al., 2020; Allotey et al., 2020; Capobianco et al., 2020; Di Mascio et al., 2020a; Dubey et al., 2020; Kasraeian et al., 2020; Khalil et al., 2020a; Matar et al., 2021; Muhidin et al., 2020; Smith et al., 2020; Trippella et al., 2020; Trocado et al., 2020; Woodworth et al., 2020; Zaigham and Andersson, 2020).

Several factors interlace in the higher CD rates among pregnant women with COVID-19: (1) lack of recommendations and guidelines at the beginning of the pandemic, when reasons such as fear of maternal–fetal transmission were sometimes the only indication for CD, (2) the already higher CD rates in China where the pandemic first started and from where the most early publications, which were mainly case reports and case series were published, (3) the assumed higher incidence of prematurity among women with COVID-19, which under certain circumstances can necessitate CD, (4) despite induction of labor being safe and successful in mechanically ventilated pregnant women, sometimes the circumstances are challenging regarding the operating theater and intensive care staff and ease of equipment provision, (5) sometimes the reported indication for CD by many authors is the belief of rapid need for termination of pregnancy due to aggravation of maternal illness and rapid need for relief of abdominal distension and improving efficiency of mechanical ventilation, and (6) also, the need for rapid termination to safely allow maternal administration of possibly teratogenic investigational antiviral medications (Berghella and Hughes, 2021; Boerma et al., 2018; Ming et al., 2019; Papapanou et al., 2021; Yang et al., 2020b).

TABLE 6.5
Frequency and Percent of Cesarean Deliveries in COVID-19 Pregnant Women.

Study (First Author, Year)	CS Rate (Percent) (%)
Allotey (2020)	54.8
Capobianco (2020)	88
Di Mascio et al. (2020a)	92.7
Dubey (2020)	72
Kasraeian (2020)	92
Khalil et al. (2020a)	48.3
Matar et al. (2021)	76.3
Muhidin (2020)	88.8
Smith (2020)	80
Trippella (2020)	75
Trocado (2020)	94
Woodworth (2020)	51.4
Zaigham (2020)	92

Preterm labor and preterm birth: The classic definition of preterm labor (PTL) is occurrence of regular uterine contractions accompanied by cervical changes before 37 weeks, while preterm birth (PTB) or preterm delivery means actual delivery before 37 weeks of gestation. Although a proportion of women with PL will go into preterm birth, this is not the condition in all cases. PTB rate in the United Sates was estimated to be 10.23% in 2019 (Hamilton et al., 2020; Howson et al., 2013; Rundell and Panchal, 2017).

PTL rates are estimated between 22.7% and 32.2%, and PTB rates in studies with ample sample sizes constitute 20.6%−25% of all deliveries. When observing studies with smaller samples rates reach as high as 63.8%. However, the PTB rates in the United States might not be affected; the CDC COVID-19 Response Pregnancy and Infant Linked Outcomes Team reports PTB rates of 12.9% (506/3912 infants). The proportion of preterm births, which are spontaneous among the overall preterm births in COVID-19 pregnant women, do not seem to be actually high and represent around 5%−6% of all preterm births. The larger fraction of preterm deliveries is in fact iatrogenic, and a study mentions being diagnosed with maternal COVID-19 to be the main indication for preterm delivery in half of the PTBs. Iatrogenic preterm deliveries with medically relevant indications could perhaps represent only 18% of births. In summary, COVID-19 mothers have higher chances of delivering preterm, although usually spontaneous PTB would not be the cause. The mechanism by which spontaneous PTL and PTB occur is not fully understood yet; however, maternal hypoxia, maternal fever, placental pathological changes, and uteroplacental insufficiency could be responsible (Allotey et al., 2020; Khalil et al., 2020a; Papapanou et al., 2021; Smith et al., 2020; Turan et al., 2020; Woodworth et al., 2020).

Miscarriage: COVID-19 does not seem to raise the rates of miscarriage in pregnant women. Unlike SARS (severe acute respiratory syndrome) and MERS (Middle East respiratory syndrome) in which miscarriage rates were approximately 25% and 18%, respectively, rates of miscarriage in SARS-CoV-2-positive women are below 2%. Woodworth et al. (2020) report pregnancy loss before 20 weeks to be as low as 0.3%. However, the paucity of data on women in early pregnancy must be kept in mind. There is some concern that placental infection and pathological changes— mentioned later in vertical transmission—could be responsible for midtrimester miscarriage. Anxiety and fear of "consequences of the novel disease on the unborn fetus" may increase women's requests for abortion (Baud et al., 2020; Berghella and Hughes, 2021; Dashraath et al., 2020; Papapanou et al., 2021).

Preeclampsia: A case-control Swedish study demonstrated higher prevalence of preeclampsia (PE) among SARS-CoV-2-positive pregnant women compared with their matched negative controls with a prevalence of

7.7% the positive group compared with 4.3% in controls. Development of preeclampsia could be related to the immunological and inflammatory changes occurring in COVID-19 and involved in the pathophysiology of PE. Moreover, a possible explanation of development of preeclampsia in COVID-19 is that after SARS-CoV-2 uses the ACE2 receptor for cell entry, it causes its downregulation—after the initial upregulation mentioned before—thus decreasing conversion of angiotensin II to angiotensin-(1−7), thus resulting in a pathophysiology, which mimics preeclampsia. In addition to this shared mechanism, COVID-19 and preeclampsia share other pathophysiology such as endothelial cell dysfunction and complement activation. In conclusion, preeclampsia could be aggravated by COVID-19 or could worsen COVID-19 (Ahlberg et al., 2020; Alrahmani and Willrich, 2018; Ferrario et al., 2005; Narang et al., 2020; Pringle et al., 2011; Risitano et al., 2020; Varga et al., 2020).

COVID-19 and preeclampsia also share common abnormalities particularly thrombocytopenia, elevated transaminases, and evidence of disseminated intravascular coagulopathy (DIC) with prolonged prothrombin time, low fibrinogen, and elevated fibrinogen degradation products, which renders differentiation between COVID-19 related laboratory abnormalities and severe preeclampsia—especially with HELLP syndrome—difficult. Other interlacing aberrations are hemolysis and lupus anticoagulant positivity. Moreover, symptoms of COVID-19 and severe preeclampsia and eclampsia could intermingle, such as headache, and cerebrovascular and neurological manifestations (Berghella and Hughes, 2021; Futterman et al., 2020; Lippi et al., 2020; Narang et al., 2020; Zhang et al., 2020).

Vertical Transmission and Neonatal COVID-19 Infection

In general, vertical transmission of a virus refers to its transmission from the mother to her offspring. This can happen either intrauterine, intrapartum, or postpartum (Fermin, 2018).

The possibility of the vertical transmission of SARS-CoV-2 has been a major concern of the medical community. Theoretically, maternal−fetal transmission is possible because the ACE2 receptor—which is now established as the receptor of SARS-CoV-2—is expressed in the placental villi, in both cytotrophoblasts and syncytiotrophoblasts, in addition to extravillous trophoblasts (Gengler et al., 2021; Hanna et al., 2020; Hikmet et al., 2020; Li et al., 2020a; Pringle et al., 2011; Valdés et al., 2006).

It is assumed that SARS-CoV-2 can be transmitted from the mother to the fetus or neonate through several probable routes: (1) intrauterine, either transplacental, through ingestion or aspiration of infected amniotic fluid, or due to maternal viremia, (2) intrapartum, through contact with maternal blood or secretions, or (3) postpartum, through breast milk. The neonate could also get infected during the postpartum period through the conventional respiratory route from an infected mother or contact. However, transient positive neonatal nasopharyngeal swabs have been collected in the first 24 h with subsequent negative swabs and negative IgM antibodies, which could be explained by contamination of the newborn's mouth or airway with maternal secretions during resuscitation (Blumberg et al., 2020; Caparros-Gonzalez et al., 2020).

To prove the possibility of maternal−fetal transmission of the SARS-CoV-2, two factors should be fulfilled: (1) evidence of the presence of the virus in amniotic fluid, umbilical cord blood, placenta, cervical secretions, or breast milk by immunohistochemistry or nucleic acid methods such as in situ hybridization, and (2) evidence of neonatal COVID-19 infection by a positive RT-PCR within the first 12−72 h (Caparros-Gonzalez et al., 2020; Schwartz et al., 2021; Schwartz et al., 2020; Walker et al., 2020).

A handful of studies have assessed infection of the amniotic fluid by SARS-CoV-2 in COVID-19 positive mothers, all of which have disproved evidence of the virus (Chen et al., 2020b; Li et al., 2020b; Liu et al., 2020b; Peng et al., 2020; Stonoga et al., 2021; Vivanti et al., 2020; Wang et al., 2020b; Xiong et al., 2020; Yang et al., 2020b; Yu et al., 2020) except for one case report from Iran (Zamaniyan et al., 2020). It seems maternal−fetal transmission via the amniotic fluid is unlikely to occur (Caparros-Gonzalez et al., 2020; Salem et al., 2021).

Multiple studies also confirmed the absence of SARS-CoV-2 RNA in umbilical cord blood (Chen et al., 2020b; Khan et al., 2020; Li et al., 2020b; Liu et al., 2020b; Patanè et al., 2020; Peng et al., 2020; Wang et al., 2020a; Yang et al., 2020a; Yang et al., 2020b; Zambrano et al., 2020b). Only two studies demonstrated viral evidence, in one of which the sample was obtained from the umbilical cord stump; however, neonatal RT-PCR swabs were negative (Kulkarni et al., 2021; Stonoga et al., 2021).

Unlike the amniotic fluid which seems—according to current evidence—not infectious to the fetus of a COVID-19 mother, the placenta could be accused of in utero maternal−fetal transmission of SARS-CoV-2. While several studies have shown negative results from placental tissues, other studies have demonstrated positive SARS-CoV-2 RNA on the placental fetal side. RNA belonging to SARS-CoV has been found in different types of cells of the placental barrier: cytotrophoblasts, syncytiotrophoblasts, villous endothelial cells, and extravillous trophoblasts. The histopathological examination of placentae from COVID-19 mothers has shown accelerated villous maturation, acute intervillositis, chronic histiocytic—lymphocytic intervillositis, necrosis of syncytiotrophoblasts, acute deciduitis and chorioamnionitis, decidua capsularis thrombotic vasculopathy, chorangiosis, funisitis, subchorial hematoma, and perivillous fibrin deposition with infarctions. Moreover, Algarroba et al. (2020) were able to visualize SARS-CoV-2 virions in the villous syncytiotrophoblasts by the electron microscope. A handful of neonates born to mothers with these changes have tested positive for SARS-CoV, and the number is still rising. Also, Kulkarni et al., 2021 report a term neonate delivered to a suspected COVID-19 mother, whose placenta tested positive for SARS-CoV. The mother although having a negative nasopharyngeal swab later on developed detectable antibodies, and the neonate developed symptomatic COVID-19. There is growing evidence that the two findings "trophoblast necrosis" and "chronic histiocytic intervillositis" when together are possible markers for placental infection with SARS-CoV with risk of fetal transmission (Algarroba et al., 2020; Baud et al., 2020; Facchetti et al., 2020; Hecht et al., 2020; Hosier et al., 2020; Hsu et al., 2021; Kirtsman et al., 2020; Kulkarni et al., 2021; Patanè et al., 2020; Penfield et al., 2020; Stonoga et al., 2021; Vivanti et al., 2020).

On the contrary, other few studies have reported negative swabs from placentae of COVID-19 women, with delivery of healthy neonates with negative swabs. Chen et al. (2020e) performed histological examination on placentae from three COVID-19 mothers and were found to show no histopathologic changes, in addition to negative RT-PCR (Chen et al., 2020d, 2020e; Li et al., 2020b; Peng et al., 2020; Schwartz, 2020; Schwartz et al., 2021; Schwartz and Morotti, 2020; Wang et al., 2020b; Xiong et al., 2020).

Regarding presence in breast milk, interpretation of results of studies may be more difficult. While several studies have negated the presence of viral evidence in breast milk (Chen et al., 2020b; Li et al., 2020b; Liu et al., 2020b; Peng et al., 2020; Wang et al., 2020a; Xiong et al., 2020) SARS-CoV-2 RNA has been isolated from multiple samples in other studies (Bastug et al., 2020; Chambers et al., 2020; Kirtsman et al., 2020; Lugli et al., 2020; Tam et al., 2021; Zhu et al., 2020).

However, the infectious potential of breast milk of COVID-19 infected mothers remains unclear till now (Caparros-Gonzalez et al., 2020; Chen et al., 2020a), as it is hard to prove whether a previously negative breastfed neonate that transforms to positive acquired it from the breast milk of its COVID-19-infected mother or via the conventional respiratory route.

Whether vertically transmitted or not, newborns born to mothers with COVID-19 have a chance of testing positive for SARS-CoV-2 in their early neonatal life by RT-PCR for throat swabs. The incidence is not common, but some reviews support that vertical transmission is possible. Table 6.6 summarizes the percent of newborns born to COVID-19-infected mothers, who tested positive for SARS-CoV-2 (Amaral et al., 2020; Ashraf et al., 2020; Bellos et al., 2021; Dhir et al., 2020; Di Toro et al., 2021; Figueiro-Filho et al., 2020; Han et al., 2020; Khalil et al., 2020a; Trocado et al., 2020; Turan et al., 2020; Walker et al., 2020; Yee et al., 2020; Yoon et al., 2020). The main drawback is that most of the publications included in these reviews are case series and case reports. Some publications have not accurately mentioned the time frame during which the neonatal nasopharyngeal RT-PCR was carried out. Also, there are some overlapping publications included in the reviews, not to mention the possibility of overlapping participants in the included publications. After conducting an analysis of systematic reviews, Papapanou et al. (2021) estimated the possibility of COVID-19 mothers to have an infected newborn to be around 2.5%.

In summary, congenital intrauterine transmission seems possible and cannot be excluded; however, neonatal COVID-9 is uncommon. The exact method of maternal—fetal transmission of SARS-CoV—if actually occurring—needs further meticulous studying (Salem et al., 2021; Schwartz et al., 2020; Sheth et al., 2020; Walker et al., 2020).

Perinatal Morbidity and Mortality

At first glance, possible adverse perinatal outcomes to fetuses and neonates of mothers with COVID-19 that would jump in mind the obstetrician are fetal distress, fetal growth restriction (FGR), stillbirth, prematurity, low birth weight (LBW), low Apgar scores, and neonatal asphyxia, with subsequent NICU (neonatal intensive care unit) admission and probable neonatal mortality. It could be wrongfully guessed that newborns infected in utero or having early neonatal COVID-19 could have worse outcomes. Actually, there seems no difference in these outcomes between infected and normal neonates; however, an evident contributing factor is

TABLE 6.6
Percent of Newborns Who Tested Positive for SARS-CoV-2 Born to COVID-19 Infected Mothers.

Study (First Author, Year)	Publications Included	Total Neonates	Tested Neonates	Positive Neonates	% of Positive to Total	% of Positive to Tested
Amaral (2020)	70	1042	–	39	3.74	–
Ashraf (2020)	21	92	86	4	4.35	4.65
Bellos et al. (2021)	60	920	–	4	0.43	–
Dhir (2020)	86	1141	–	58	5.08	–
Di Toro et al. (2021)	5	444	–	19	4.28	–
Figueiro-Filho (2020)	8	1116	–	18	1.61	–
Han (2020)	30	559	–	21	3.76	–
Khalil et al. (2020a)	9	751	–	19	2.53	–
Trocado (2020)	8	51	–	1	1.96	–
Turan (2020)	63	479	405	8	1.67	1.98
Walker (2020)	49	666	–	28	4.20	–
Yee (2020)	11	338	154	5	1.48	3.25
Yoon (2020)	28	201	167	4	1.99	2.40

the severity of the maternal disease, with worse perinatal outcomes among severe and critically ill COVID-19 mothers. Other risk factors include gestational age at diagnosis, LBW, and need for maternal ventilatory support. To this date, there are no cases of neonatal mortalities due to early neonatal diagnosis of COVID-19 in newborns of COVID-19 mothers. Perinatal outcomes in hospitalized pregnant women are generally worse, being those with severe or critical COVID-19 (Allotey et al., 2020; Di Mascio et al., 2020b; Papapanou et al., 2021; Turan et al., 2020).

Fetal distress: The occurrence of fetal distress as one of the outcomes of neonates born to pregnant women with COVID-19 has shown a wide range of reporting. Systematic reviews have described frequencies ranging from 8.5% up to 61%. One of the largest neonatal samples is that reported by Allotey et al. (2020), in which 8.5% (25/293) of neonates had fetal distress. On the other end of the wide range, Smith et al. (2020) report a much higher frequency of fetal distress of 61% (11/18 neonates). However, the latter systematic review includes a smaller number of pregnancies, all from case reports and series, solely from China, and very early in the pandemic experience (Allotey et al., 2020; Papapanou et al., 2021; Smith et al., 2020). Also,

discrepancies could be attributed to different definitions used for fetal distress along different publications.

Cardiotocographic changes: A Spanish study retrospectively analyzed cardiotocograph (CTG) tracings of 12 pregnant women with COVID-19, 10 of which had abnormalities in their tracings. Abnormalities included elevated fetal heart rate baseline, absent accelerations, occurrence of late decelerations, and abnormal variability. The authors hypothesized that these alterations could be attributed to fetal response to maternal fever, transplacental passage of inflammatory and immunological mediators of the cytokine storm, or possible uteroplacental insufficiency and FGR (Gracia-Perez-Bonfils et al., 2020).

Fetal growth restriction and small for gestational age: Although FGR seems as an important outcome worth studying, it is still underreported. Available reviews report rates between zero and 9% of neonates born to COVID-19 mothers (Di Mascio et al., 2020a; Diriba et al., 2020; Segars et al., 2020). Restricted fetal growth could be explained by uteroplacental insufficiency, which could be due to severe maternal illness, maternal hypoxia, placental COVID-19-induced pathological changes, or maternal hypercoagulable state, leading to thrombosis and infarction in the placenta and

TABLE 6.7
Frequency and Percent of Newborns With Low Birth Weight (LBW) Born to COVID-19-Infected Mothers.

Study (First Author, Year)	Neonates with LBW (Number)	Total Neonates (Number)	Neonates with LBW (Percent) (%)
Chang (2020)	9	92	9.8
Dubey (2020)	41	548	7.5
Elshafeey (2020)	20	256	7.8
Juan (2020)	8	221	3.6
Muhidin (2020)	7	89	7.9
Smith (2020)	9	21	42.8

umbilical vessels (Gracia-Perez-Bonfils et al., 2020; Schwartz et al., 2020a). Woodworth et al. (2020) state in their report in November 2020 that out of 3486 live births of women with COVID-19 with known weight and gestational age at birth, 198 (5.7%) were small for gestational age.

Stillbirth: Rates of stillbirth in COVID-19 pregnant women are calculated to be as low as 0.6% by Allotey et al. (2020) and as high as 2.4% by Di Mascio et al. (2020a). Hospitalized pregnant women have a higher stillbirth rate, which could reach 3% (Panagiotakopoulos et al., 2020). Possible explanations of stillbirths in COVID-19 patients could be (1) the aforementioned uteroplacental changes, maternal hypoxia, placental and umbilical thrombosis, and infarctions, (2) the transient elevation of lupus anticoagulant observed in COVID-19 patients, (3) the severity of maternal illness, and (4) inadequate antenatal care, due to lockdowns, lack of face-to-face services, or fear of patients to visit healthcare facilities during the pandemic. Stillbirths being probably explained by the aforementioned uteroplacental changes, which could be more advanced as the maternal disease is more severe, hence the higher still birth rates amidst hospitalized pregnant women (Bowles et al., 2020; Devreese et al., 2020; Helms et al., 2020; KC et al., 2020; Khalil et al., 2020b; Reyes Gil et al., 2020).

Congenital anomalies/birth defects: There is no increase in frequency of birth defects in newborns of women with COVID-19 during pregnancy compared with non-COVID-19 mothers. Woodworth et al. (2020) report a frequency of 28 neonates with birth defects out of 4447 live births (0.6%) to mothers with COVID-19. However, till now, data on patients who were infected in their first trimester are still limited.

LBWs: The World Health Organization (2014) classically defines LBW as "weight at birth less than 2500 g."

Like fetal distress, LBW rates show a wide discrepancy between currently available systematic reviews with rates ranging from around 3.6%–42.8%. Reviews with a small number of neonates are mainly those with high rates of LBW, so perhaps the small sample of the represented neonatal population could be the cause. Dubey et al. (2020) in their analysis report an LBW rate of 7.5% (41/548); however, when publications are divided into two main categories, case series and case reports, a discrepancy is seen with LBW rates of 5.3% (27/505) and 32.6% (14/43), respectively. The main contributing factor to LBWs of these neonates is prematurity, as suggested by Smith et al. (2020). Table 6.7 summarizes rates of LBW among neonates born to mothers with COVID-19 (Chang et al., 2020; Dubey et al., 2020; Elshafeey et al., 2020; Juan et al., 2020; Muhidin et al., 2020; Smith et al., 2020; Trocado et al., 2020).

Low/abnormal Apgar scores: Since first proposed by Virginia Apgar in 1953, the Apgar scoring system—now well established—has been a quick and efficient quantitative way for early assessment of neonates following delivery. Hence, one of the aspects to assess the probable effect of any maternal illness on pregnancy outcomes—especially neonatal outcomes—is to evaluate its effect on the Apgar score (Apgar, 1953, 2015; Finster and Wood, 2005). Since Apgar scores are generally lower in cases of fetal distress and prematurity, it is an obvious finding that when critically ill mothers are delivered prematurely due to fetal distress, their neonates would have abnormal scores (Pettirosso et al., 2020; Turan et al., 2020). As other neonatal outcomes, still some controversy exists; with most figures showing no abnormalities in Apgar scores of neonates delivered to diseased mothers and average scores of 8.8 and 9.5 at 1 and 5 min, respectively (Ashraf et al., 2020; Chang et al., 2020; Juan et al., 2020; Matar et al., 2021; Mustafa and Selim, 2020; Smith et al., 2020; Trocado et al., 2020). An even larger systematic review including 500 neonates assessed with the Apgar scoring system at 5 min shows only 11 (2.2%) had scores less than 7 (Allotey et al., 2020). Collectively, Apgar scores seem to be affected in cases of fetal distress, prematurity, and severe/critical maternal disease.

Neonatal asphyxia: Factors predisposing to neonatal asphyxia include maternal hypoxia due to severe illness, fetal distress, and prematurity. Rates are reported to be around 0.6% and 1.8% of live born neonates in systematic reviews with reasonable number of neonates included (Juan et al., 2020; Yoon et al., 2020). Some extremes are also reported; several systematic reviews have found no cases at all (Di Mascio et al., 2020a; Matar et al., 2021; Trocado et al., 2020), while others show rates of 7.7% and 13% (Capobianco et al., 2020; Trippella et al., 2020). In the aforementioned reviews, neonatal asphyxia does not seem to be caused by neonatal COVID-19; Yoon et al. (2020) reported their 168 neonates to have one case of asphyxia out of their four SARS-CoV-2 positive neonates, with a total rate of 1.8% (3/168 neonates).

NICU admission: Overall, causes of ICU admissions of neonates born to COVID-19 mothers mostly are due to either prematurity, or the sole need for neonatal isolation—especially early in the pandemic when very little was understood regarding the new disease—while neonatal SARS-CoV-2 infection per se is not the actual cause necessitating NICU admission (Papapanou et al., 2021; Turan et al., 2020; Woodworth et al., 2020). Therefore, reports widely range in rates of NICU admissions. The largest available report till now is that of Woodworth et al. (2020), in which 279 term live-born neonates were admitted to the NICU out of 2995 with an admission rate of 9.3%, followed by the estimation of Allotey et al. (2020) with 1348 neonates of which 368 (27.3%) were admitted. Prematurity is the commonest cause of NICU admission—regardless of the SARS-CoV-2 neonatal status—where preterm SARS-CoV-2-negative neonates accounted for 96.3% (52/54) of the total NICU admissions in the systematic review of Turan et al. (2020). Dhir et al. (2020) estimated the admission rate solely among SARS-CoV-2-positive neonates and was 38% (22/58 positive neonates). Overall, admission rate to the NICU ranges between 3.1% and 76.9%; the large disparity is due to presence of studies with very small numbers of neonates and due to lack of universal standardized indications for NICU admission in neonates born to SARS-CoV-2 positive mothers, not to mention that a lot of studies do not mention the cause for NICU admission (Allotey et al., 2020; Della Gatta et al., 2020; Dhir

et al., 2020; Di Mascio et al., 2020a; Diriba et al., 2020; Elshafeey et al., 2020; Huntley et al., 2020; Juan et al., 2020; Matar et al., 2021; Smith et al., 2020; Turan et al., 2020; Zaigham and Andersson, 2020).

Neonatal mortality: Neonatal mortalities in cases maternal COVID-19 infection do not exceed 3.2%. Conclusions from available systematic reviews and metaanalyses suggest that mortalities are not related to neonatal SARS-CoV-2 positivity. Risk factors that could contribute to neonatal mortality are prematurity—on the top of the list—and the severity of maternal condition (Di Mascio et al., 2020b; Hessami et al., 2020). Table 6.8 provides neonatal mortalities according to different studies (Allotey et al., 2020; Capobianco et al., 2020; Chang et al., 2020; Juan et al., 2020; Kasraeian et al., 2020; Khalil et al., 2020a; Matar et al., 2021; Muhidin et al., 2020; Trippella et al., 2020; Trocado et al., 2020; Zaigham and Andersson, 2020).

In summary, fetal and neonatal outcomes appear to be favorable apart from higher incidence of prematurity, and higher NICU admissions, the latter possibly explained by prematurity and extraprecautionary measures (Papapanou et al., 2021).

Rare Reported Outcomes

Preeclampsia-like syndrome: Hosier et al. (2020) have described a 22-week-pregnant lady who contracted SARS-CoV-2 during pregnancy who developed hypertension, proteinuria, elevated transaminases, thrombocytopenia, and hypofibrinogenemia consistent with DIC, and hence was diagnosed as severe PE and terminated. Histopathological examination of the placenta revealed the COVID-19-related changes of histiocytic intervillositis and fibrin deposition, but lacked classic PE vasculopathy. Mendoza et al. (2020) name this condition "preeclampsia-like syndrome" and describe it as a PE-like condition developing in pregnant women infected with COVID-19 and that can resolve after resolution of manifestations of COVID-19. In their prospective study, 42 SARS-CoV-2 pregnant ladies were classified—as per presence of severe pneumonia—as severe and nonsevere COVID-19, with 8 and 34 ladies, respectively. Any lady suspected to have PE was evaluated for uterine artery pulsatility index (UtAPI) and sFlt-1/PlGF (soluble fms-like tyrosine kinase-1/placental growth factor) ratio. Of the eight with severe COVID-19, five developed PE. Abnormal sFlt-1/PlGF and UtAPI, which are predictors for PE, were confirmed only in one case, while the other four had normal parameters. One case continued her pregnancy after recovery from her severe pneumonia and her preeclampsia-like syndrome resolved (Hosier et al., 2020; Mendoza et al., 2020; Zeisler et al., 2016).

TABLE 6.8
Frequency and Percent of Neonatal Mortality in Neonates Born to COVID-19 Infected Mothers.

Review Study (First Author, Year)	Neonatal Death (Number)	Total Neonates (Number)	Neonatal Death (Percent) (%)
Allotey (2020)	6	1728	0.3
Capobianco (2020)	2	108	1.9
Chang (2020)	0	92	0.0
Juan (2020)	1	221	0.5
Kasraeian (2020)	0	86	0.0
Khalil et al. (2020a)	4	668	0.6
Matar et al. (2021)	3	94	3.2
Muhidin (2020)	2	89	2.2
Trippella (2020)	1	248	0.4
Trocado (2020)	1	51	2.0
Zaigham (2020)	1	87	1.1

Transient fetal skin edema: Garcia-Manau et al. (2020) describe a rare finding of two pregnant cases diagnosed with COVID-19: the first was severe and required ICU admission and mechanical ventilation, while the other one had mild symptoms, who were diagnosed by positive RT-PCR nasopharyngeal and oropharyngeal swabs for SARS-CoV-2. On fetal ultrasound examination, both revealed isolated fetal skin edema, without any other manifestations of fetal hydrops—fetal ascites, hydrothorax, placentomegaly. No other anomalies were evident, with normal fetal heart rate, echocardiography, and Doppler indices. Maternal serological tests for TORCH infections and antibodies were negative, and amniocentesis was also negative for known TORCH infections in addition to negative SARS-CoV-2 RT-PCR. Fetal skin edema was self-limited and resolved spontaneously on follow-up scans. Possible explanations could be congenital SARS-CoV-2 infection, effect of the maternal general condition, viral-induced immunological response, or mere coincidence.

REFERENCES

Afshar, Y., Gaw, S.L., Flaherman, V.J., Chambers, B.D., Krakow, D., Berghella, V., Shamshirsaz, A.A., Boatin, A.A., Aldrovandi, G., Greiner, A., Riley, L., Boscardin, W.J., Jamieson, D.J., Jacoby, V.L., 2020. Clinical presentation of coronavirus disease 2019 (COVID-19) in pregnant and recently pregnant people. Obstet. Gynecol. 136, 1117−1125.

Ahlberg, M., Neovius, M., Saltvedt, S., Söderling, J., Pettersson, K., Brandkvist, C., Stephansson, O., 2020. Association of SARS-CoV-2 test status and pregnancy outcomes. J. Am. Med. Assoc. 324, 1782.

Algarroba, G.N., Rekawek, P., Vahanian, S.A., Khullar, P., Palaia, T., Peltier, M.R., Chavez, M.R., Vintzileos, A.M., 2020. Visualization of severe acute respiratory syndrome coronavirus 2 invading the human placenta using electron microscopy. Am. J. Obstet. Gynecol. 223, 275−278.

Allotey, J., Stallings, E., Bonet, M., Yap, M., Chatterjee, S., Kew, T., Debenham, L., Llavall, A.C., Dixit, A., Zhou, D., Balaji, R., Lee, S.I., Qiu, X., Yuan, M., Coomar, D., Van Wely, M., Van Leeuwen, E., Kostova, E., Kunst, H., Khalil, A., Tiberi, S., Brizuela, V., Broutet, N., Kara, E., Kim, C.R., Thorson, A., Oladapo, O.T., Mofenson, L., Zamora, J., Thangaratinam, S., 2020. Clinical manifestations, risk factors, and maternal and perinatal outcomes of coronavirus disease 2019 in pregnancy: living systematic review and meta-analysis. BMJ 370, m3320.

Alrahmani, L., Willrich, M.A.V., 2018. The complement alternative pathway and preeclampsia. Curr. Hypertens. Rep. 20, 1−8.

Amaral, W.N.D., Moraes, C.L.D., Rodrigues, A.P.D.S., Noll, M., Arruda, J.T., Mendonça, C.R., 2020. Maternal coronavirus infections and neonates born to mothers with SARS-CoV-2: a systematic review. Healthcare 8, 511.

Apgar, V., 1953. A proposal for a new method of evaluation of the newborn infant. Curr. Res. Anesth. Analg. 32, 260−267.

Apgar, V., 2015. A proposal for a new method of evaluation of the newborn infant. Anesth. Analg. 120, 1056−1059.

Ashraf, M.A., Keshavarz, P., Hosseinpour, P., Erfani, A., Roshanshad, A., Pourdast, A., Nowrouzi-Sohrabi, P., Chaichian, S., Poordast, T., 2020. Coronavirus disease 2019 (COVID-19): a systematic review of pregnancy and the possibility of vertical transmission. J. Reproduction Infertil. 21, 157−168.

Badr, D.A., Mattern, J., Carlin, A., Cordier, A.-G., Maillart, E., El Hachem, L., El Kenz, H., Andronikof, M., De Bels, D., Damoisel, C., Preseau, T., Vignes, D., Cannie, M.M., Vauloup-Fellous, C., Fils, J.-F., Benachi, A., Jani, J.C., Vivanti, A.J., 2020. Are clinical outcomes worse for pregnant women at ≥20 weeks' gestation infected with coronavirus disease 2019? A multicenter case-control study with propensity score matching. Am. J. Obstet. Gynecol. 223, 764–768.

Bastug, A., Hanifehnezhad, A., Tayman, C., Ozkul, A., Ozbay, O., Kazancioglu, S., Bodur, H., 2020. Virolactia in an asymptomatic mother with COVID-19. Breastfeed. Med. 15, 488–491.

Baud, D., Greub, G., Favre, G., Gengler, C., Jaton, K., Dubruc, E., Pomar, L., 2. Second-trimester miscarriage in a pregnant woman with SARS-CoV-2 infection. J. Am. Med. Assoc. 323, 2198–2200.

Begbie, M., Notley, C., Tinlin, S., Sawyer, L., Lillicrap, D., 2000. The factor VIII acute phase response requires the participation of NFkappaB and C/EBP. Thromb. Haemostasis 84, 216–222.

Bellos, I., Pandita, A., Panza, R., 2021. Maternal and perinatal outcomes in pregnant women infected by SARS-CoV-2: a meta-analysis. Eur. J. Obstet. Gynecol. Reprod. Biol. 256, 194–204.

Berghella, V., Hughes, B., 2021. Coronavirus Disease 2019 (COVID-19): Pregnancy Issues and Antenatal Care. UpToDate. URL: https://www.uptodate.com/contents/coronavirus-disease-2019-covid-19-pregnancy-issues-and-antenatal-care#H1012270637. (Accessed 20 January 2021).

Bhattacharjee, S., Banerjee, M., 2020. Immune thrombocytopenia secondary to COVID-19: a systematic review. SN Compr. Clin. Med. 2, 2048–2058.

Blumberg, D.A., Underwood, M.A., Hedriana, H.L., Lakshminrusimha, S., 2020. Vertical transmission of SARS-CoV-2: what is the optimal definition? Am. J. Perinatol. 37, 769–772.

Boerma, T., Ronsmans, C., Melesse, D.Y., Barros, A.J.D., Barros, F.C., Juan, L., Moller, A.-B., Say, L., Hosseinpoor, A.R., Yi, M., de Lyra Rabello Neto, D., Temmerman, M., 2018. Global epidemiology of use of and disparities in caesarean sections. Lancet 392, 1341–1348.

Bowles, L., Platton, S., Yartey, N., Dave, M., Lee, K., Hart, D.P., MacDonald, V., Green, L., Sivapalaratnam, S., Pasi, K.J., MacCallum, P., 2020. Lupus anticoagulant and abnormal coagulation tests in patients with Covid-19. N. Engl. J. Med. 383, 288–290.

Breslin, N., Baptiste, C., Gyamfi-Bannerman, C., Miller, R., Martinez, R., Bernstein, K., Ring, L., Landau, R., Purisch, S., Friedman, A.M., Fuchs, K., Sutton, D., Andrikopoulou, M., Rupley, D., Sheen, J.-J., Aubey, J., Zork, N., Moroz, L., Mourad, M., Wapner, R., Simpson, L.L., D'Alton, M.E., Goffman, D., 2020. Coronavirus disease 2019 infection among asymptomatic and symptomatic pregnant women: two weeks of confirmed presentations to an affiliated pair of New York city hospitals. Am. J. Obstet. Gynecol. MFM 2, 100118.

Campbell, L.A., Klocke, R.A., 2001. Implications for the pregnant patient. Am. J. Respir. Crit. Care Med. 163, 1051–1054.

Caparros-Gonzalez, R.A., Pérez-Morente, M.A., Hueso-Montoro, C., Álvarez-Serrano, M.A., de la Torre-Luque, A., 2020. Congenital, intrapartum and postnatal maternal-fetal-neonatal sars-cov-2 infections: a narrative review. Nutrients 12, 1–15.

Capobianco, G., Saderi, L., Aliberti, S., Mondoni, M., Piana, A., Dessole, F., Dessole, M., Cherchi, P.L., Dessole, S., Sotgiu, G., 2020. COVID-19 in pregnant women: a systematic review and meta-analysis. Eur. J. Obstet. Gynecol. Reprod. Biol. 215, 153–163.

Caputo, N.D., Strayer, R.J., Levitan, R., 2020. Early self-proning in awake, non-intubated patients in the emergency department: a single ED's experience during the COVID-19 pandemic. Acad. Emerg. Med. 27, 375–378.

Centers for Disease Control and Prevention, 2020. COVID-19 Pandemic Planning Scenarios. URL: https://www.cdc.gov/coronavirus/2019-ncov/hcp/planning-scenarios.html. (Accessed 26 January 2021).

Chambers, C., Krogstad, P., Bertrand, K., Contreras, D., Tobin, N.H., Bode, L., Aldrovandi, G., 2020. Evaluation for SARS-CoV-2 in breast milk from 18 infected women. J. Am. Med. Assoc. 324, 1347.

Chang, T.H., Wu, J.L., Chang, L.Y., 2020. Clinical characteristics and diagnostic challenges of pediatric COVID-19: a systematic review and meta-analysis. J. Formos. Med. Assoc. 119, 982–989.

Chen, D., Yang, H., Cao, Y., Cheng, W., Duan, T., Fan, C., Fan, S., Feng, L., Gao, Y., He, F., He, J., Hu, Y., Jiang, Y., Li, Y., Li, J., Li, X., Li, X., Lin, K., Liu, C., Liu, J., Liu, X., Pan, X., Pang, Q., Pu, M., Qi, H., Shi, C., Sun, Y., Sun, J., Wang, X., Wang, Y., Wang, Z., Wang, Z., Wang, C., Wu, S., Xin, H., Yan, J., Zhao, Y., Zheng, J., Zhou, Y., Zou, L., Zeng, Y., Zhang, Y., Guan, X., 2020a. Expert consensus for managing pregnant women and neonates born to mothers with suspected or confirmed novel coronavirus (COVID-19) infection. Int. J. Gynecol. Obstet. 149, 130–136.

Chen, H., Guo, J., Wang, C., Luo, F., Yu, X., Zhang, W., Li, J., Zhao, D., Xu, D., Gong, Q., Liao, J., Yang, H., Hou, W., Zhang, Y., 2020b. Clinical characteristics and intrauterine vertical transmission potential of COVID-19 infection in nine pregnant women: a retrospective review of medical records. Lancet 395, 809–815.

Chen, N., Zhou, M., Dong, X., Qu, J., Gong, F., Han, Y., Qiu, Y., Wang, J., Liu, Y., Wei, Y., Xia, J., Yu, T., Zhang, X., Zhang, L., 2020c. Epidemiological and clinical characteristics of 99 cases of 2019 novel coronavirus pneumonia in Wuhan, China: a descriptive study. Lancet 395, 507–513.

Chen, S., Huang, B., Luo, D.J., Li, X., Yang, F., Zhao, Y., Nie, X., Huang, B.X., 2020d. Pregnancy with new coronavirus infection: a clinical characteristics and placental pathological analysis of three cases. Zhonghua Bing Li Xue Za Zhi 49, 418–423.

Chen, S., Liao, E., Cao, D., Gao, Y., Sun, G., Shao, Y., 2020e. Clinical analysis of pregnant women with 2019 novel coronavirus pneumonia. J. Med. Virol. 92, 1556–1561.

Dashraath, P., Wong, J.L.J., Lim, M.X.K., Lim, L.M., Li, S., Biswas, A., Choolani, M., Mattar, C., Su, L.L., 2020. Coronavirus disease 2019 (COVID-19) pandemic and pregnancy. Am. J. Obstet. Gynecol. 222, 521–531.

de Lorenzo, A., Kasal, D.A.B., Tura, B.R., da Cruz Lamas, C., Rey, H.C.V., 2020. Acute cardiac injury in patients with COVID-19. Am. J. Cardiovasc. Dis. 10, 28–33.

Delahoy, M.J., Whitaker, M., O'Halloran, A., Chai, S.J., Kirley, P.D., Alden, N., Kawasaki, B., Meek, J., Yousey-Hindes, K., Anderson, E.J., Openo, K.P., Monroe, M.L., Ryan, P.A., Fox, K., Kim, S., Lynfield, R., Siebman, S., Davis, S.S., Sosin, D.M., Barney, G., Muse, A., Bennett, N.M., Felsen, C.B., Billing, L.M., Shiltz, J., Sutton, M., West, N., Schaffner, W., Talbot, H.K., George, A., Spencer, M., Ellington, S., Galang, R.R., Gilboa, S.M., Tong, V.T., Piasecki, A., Brammer, L., Fry, A.M., Hall, A.J., Wortham, J.M., Kim, L., Garg, S., Apostol, M., Brooks, S., Coates, A., Frank, L., Heidenga, B., Hundal, K., Nadle, J., Quach, S., Roland, J., Rosales, M., Armistead, I., Herlihy, R., McLafferty, H., Misiorski, A., Parisi, C., Olson, D., Lyons, C., Maslar, A., Clogher, P., Blythe, D., Brooks, A., Park, R., Wilson, M., Bye, E., Como-Sabetti, K., Danila, R., Sullivan, M., Angeles, K.M., Christian, M., Eisenberg, N., Habrun, C., Hancock, E.B., Khanlian, S.A., Novi, M., Salazar-Sanchez, Y., Dufort, E., Spina, N., Owusu-Dommey, A., Markus, T., Chatelain, R., McCullough, L., Ortega, J., Price, A., Swain, A., Kambhampati, A., Meador, S., 2020. Characteristics and maternal and birth outcomes of hospitalized pregnant women with laboratory-confirmed COVID-19 — COVID-NET, 13 states, March 1–August 22, 2020. Morb. Mortal. Wkly. Rep. 69, 1347–1354.

Della Gatta, A.N., Rizzo, R., Pilu, G., Simonazzi, G., 2020. Coronavirus disease 2019 during pregnancy: a systematic review of reported cases. Am. J. Obstet. Gynecol. 223, 36–41.

Devreese, K.M.J., Linskens, E.A., Benoit, D., Peperstraete, H., 2020. Antiphospholipid antibodies in patients with COVID-19: a relevant observation? J. Thromb. Haemostasis 18, 2191–2201.

Dhir, S.K., Kumar, J., Meena, J., Kumar, P., 2020. Clinical features and outcome of SARS-CoV-2 infection in neonates: a systematic review. J. Trop. Pediatr. 1–14.

Di Mascio, D., Khalil, A., Saccone, G., Rizzo, G., Buca, D., Liberati, M., Vecchiet, J., Nappi, L., Scambia, G., Berghella, V., D'Antonio, F., 2020a. Outcome of coronavirus spectrum infections (SARS, MERS, COVID-19) during pregnancy: a systematic review and meta-analysis. Am. J. Obstet. Gynecol. MFM 2, 100107.

Di Mascio, D., Sen, C., Saccone, G., Galindo, A., Grünebaum, A., Yoshimatsu, J., Stanojevic, M., Kurjak, A., Chervenak, F., 2020b. Risk factors associated with adverse fetal outcomes in pregnancies affected by Coronavirus disease 2019 (COVID-19): a secondary analysis of the WAPM study on COVID-19. J. Perinat. Med. 48, 950–958.

Di Toro, F., Gjoka, M., Di Lorenzo, G., De Santo, D., De Seta, F., Maso, G., Risso, F.M., Romano, F., Wiesenfeld, U., Levi-D'Ancona, R., Ronfani, L., Ricci, G., 2021. Impact of COVID-19 on maternal and neonatal outcomes: a systematic review and meta-analysis. Clin. Microbiol. Infect. 27, 36–46.

Diriba, K., Awulachew, E., Getu, E., 2020. The effect of coronavirus infection (SARS-CoV-2, MERS-CoV, and SARS-CoV) during pregnancy and the possibility of vertical maternal-fetal transmission: a systematic review and meta-analysis. Eur. J. Med. Res. 25, 39.

Dubey, P., Reddy, S.Y., Manuel, S., Dwivedi, A.K., 2020. Maternal and neonatal characteristics and outcomes among COVID-19 infected women: an updated systematic review and meta-analysis. Eur. J. Obstet. Gynecol. Reprod. Biol. 252, 490–501.

Ellington, S., Strid, P., Tong, V.T., Woodworth, K., Galang, R.R., Zambrano, L.D., Nahabedian, J., Anderson, K., Gilboa, S.M., 2020. Characteristics of women of reproductive age with laboratory-confirmed SARS-CoV-2 infection by pregnancy status — United States, January 22–June 7, 2020. Morb. Mortal. Wkly. Rep. 69, 769–775.

Elshafeey, F., Magdi, R., Hindi, N., Elshebiny, M., Farrag, N., Mahdy, S., Sabbour, M., Gebril, S., Nasser, M., Kamel, M., Amir, A., Maher Emara, M., Nabhan, A., 2020. A systematic scoping review of COVID-19 during pregnancy and childbirth. Int. J. Gynecol. Obstet. 150, 47–52.

European Medicines Agency, 2020. Update on Remdesivir - EMA Will Evaluate New Data from Solidarity Trial. URL: https://www.ema.europa.eu/en/news/update-remdesivir-ema-will-evaluate-new-data-solidarity-trial. (Accessed 28 January 2021).

Facchetti, F., Bugatti, M., Drera, E., Tripodo, C., Sartori, E., Cancila, V., Papaccio, M., Castellani, R., Casola, S., Boniotti, M.B., Cavadini, P., Lavazza, A., 2020. SARS-CoV2 vertical transmission with adverse effects on the newborn revealed through integrated immunohistochemical, electron microscopy and molecular analyses of placenta. EBioMedicine 59, 102951.

Farkash, E.A., Wilson, A.M., Jentzen, J.M., 2020. Ultrastructural evidence for direct renal infection with SARS-CoV-2. J. Am. Soc. Nephrol. 31, 1683–1687.

Fermin, G., 2018. Host range, host–virus interactions, and virus transmission. In: Tennant, P., Fermin, G., Foster, J.E. (Eds.), Viruses: Molecular Biology, Host Interactions, and Applications to Biotechnology. Academic Press, Cambridge, pp. 101–134.

Ferrario, C.M., Trask, A.J., Jessup, J.A., 2005. Advances in biochemical and functional roles of angiotensin-converting enzyme 2 and angiotensin-(1-7) in regulation of cardiovascular function. Am. J. Physiol. Heart Circ. Physiol. 289, H2281–H2290.

Figueiro-Filho, E.A., Yudin, M., Farine, D., 2020. COVID-19 during pregnancy: an overview of maternal characteristics, clinical symptoms, maternal and neonatal outcomes of 10,996 cases described in 15 countries. J. Perinat. Med. 48, 900–911.

Finster, M., Wood, M., 2005. The apgar score has survived the test of time. Anesthesiology 102, 855–857.

Futterman, I., Toaff, M., Navi, L., Clare, C.A., 2020. COVID-19 and HELLP: overlapping clinical pictures in two gravid patients. AJP Rep. 10, e179–e182.

Garcia-Manau, P., Garcia-Ruiz, I., Rodo, C., Sulleiro, E., Maiz, N., Catalan, M., Fernández-Hidalgo, N., Balcells, J., Antón, A., Carreras, E., Suy, A., 2020. Fetal transient skin edema in two pregnant women with coronavirus disease 2019 (COVID-19). Obstet. Gynecol. 136, 1016–1020.

Gengler, C., Dubruc, E., Favre, G., Greub, G., de Leval, L., Baud, D., 2021. SARS-CoV-2 ACE-receptor detection in the placenta throughout pregnancy. Clin. Microbiol. Infect. 27, 489–490.

Gracia-Perez-Bonfils, A., Martinez-Perez, O., Llurba, E., Chandraharan, E., 2020. Fetal heart rate changes on the cardiotocograph trace secondary to maternal COVID-19 infection. Eur. J. Obstet. Gynecol. Reprod. Biol. 252, 286–293.

Guo, T., Fan, Y., Chen, M., Wu, X., Zhang, L., He, T., Wang, H., Wan, J., Wang, X., Lu, Z., 2020. Cardiovascular implications of fatal outcomes of patients with coronavirus disease 2019 (COVID-19). JAMA Cardiol. 5, 811–818.

Hamilton, B.E., Martin, J.A., Osterman, M.J.K., Driscoll, A.K., Rossen, L.M., 2020. Births: Provisional Data for 2019. Natl. Cent. Heal. Stat., Hyattsville, MD. URL: https://www.cdc.gov/nchs/data/vsrr/vsrr-8-508.pdf. (Accessed 24 January 2021).

Han, Y., Ma, H., Suo, M., Han, F., Wang, F., Ji, J., Ji, J., Yang, H., 2020. Clinical manifestation, outcomes in pregnant women with COVID-19 and the possibility of vertical transmission: a systematic review of the current data. J. Perinat. Med. 48, 912–924.

Hanna, N., Hanna, M., Sharma, S., 2020. Is pregnancy an immunological contributor to severe or controlled COVID-19 disease? Am. J. Reprod. Immunol. 84, e13317.

Hantoushzadeh, S., Shamshirsaz, A.A., Aleyasin, A., Seferovic, M.D., Aski, S.K., Arian, S.E., Pooransari, P., Ghotbizadeh, F., Aalipour, S., Soleimani, Z., Naemi, M., Molaei, B., Ahangari, R., Salehi, M., Oskoei, A.D., Pirozan, P., Darkhaneh, R.F., Laki, M.G., Farani, A.K., Atrak, S., Miri, M.M., Kouchek, M., Shojaei, S., Hadavand, F., Keikha, F., Hosseini, M.S., Borna, S., Ariana, S., Shariat, M., Fatemi, A., Nouri, B., Nekooghadam, S.M., Aagaard, K., 2020. Maternal death due to COVID-19. Am. J. Obstet. Gynecol. 223, 109.e1–109.e16.

Hecht, J.L., Quade, B., Deshpande, V., Mino-Kenudson, M., Ting, D.T., Desai, N., Dygulska, B., Heyman, T., Salafia, C., Shen, D., Bates, S.V., Roberts, D.J., 2020. SARS-CoV-2 can infect the placenta and is not associated with specific placental histopathology: a series of 19 placentas from COVID-19+ mothers. Mod. Pathol. 33, 2092–2103.

Helms, J., Tacquard, C., Severac, F., Leonard-Lorant, I., Ohana, M., Delabranche, X., Merdji, H., Clere-Jehl, R., Schenck, M., Fagot Gandet, F., Fafi-Kremer, S., Castelain, V., Schneider, F., Grunebaum, L., Anglés-Cano, E., Sattler, L., Mertes, P.-M., Meziani, F., CRICS TRIGGERSEP Group (Clinical Research in Intensive Care and Sepsis Trial Group for Global Evaluation and Research in Sepsis), 2020. High risk of thrombosis in patients with severe SARS-CoV-2 infection: a multicenter prospective cohort study. Intensive Care Med. 46, 1089–1098.

Hendren, N.S., Drazner, M.H., Bozkurt, B., Cooper, L.T., 2020. Description and proposed management of the acute COVID-19 cardiovascular syndrome. Circulation 141, 1903–1914.

Hessami, K., Homayoon, N., Hashemi, A., Vafaei, H., Kasraeian, M., Asadi, N., 2020. COVID-19 and maternal, fetal and neonatal mortality: a systematic review. J. Matern. Fetal Neonatal Med. On line ahead of print, 1–6.

Hikmet, F., Méar, L., Edvinsson, Å., Micke, P., Uhlén, M., Lindskog, C., 2020. The protein expression profile of ACE2 in human tissues. Mol. Syst. Biol. 16, e9610.

Hosier, H., Farhadian, S.F., Morotti, R.A., Deshmukh, U., Lu-Culligan, A., Campbell, K.H., Yasumoto, Y., Vogels, C.B.F., Casanovas-Massana, A., Vijayakumar, P., Geng, B., Odio, C.D., Fournier, J., Brito, A.F., Fauver, J.R., Liu, F., Alpert, T., Tal, R., Szigeti-Buck, K., Perincheri, S., Larsen, C., Gariepy, A.M., Aguilar, G., Fardelmann, K.L., Harigopal, M., Taylor, H.S., Pettker, C.M., Wyllie, A.L., Dela Cruz, C., Ring, A.M., Grubaugh, N.D., Ko, A.I., Horvath, T.L., Iwasaki, A., Reddy, U.M., Lipkind, H.S., 2020. SARS-CoV-2 infection of the placenta. J. Clin. Invest. 130, 4947–4953.

Howson, C.P., Kinney, M.V., McDougall, L., Lawn, J.E., 2013. Born too soon: preterm birth matters. Reprod. Health 10, S1.

Hsu, A.L., Guan, M., Johannesen, E., Stephens, A.J., Khaleel, N., Kagan, N., Tuhlei, B.C., Wan, X.F., 2021. Placental SARS-CoV-2 in a pregnant woman with mild COVID-19 disease. J. Med. Virol. 93, 1038–1044.

Huntley, B.J.F., Huntley, E.S., Di Mascio, D., Chen, T., Berghella, V., Chauhan, S.P., 2020. Rates of maternal and perinatal mortality and vertical transmission in pregnancies complicated by severe acute respiratory syndrome coronavirus 2 (SARS-Co-V-2) infection. Obstet. Gynecol. 136, 303–312.

James, A.H., 2010. Pregnancy and thrombotic risk. Crit. Care Med. 38, S57–S63.

Jose, R.J., Manuel, A., 2020. COVID-19 cytokine storm: the interplay between inflammation and coagulation. Lancet Respir. Med. 8, e46–e47.

Juan, J., Gil, M.M., Rong, Z., Zhang, Y., Yang, H., Poon, L.C., 2020. Effects of coronavirus disease 2019 (COVID-19) on maternal, perinatal and neonatal outcomes: a systematic review. Ultrasound Obstet. Gynecol. 56, 15–27.

Juusela, A., Nazir, M., Gimovsky, M., 2020. Two cases of coronavirus 2019–related cardiomyopathy in pregnancy. Am. J. Obstet. Gynecol. MFM 2, 100113.

Kasraeian, M., Zare, M., Vafaei, H., Asadi, N., Faraji, A., Bazrafshan, K., Roozmeh, S., 2020. COVID-19 pneumonia and pregnancy; a systematic review and meta-analysis. J. Matern. Fetal Neonatal Med. Online ahead of print, 1–8.

KC, A., Gurung, R., Kinney, M.V., Sunny, A.K., Moinuddin, M., Basnet, O., Paudel, P., Bhattarai, P., Subedi, K., Shrestha, M.P., Lawn, J.E., Mälqvist, M., 2020. Effect of the COVID-19 pandemic response on intrapartum care, stillbirth, and neonatal mortality outcomes in Nepal: a prospective observational study. Lancet Glob. Health 8, e1273–e1281.

Khalil, A., Kalafat, E., Benlioglu, C., O'Brien, P., Morris, E., Draycott, T., Thangaratinam, S., Le Doare, K., Heath, P., Ladhani, S., von Dadelszen, P., Magee, L.A., 2020a. SARS-CoV-2 infection in pregnancy: a systematic review and meta-analysis of clinical features and pregnancy outcomes. EClinicalMedicine 25, 100446.

Khalil, A., Von Dadelszen, P., Draycott, T., Ugwumadu, A., O'Brien, P., Magee, L., 2020b. Change in the incidence of stillbirth and preterm delivery during the COVID-19 pandemic. J. Am. Med. Assoc. 324, 705–706.

Khan, S., Peng, L., Siddique, R., Nabi, G., Nawsherwan, Xue, M., Liu, J., Han, G., 2020. Impact of COVID-19 infection on pregnancy outcomes and the risk of maternal-to-neonatal intrapartum transmission of COVID-19 during natural birth. Infect. Control Hosp. Epidemiol. 41, 748–750.

Khoury, R., Bernstein, P.S., Debolt, C., Stone, J., Sutton, D.M., Simpson, L.L., Limaye, M.A., Roman, A.S., Fazzari, M., Penfield, C.A., Ferrara, L., Lambert, C., Nathan, L., Wright, R., Bianco, A., Wagner, B., Goffman, D., Gyamfi-Bannerman, C., Schweizer, W.E., Avila, K., Khaksari, B., Proehl, M., Heitor, F., Monro, J., Keefe, D.L., D'Alton, M.E., Brodman, M., Makhija, S.K., Dolan, S.M., 2020. Characteristics and outcomes of 241 births to women with severe acute respiratory syndrome coronavirus 2 (SARS-CoV-2) infection at five New York City medical centers. Obstet. Gynecol. 136, 273–282.

Kirtsman, M., Diambomba, Y., Poutanen, S.M., Malinowski, A.K., Vlachodimitropoulou, E., Parks, W.T., Erdman, L., Morris, S.K., Shah, P.S., 2020. Probable congenital SARS-CoV-2 infection in a neonate born to a woman with active SARS-CoV-2 infection. Can. Med. Assoc. J. 192, E647–E650.

Knight, M., Bunch, K., Vousden, N., Morris, E., Simpson, N., Gale, C., O'Brien, P., Quigley, M., Brocklehurst, P., Kurinczuk, J.J., 2020. Characteristics and outcomes of pregnant women admitted to hospital with confirmed SARS-CoV-2 infection in UK: national population based cohort study. BMJ 369, m2107.

Kulkarni, R., Rajput, U., Dawre, R., Valvi, C., Nagpal, R., Magdum, N., Vankar, H., Sonkawade, N., Das, A., Vartak, S., Joshi, S., Varma, S., Karyakarte, R., Bhosale, R., Kinikar, A., 2021. Early-onset symptomatic neonatal COVID-19 infection with high probability of vertical transmission. Infection 49, 339–343.

Li, M., Chen, L., Zhang, J., Xiong, C., Li, X., 2020a. The SARS-CoV-2 receptor ACE2 expression of maternal-fetal interface and fetal organs by single-cell transcriptome study. PloS One 15, e0230295.

Li, Y., Zhao, R., Zheng, S., Chen, X., Wang, J., Sheng, X., Zhou, J., Cai, H., Fang, Q., Yu, F., Fan, J., Xu, K., Chen, Y., Sheng, J., 2020b. Lack of vertical transmission of severe acute respiratory syndrome coronavirus 2, China. Emerg. Infect. Dis. 26, 1335–1336.

Libby, P., Lüscher, T., 2020. COVID-19 is, in the end, an endothelial disease. Eur. Heart J. 41, 3038–3044.

Lin, L., Wang, X., Ren, J., Sun, Y., Yu, R., Li, K., Zheng, L., Yang, J., 2020. Risk factors and prognosis for COVID-19-induced acute kidney injury: a meta-analysis. BMJ Open 10, e042573.

Lippi, G., Plebani, M., Henry, B.M., 2020. Thrombocytopenia is associated with severe coronavirus disease 2019 (COVID-19) infections: a meta-analysis. Clin. Chim. Acta 506, 145–148.

Liu, H., Liu, F., Li, J., Zhang, T., Wang, D., Lan, W., 2020a. Clinical and CT imaging features of the COVID-19 pneumonia: focus on pregnant women and children. J. Infect. 80, e7–e13.

Liu, W., Wang, J., Li, W., Zhou, Z., Liu, S., Rong, Z., 2020b. Clinical characteristics of 19 neonates born to mothers with COVID-19. Front. Med. 14, 193–198.

Long, B., Brady, W.J., Koyfman, A., Gottlieb, M., 2020. Cardiovascular complications in COVID-19. Am. J. Emerg. Med. 38, 1504–1507.

Lowenstein, C.J., Solomon, S.D., 2020. Severe COVID-19 is a microvascular disease. Circulation 142, 1609–1611.

Lugli, L., Bedetti, L., Lucaccioni, L., Gennari, W., Leone, C., Ancora, G., Berardi, A., 2020. An uninfected preterm newborn inadvertently fed SARS-CoV-2–positive breast milk. Pediatrics 146 e2020004960.

Maier, C.L., Truong, A.D., Auld, S.C., Polly, D.M., Tanksley, C.L., Duncan, A., 2020. COVID-19-associated hyperviscosity: a link between inflammation and thrombophilia? Lancet 395, 1758–1759.

Matar, R., Alrahmani, L., Monzer, N., Debiane, L.G., Berbari, E., Fares, J., Fitzpatrick, F., Murad, M.H., 2021. Clinical presentation and outcomes of pregnant women with coronavirus disease 2019: a systematic review and meta-analysis. Clin. Infect. Dis. 72, 521–533.

Mendoza, M., Garcia-Ruiz, I., Maiz, N., Rodo, C., Garcia-Manau, P., Serrano, B., Lopez-Martinez, R., Balcells, J., Fernandez-Hidalgo, N., Carreras, E., Suy, A., 2020. Preeclampsia-like syndrome induced by severe COVID-19: a prospective observational study. BJOG Int. J. Obstet. Gynaecol. 127, 1374–1380.

Mercedes, B.R., Serwat, A., Naffaa, L., Ramirez, N., Khalid, F., Steward, S.B., Feliz, O.G.C., Kassab, M.B., Karout, L., 2021. New-onset myocardial injury in pregnant patients with coronavirus disease 2019: a case series of 15 patients. Am. J. Obstet. Gynecol. 224, 387.e1–387.e9.

Ming, Y., Li, M., Dai, F., Huang, R., Zhang, J., Zhang, L., Qin, M., Zhu, L., Yu, H., Zhang, J., 2019. Dissecting the current caesarean section rate in Shanghai, China. Sci. Rep. 9, 2080.

Mitchell, A., Chiwele, I., 2021. Coronavirus Disease 2019 (COVID-19) | BMJ Best Practice. BMC Publ. Gr. URL: https://bestpractice.bmj.com/topics/en-gb/3000201. (Accessed 26 January 2021).

Mor, G., Cardenas, I., 2010. The immune system in pregnancy: a unique complexity. Am. J. Reprod. Immunol. 63, 425–433.

Muhidin, S., Behboodi Moghadam, Z., Vizheh, M., 2020. Analysis of maternal coronavirus infections and neonates born to mothers with 2019-nCoV; a systematic review. Arch. Acad. Emerg. Med. 8, e49.

Mustafa, N.M., Selim, L.A., 2020. Characterisation of COVID-19 pandemic in paediatric age group: a systematic review and meta-analysis. J. Clin. Virol. 128, 104395.

Naicker, S., Yang, C.W., Hwang, S.J., Liu, B.C., Chen, J.H., Jha, V., 2020. The Novel Coronavirus 2019 epidemic and kidneys. Kidney Int. 97, 824–828.

Nakamura-Pereira, M., Andreucci, C., Menezes, M., Knobel, R., Takemoto, M.L.S., 2020. Worldwide maternal deaths due to COVID-19: a brief review. Int. J. Gynecol. Obstet. 151, 148–150.

Narang, K., Enninga, E.A.L., Gunaratne, M.D.S.K., Ibirogba, E.R., Trad, A.T.A., Elrefaei, A., Theiler, R.N., Ruano, R., Szymanski, L.M., Chakraborty, R., Garovic, V.D., 2020. SARS-CoV-2 infection and COVID-19 during pregnancy: a multidisciplinary review. Mayo Clin. Proc. 95, 1750–1765.

National Institute for Health and Care Excellence, 2020. COVID-19 Rapid Guideline: Acute Kidney Injury in Hospital. URL: https://www.nice.org.uk/guidance/ng175/resources/covid19-rapid-guideline-acute-kidney-injury-in-hospital-pdf-66141962895301. (Accessed 28 January 2021).

National Institutes of Health, 2020. Coronavirus Disease 2019 (COVID-19) Treatment Guidelines. NIH. URL: https://www.covid19treatmentguidelines.nih.gov/whats-new/. (Accessed 26 January 2021).

Nesr, G., Garnett, C., Bailey, C., Arami, S., 2020. Immune thrombocytopenia flare with mild COVID-19 infection in pregnancy: a case report. Br. J. Haematol. 190 bjh.16928.

Pacheco, L.D., Saad, A., 2018. Ventilator management in critical illness. In: Phelan, J.P., Pacheco, L.D., Foley, M.R., Saade, G.R., Dildy, G.A., Belfort, M.A. (Eds.), Critical Care Obstetrics. John Wiley & Sons, Ltd., Chichester, UK, pp. 215–248.

Panagiotakopoulos, L., Myers, T.R., Gee, J., Lipkind, H.S., Kharbanda, E.O., Ryan, D.S., Williams, J.T.B., Naleway, A.L., Klein, N.P., Hambidge, S.J., Jacobsen, S.J., Glanz, J.M., Jackson, L.A., Shimabukuro, T.T., Weintraub, E.S., 2020. SARS-CoV-2 infection among hospitalized pregnant women: reasons for admission and pregnancy characteristics — eight U.S. Health care centers, March 1–May 30, 2020. Morb. Mortal. Wkly. Rep. 69, 1355–1359.

Panigada, M., Bottino, N., Tagliabue, P., Grasselli, G., Novembrino, C., Chantarangkul, V., Pesenti, A., Peyvandi, F., Tripodi, A., 2020. Hypercoagulability of COVID-19 patients in intensive care unit: a report of thromboelastography findings and other parameters of hemostasis. J. Thromb. Haemostasis 18, 1738–1742.

Papapanou, M., Papaioannou, M., Petta, A., Routsi, E., Farmaki, M., Vlahos, N., Siristatidis, C., 2021. Maternal and neonatal characteristics and outcomes of COVID-19 in pregnancy: an overview of systematic reviews. Int. J. Environ. Res. Publ. Health 18, 596.

Patanè, L., Morotti, D., Giunta, M.R., Sigismondi, C., Piccoli, M.G., Frigerio, L., Mangili, G., Arosio, M., Cornolti, G., 2020. Vertical transmission of coronavirus disease 2019: severe acute respiratory syndrome coronavirus 2 RNA on the fetal side of the placenta in pregnancies with coronavirus disease 2019–positive mothers and neonates at birth. Am. J. Obstet. Gynecol. MFM 2, 100145.

Penfield, C.A., Brubaker, S.G., Limaye, M.A., Lighter, J., Ratner, A.J., Thomas, K.M., Meyer, J.A., Roman, A.S., 2020. Detection of severe acute respiratory syndrome coronavirus 2 in placental and fetal membrane samples. Am. J. Obstet. Gynecol. MFM 2, 100133.

Peng, Z., Wang, J., Mo, Y., Duan, W., Xiang, G., Yi, M., Bao, L., Shi, Y., 2020. Unlikely SARS-CoV-2 vertical transmission from mother to child: a case report. J. Infect. Publ. Health 13, 818–820.

Pettirosso, E., Giles, M., Cole, S., Rees, M., 2020. COVID-19 and pregnancy: a review of clinical characteristics, obstetric outcomes and vertical transmission. Aust. N. Z. J. Obstet. Gynaecol. 60, 640–659.

Pierce-Williams, R.A.M., Burd, J., Felder, L., Khoury, R., Bernstein, P.S., Avila, K., Penfield, C.A., Roman, A.S., DeBolt, C.A., Stone, J.L., Bianco, A., Kern-Goldberger, A.R., Hirshberg, A., Srinivas, S.K., Jayakumaran, J.S., Brandt, J.S., Anastasio, H., Birsner, M., O'Brien, D.S., Sedev, H.M., Dolin, C.D., Schnettler, W.T., Suhag, A., Ahluwalia, S., Navathe, R.S., Khalifeh, A., Anderson, K., Berghella, V., 2020. Clinical course of severe and critical coronavirus disease 2019 in hospitalized pregnancies: a United States cohort study. Am. J. Obstet. Gynecol. MFM 2, 100134.

Poon, L.C., Yang, H., Kapur, A., Melamed, N., Dao, B., Divakar, H., McIntyre, H.D., Kihara, A.B., Ayres-de-Campos, D., Ferrazzi, E.M., Di Renzo, G.C., Hod, M., 2020. Global interim guidance on coronavirus disease 2019 (COVID-19) during pregnancy and puerperium from FIGO and allied partners: information for healthcare professionals. Int. J. Gynecol. Obstet. 149, 273–286.

Prasitlumkum, N., Chokesuwattanaskul, R., Thongprayoon, C., Bathini, T., Vallabhajosyula, S., Cheungpasitporn, W., 2020. Incidence of myocardial injury in COVID-19-infected patients: a systematic review and meta-analysis. Diseases 8, 40.

Pringle, K.G., Tadros, M.A., Callister, R.J., Lumbers, E.R., 2011. The expression and localization of the human placental prorenin/renin-angiotensin system throughout pregnancy: roles in trophoblast invasion and angiogenesis? Placenta 32, 956–962.

Ranucci, M., Ballotta, A., Di Dedda, U., Bayshnikova, E., Dei Poli, M., Resta, M., Falco, M., Albano, G., Menicanti, L., 2020. The procoagulant pattern of patients with COVID-19 acute respiratory distress syndrome. J. Thromb. Haemostasis 18, 1747–1751.

Reyes Gil, M., Barouqa, M., Szymanski, J., Gonzalez-Lugo, J.D., Rahman, S., Billett, H.H., 2020. Assessment of lupus anticoagulant positivity in patients with coronavirus disease 2019 (COVID-19). JAMA Netw. Open 3, e2017539.

Righini, M., Perrier, A., De Moerloose, P., Bounameaux, H., 2008. D-Dimer for venous thromboembolism diagnosis: 20 years later. J. Thromb. Haemostasis 6, 1059–1071.

Risitano, A.M., Mastellos, D.C., Huber-Lang, M., Yancopoulou, D., Garlanda, C., Ciceri, F., Lambris, J.D.,

2020. Complement as a target in COVID-19? Nat. Rev. Immunol. 20, 343–344.

Ruan, Q., Yang, K., Wang, W., Jiang, L., Song, J., 2020. Clinical predictors of mortality due to COVID-19 based on an analysis of data of 150 patients from Wuhan, China. Intensive Care Med. 46, 846–848.

Rundell, K., Panchal, B., 2017. Preterm Labor: Prevention and Management. Am. Fam. Physician 95, 366–372.

Salem, D., Katranji, F., Bakdash, T., 2021. COVID-19 infection in pregnant women: review of maternal and fetal outcomes. Int. J. Gynaecol. Obstet. 152, 291–298.

Schwartz, D.A., 2020. An analysis of 38 pregnant women with COVID-19, their newborn infants, and maternal-fetal transmission of SARS-CoV-2: maternal coronavirus infections and pregnancy outcomes. Arch. Pathol. Lab Med. 144, 799–805.

Schwartz, D.A., Morotti, D., 2020. Placental pathology of COVID-19 with and without fetal and neonatal infection: trophoblast necrosis and chronic histiocytic intervillositis as risk factors for transplacental transmission of SARS-CoV-2. Viruses 12, 1308.

Schwartz, D.A., Baldewijns, M., Benachi, A., Bugatti, M., Collins, R.R.J., De Luca, D., Facchetti, F., Linn, R.L., Marcelis, L., Morotti, D., Morotti, R., Parks, W.T., Patanè, L., Prevot, S., Pulinx, B., Rajaram, V., Strybol, D., Thomas, K., Vivanti, A.J., 2021. Chronic histiocytic intervillositis with trophoblast necrosis are risk factors associated with placental infection from coronavirus disease 19 (COVID-19) and intrauterine maternal-fetal systemic acute respiratory coronavirus 2 (SARSCoV-2) transmission in liveborn and stillborn infants. Arch. Pathol. Lab Med. 145, 517–528.

Schwartz, D.A., Morotti, D., Beigi, B., Moshfegh, F., Zafaranloo, N., Patanè, L., 2020. Confirming vertical fetal infection with coronavirus disease 2019: neonatal and pathology criteria for early onset and transplacental transmission of severe acute respiratory syndrome coronavirus 2 from infected pregnant mothers. Arch. Pathol. Lab Med. 144, 1451–1456.

Segars, J., Katler, Q., McQueen, D.B., Kotlyar, A., Glenn, T., Knight, Z., Feinberg, E.C., Taylor, H.S., Toner, J.P., Kawwass, J.F., 2020. Prior and novel coronaviruses, Coronavirus disease 2019 (COVID-19), and human reproduction: what is known? Fertil. Steril. 113, 1140–1149.

Sheth, S., Shah, N., Bhandari, V., 2020. Outcomes in COVID-19 positive neonates and possibility of viral vertical transmission: a narrative review. Am. J. Perinatol. 37, 1208–1216.

Smith, V., Seo, D., Warty, R., Payne, O., Salih, M., Chin, K.L., Ofori-Asenso, R., Krishnan, S., da Silva Costa, F., Vollenhoven, B., Wallace, E., 2020. Maternal and neonatal outcomes associated with COVID-19 infection: a systematic review. PloS One 15, e0234187.

Soma-Pillay, P., Nelson-Piercy, C., Tolppanen, H., Mebazaa, A., 2016. Physiological changes in pregnancy. Cardiovasc. J. Afr. 27, 89–94.

Stephens, A.J., Barton, J.R., Bentum, N.A.A., Blackwell, S.C., Sibai, B.M., 2020. General guidelines in the management of an obstetrical patient on the labor and delivery unit during the COVID-19 pandemic. Am. J. Perinatol. 37, 829–836.

Stonoga, E.T.S., de Almeida Lanzoni, L., Rebutini, P.Z., Permegiani de Oliveira, A.L., Chiste, J.A., Fugaça, C.A., Prá, D.M.M., Percicote, A.P., Rossoni, A., Nogueira, M.B., de Noronha, L., Raboni, S.M., 2021. Intrauterine transmission of SARS-CoV-2. Emerg. Infect. Dis. 27, 638–641.

Syeda, S., Baptiste, C., Breslin, N., Gyamfi-Bannerman, C., Miller, R., 2020. The clinical course of COVID in pregnancy. Semin. Perinatol. 44, 151284.

Szecsi, P., Jørgensen, M., Klajnbard, A., Andersen, M., Colov, N., Stender, S., 2010. Haemostatic reference intervals in pregnancy. Thromb. Haemostasis 103, 718–727.

Taghizadieh, A., Mikaeili, H., Ahmadi, M., Valizadeh, H., 2020. Acute kidney injury in pregnant women following SARS-CoV-2 infection: a case report from Iran. Respir. Med. Case Rep. 30, 101090.

Takemoto, M.L.S., Menezes, M.O., Andreucci, C.B., Knobel, R., Sousa, L.A.R., Katz, L., Fonseca, E.B., Magalhães, C.G., Oliveira, W.K., Rezende-Filho, J., Melo, A.S.O., Amorim, M.M.R., 2020. Maternal mortality and COVID-19. J. Matern. Fetal Neonatal Med. 136, 313–316.

Tang, M.W., Nur, E., Biemond, B.J., 2020a. Immune thrombocytopenia due to COVID-19 during pregnancy. Am. J. Hematol. 95, E191–E192.

Tang, N., Li, D., Wang, X., Sun, Z., 2020b. Abnormal coagulation parameters are associated with poor prognosis in patients with novel coronavirus pneumonia. J. Thromb. Haemostasis 18, 844–847.

Tam, P.C.K., Ly, K.M., Kernich, M.L., Spurrier, N., Lawrence, D., Gordon, D.L., Tucker, E.C., 2021. Detectable severe acute respiratory syndrome coronavirus 2 (SARS-CoV-2) in human breast milk of a mildly symptomatic patient with coronavirus disease 2019 (COVID-19). Clin. Infect. Dis. 72, 128–130.

Teuwen, L.A., Geldhof, V., Pasut, A., Carmeliet, P., 2020. COVID-19: the vasculature unleashed. Nat. Rev. Immunol. 20, 389–391.

Tolcher, M.C., McKinney, J.R., Eppes, C.S., Muigai, D., Shamshirsaz, A., Guntupalli, K.K., Nates, J.L., 2020. Prone positioning for pregnant women with hypoxemia due to coronavirus disease 2019 (COVID-19). Obstet. Gynecol. 136, 259–261.

Trippella, G., Ciarcià, M., Ferrari, M., Buzzatti, C., Maccora, I., Azzari, C., Dani, C., Galli, L., Chiappini, E., 2020. COVID-19 in pregnant women and neonates: a systematic review of the literature with quality assessment of the studies. Pathogens 9, 485.

Trocado, V., Silvestre-Machado, J., Azevedo, L., Miranda, A., Nogueira-Silva, C., 2020. Pregnancy and COVID-19: a systematic review of maternal, obstetric and neonatal outcomes. J. Matern. Fetal Neonatal Med. Online ahead of print, 1–13.

Turan, O., Hakim, A., Dashraath, P., Jeslyn, W.J.L., Wright, A., Abdul-Kadir, R., 2020. Clinical characteristics, prognostic factors, and maternal and neonatal outcomes of SARS-CoV-2 infection among hospitalized pregnant women: a systematic review. Int. J. Gynecol. Obstet. 151, 7–16.

Valdés, G., Neves, L.A.A., Anton, L., Corthorn, J., Chacón, C., Germain, A.M., Merrill, D.C., Ferrario, C.M., Sarao, R., Penninger, J., Brosnihan, K.B., 2006. Distribution of angiotensin-(1-7) and ACE2 in human placentas of normal and pathological pregnancies. Placenta 27, 200–207.

Varga, Z., Flammer, A.J., Steiger, P., Haberecker, M., Andermatt, R., Zinkernagel, A.S., Mehra, M.R., Schuepbach, R.A., Ruschitzka, F., Moch, H., 2020. Endothelial cell infection and endotheliitis in COVID-19. Lancet 395, 1417–1418.

Verity, R., Okell, L.C., Dorigatti, I., Winskill, P., Whittaker, C., Imai, N., Cuomo-Dannenburg, G., Thompson, H., Walker, P.G.T., Fu, H., Dighe, A., Griffin, J.T., Baguelin, M., Bhatia, S., Boonyasiri, A., Cori, A., Cucunubá, Z., FitzJohn, R., Gaythorpe, K., Green, W., Hamlet, A., Hinsley, W., Laydon, D., Nedjati-Gilani, G., Riley, S., van Elsland, S., Volz, E., Wang, H., Wang, Y., Xi, X., Donnelly, C.A., Ghani, A.C., Ferguson, N.M., 2020. Estimates of the severity of coronavirus disease 2019: a model-based analysis. Lancet Infect. Dis. 20, 669–677.

Vivanti, A.J., Vauloup-Fellous, C., Prevot, S., Zupan, V., Suffee, C., Do Cao, J., Benachi, A., De Luca, D., 2020. Transplacental transmission of SARS-CoV-2 infection. Nat. Commun. 11, 2069–2076.

Walker, K.F., O'Donoghue, K., Grace, N., Dorling, J., Comeau, J.L., Li, W., Thornton, J.G., 2020. Maternal transmission of SARS-COV-2 to the neonate, and possible routes for such transmission: a systematic review and critical analysis. BJOG Int. J. Obstet. Gynaecol. 127, 1324–1336.

Wang, S., Guo, L., Chen, L., Liu, W., Cao, Y., Zhang, J., Feng, L., 2020a. A case report of neonatal 2019 coronavirus disease in China. Clin. Infect. Dis. 71, 853–857.

Wang, X., Zhou, Z., Zhang, J., Zhu, F., Tang, Y., Shen, X., 2020b. A case of 2019 novel coronavirus in a pregnant woman with preterm delivery. Clin. Infect. Dis. 71, 844–846.

Webster, C.M., Smith, K.A., Manuck, T.A., 2020. Extracorporeal membrane oxygenation in pregnant and postpartum women: a ten-year case series. Am. J. Obstet. Gynecol. MFM 2, 100108.

Williams, D., 2003. Pregnancy: a stress test for life. Curr. Opin. Obstet. Gynecol. 15, 465–471.

Woodworth, K.R., Olsen, E.O., Neelam, V., Lewis, E.L., Galang, R.R., Oduyebo, T., Aveni, K., Yazdy, M.M., Harvey, E., Longcore, N.D., Barton, J., Fussman, C., Siebman, S., Lush, M., Patrick, P.H., Halai, U.-A., Valencia-Prado, M., Orkis, L., Sowunmi, S., Schlosser, L., Khuwaja, S., Read, J.S., Hall, A.J., Meaney-Delman, D., Ellington, S.R., Gilboa, S.M., Tong, V.T., Delaney, A., Hsia, J., King, K., Perez, M., Reynolds, M., Riser, A., Rivera, M., Sancken, C., Sims, J., Smoots, A., Snead, M., Strid, P., Yowe-Conley, T., Zambrano, L., Zapata, L., Manning, S., Burkel, V., Akosa, A., Bennett, C., Griffin, I., Nahabedian, J., Newton, S., Roth, N.M., Shinde, N., Whitehouse, E., Chang, D., Fox, C., Mohamoud, Y., Whitehill, F., 2020. Birth and infant outcomes following laboratory-confirmed SARS-CoV-2 infection in pregnancy — SET-NET, 16 jurisdictions, March 29–October 14, 2020. Morb. Mortal. Wkly. Rep. 69, 1635–1640.

World Health Organization, 2014. Global Nutrition Targets 2025: Low Birth Weight Policy Brief. WHO. URL: http://www.who.int/nutrition/publications/globaltargets2025_policybrief_lbw/en/. (Accessed 22 January 2021).

World Health Organization, 2020a. Clinical Management of COVID-19: Interim Guidance. WHO. URL: https://www.who.int/publications/i/item/clinical-management-of-covid-19. (Accessed 26 January 2021).

World Health Organization, 2020b. Clinical Management of Severe Acute Respiratory Infection (SARI) when COVID-19 Disease Is Suspected: Interim Guidance, 13 March 2020. URL: https://apps.who.int/iris/handle/10665/331446.

Xie, Y., Wang, Z., Liao, H., Marley, G., Wu, D., Tang, W., 2020. Epidemiologic, clinical, and laboratory findings of the COVID-19 in the current pandemic: systematic review and meta-analysis. BMC Infect. Dis. 20, 640.

Xiong, X., Wei, H., Zhang, Z., Chang, J., Ma, X., Gao, X., Chen, Q., Pang, Q., 2020. Vaginal delivery report of a healthy neonate born to a convalescent mother with COVID-19. J. Med. Virol. 92, 1657–1659.

Yang, H., Sun, G., Tang, F., Peng, M., Gao, Y., Peng, J., Xie, H., Zhao, Y., Jin, Z., 2020a. Clinical features and outcomes of pregnant women suspected of coronavirus disease 2019. J. Infect. 81, e40–e44.

Yang, P., Wang, X., Liu, P., Wei, C., He, B., Zheng, J., Zhao, D., 2020b. Clinical characteristics and risk assessment of newborns born to mothers with COVID-19. J. Clin. Virol. 127, 104356.

Yee, J., Kim, W., Han, J.M., Yoon, H.Y., Lee, N., Lee, K.E., Gwak, H.S., 2020. Clinical manifestations and perinatal outcomes of pregnant women with COVID-19: a systematic review and meta-analysis. Sci. Rep. 10, 18126.

Yoon, S.H., Kang, J.M., Ahn, J.G., 2020. Clinical outcomes of 201 neonates born to mothers with COVID-19: a systematic review. Eur. Rev. Med. Pharmacol. Sci. 24, 7804–7815.

Yu, N., Li, W., Kang, Q., Zeng, W., Feng, L., Wu, J., 2020. No SARS-CoV-2 detected in amniotic fluid in mid-pregnancy. Lancet Infect. Dis. 20, 1364.

Zaigham, M., Andersson, O., 2020. Maternal and perinatal outcomes with COVID-19: a systematic review of 108 pregnancies. Acta Obstet. Gynecol. Scand. 99, 823–829.

Zaim, S., Chong, J.H., Sankaranarayanan, V., Harky, A., 2020. COVID-19 and multiorgan response. Curr. Probl. Cardiol. 45, 100618.

Zamaniyan, M., Ebadi, A., Aghajanpoor, S., Rahmani, Z., Haghshenas, M., Azizi, S., 2020. Preterm delivery, maternal death, and vertical transmission in a pregnant woman with COVID-19 infection. Prenat. Diagn. 40, 1759–1761.

Zambrano, L.D., Ellington, S., Strid, P., Galang, R.R., Oduyebo, T., Tong, V.T., Woodworth, K.R., Nahabedian, J.F., Azziz-Baumgartner, E., Gilboa, S.M., Meaney-Delman, D., Akosa, A., Bennett, C., Burkel, V., Chang, D., Delaney, A., Fox, C., Griffin, I., Hsia, J., Krause, K., Lewis, E., Manning, S., Mohamoud, Y., Newton, S., Neelam, V., Olsen, E.O., Perez, M., Reynolds, M., Riser, A., Rivera, M., Roth, N.M., Sancken, C., Shinde, N., Smoots, A., Snead, M., Wallace, B., Whitehill, F., Whitehouse, E., Zapata, L.,

2020a. Update: characteristics of symptomatic women of reproductive age with laboratory-confirmed SARS-CoV-2 infection by pregnancy status — United States, January 22—October 3, 2020. Morb. Mortal. Wkly. Rep. 69, 1641—1647.

Zambrano, L.I., Fuentes-Barahona, I.C., Bejarano-Torres, D.A., Bustillo, C., Gonzales, G., Vallecillo-Chinchilla, G., Sanchez-Martínez, F.E., Valle-Reconco, J.A., Sierra, M., Bonilla-Aldana, D.K., Cardona-Ospina, J.A., Rodríguez-Morales, A.J., 2020b. A pregnant woman with COVID-19 in Central America. Trav. Med. Infect. Dis. 36, 101639.

Zeisler, H., Llurba, E., Chantraine, F., Vatish, M., Staff, A.C., Sennström, M., Olovsson, M., Brennecke, S.P., Stepan, H., Allegranza, D., Dilba, P., Schoedl, M., Hund, M., Verlohren, S., 2016. Predictive value of the sFlt-1:PlGF ratio in women with suspected preeclampsia. N. Engl. J. Med. 374, 13—22.

Zhang, C., Shi, L., Wang, F.S., 2020. Liver injury in COVID-19: management and challenges. Lancet Gastroenterol. Hepatol. 5, 428—430.

Zhou, F., Yu, T., Du, R., Fan, G., Liu, Y., Liu, Z., Xiang, J., Wang, Y., Song, B., Gu, X., Guan, L., Wei, Y., Li, H., Wu, X., Xu, J., Tu, S., Zhang, Y., Chen, H., Cao, B., 2020. Clinical course and risk factors for mortality of adult inpatients with COVID-19 in Wuhan, China: a retrospective cohort study. Lancet 395, 1054—1062.

Zhu, C., Liu, W., Su, H., Li, S., Shereen, M.A., Lv, Z., Niu, Z., Li, D., Liu, F., Luo, Z., Xia, Y., 2020. Breastfeeding risk from detectable severe acute respiratory syndrome coronavirus 2 in breastmilk. J. Infect. 81, 452—482.

Author Index

A

Abbasian, L., 106
Abbassi-Ghanavati, M., 24
Abiona, O., 6−8
Abul, H., 25−26
Abu-Raya, B., 25−26
Ádány, R., 24
Agarwal, A., 108
Aguilar, G., 103
Ai, P., 65
Ai, T., 44, 53
Ai, Z., 65
Akosa, A., 146
Alagaili, A.N., 28
Alborghetti, L., 105
Alden, N., 146
Aldeyab, M.A., 94
Al Dhahiry, S.H., 3
Aleva, R.M., 94, 95t
Alexander, P.E., 102
Alexandre, J., 96
Aley, P.K., 65
Al-Hakeem, R.F., 2
Alhazzani, W., 102
Alikhan, M.F., 4
AL Johani, S.M., 28
Allotey, J., 52, 146−147, 147t, 149,
    150t, 154−156
AL-Mozaini, M.A., 28
Alotaibi, B., 93, 114−115
Alpern, J.D., 103
Al-Rabeeah, A.A., 2
AL-Rayes, S., 25−26
Alshukairi, A.N., 28
Ambrós, A., 103
Andersen, M., 146
Anderson, E.J., 146
Anderson, K., 146
Angus, B., 65
Annan, A., 3
Annweiler, C., 43
Antao, O.Q., 27
Antón, A., 157
Apostol, M., 146
Arabi, Y., 93, 114−115
Areia, A.L., 25
Árnyas, E., 24
Artis, D., 24−25
Aryal, K., 102
Asadi, N., 146
Auld, S.C., 148−149

Azhar, E.I., 93, 114−115
Azziz-Baumgartner, E., 146

B

Badr, D.A., 146−147, 147t, 149, 150t,
    154−156
Baek, J.Y., 28
Baharoon, S.A., 28
Baier, M., 29−30
Bai, H., 97, 101
Bailey, K.R., 101−102
Baillie, J.K., 107
Baker, M.G., 5
Bakker, W.W., 24
Balaji, R., 146−147, 147t, 149, 150t,
    154−156
Balcells, J., 157
Baldwin, C., 69
Baldwin, H.J., 3
Baric, R.S., 6, 9−10
Barnathan, E.S., 92
Barney, G., 146
Barzon, L., 11
Bashir, N., 2−3
Basso, O., 27−28
Bayry, J., 110
Beaver, J.T., 27
Beck, C.R., 107
Becker, S., 65
Belij-Rammerstorfer, S., 65
Bell, B.P., 67
Bell, K., 4
Belo, L., 24
Bennett, C., 146
Bennett, N.M., 146
Benson, C.L., 24
Berg, G., 25−26
Bergsbaken, T., 40
Bernstein, P.S., 146
Beruto, M.V., 109
Bestebroer, T.M., 2
Betters, D., 24
Bhatnagar, T., 108
Biemond, B.J., 148−149
Bikdeli, B., 97
Billing, L.M., 146
Biondi-Zoccai, G., 97
Birch, T., 13
Bissett, B., 69
Bleicker, T., 28
Boden, I., 69

Boersma, W.G., 94, 95t
Bojkovab, D., 93
Bokhari, A., 28
Bolt, A., 24
Bonet, M., 146−147, 147t, 149, 150t,
    154−156
Bonilla, F., 110
Bonten, M.M.J., 94, 95t
Borghuis, T., 24
Borisova, A.V., 94
Borsa, N.G., 24
Bostwick, J.M., 104
Bottino, N., 148−149
Boudjelal, M., 28
Bounameaux, H., 146
Bramante, C., 114
Brammer, L., 146
Bresnitz, E., 13
Bricker, K.M., 27
Brochard, L., 70
Brocklehurst, P., 146−147
Brooks, S., 146
Brotman, D.J., 104
Brown, I, 11
Brown, M.J., 27
Brown, T.S., 97
Brünink, S., 28
Buckeridge, D.L., 27−28
Bunch, K., 146−147
Burgos Pratx, L.D., 109
Burkel, V., 146
Burows, R.F., 24
Burrows, E.A., 24
Burwick, R.M., 24
Buss, C., 24
Butterfield, M., 5
Buycka, S.C.C., 93
Byambasuren, O., 4

C

Cai, Y., 93
Cao, B., 93, 102
Cao, W., 65
Caplan, A.I., 112
Cardenas, I., 26−27
Cardona, M., 4
Car, J., 69
Carlin, A., 146−147, 147t, 149, 150t,
    154−156
Carmeliet, P., 41, 148−149
Cascella, M., 40

Note: Page numbers followed by "t" indicate tables.

Caslake, M., 24
Castilla, J., 27
Catalan, M., 157
Chai, S.J., 146
Chakraborty, R., 146−147
Chamberland, M., 67
Chang, D., 146
Chang, G., 28
Chang, H.H., 27
Chang, R.-Q., 25
Chan, J.F.W., 2
Chan, P.K., 13
Chantarangkul, V., 148−149
Chatterjee, P., 108
Chatterjee, S., 146−147, 147t, 149, 150t, 154−156
Chauhan, S., 2
Chaw, L., 4
Cheng, F.W., 13
Chen, H., 2−3, 6−8, 10, 15
Chen, H.D., 6
Chen, J., 65
Chen, S., 6−8
Chen, V., 106
Chen, X., 93, 97, 101
Chen, Y.M., 2
Cheung, H.M., 13
Chinen, J., 110
Chinn, I.K., 110
Cho, S.Y., 28
Chu, D.K., 28
Chu, H., 2, 49
Chuich, T., 97
Ciavarella, C., 11
Cinatlb, J., 93
Clark, J., 4
Cleary, P., 107
Cohen, C., 27
Cohen, P.A., 44, 46
Colonna, M., 24−25
Colov, N., 146
Conlon, G., 94
Cookson, B.T., 40
Cooney, J., 24
Corbett, K.S., 6−8
Cordier, A.-G., 146−147, 147t, 149, 150t, 154−156
Corman, V.M., 3−4, 11, 28
Coulam, C.B., 25−26
Coumans, B., 24
Coy, K., 13
Creanga, A.A., 40−41
Cui, J., 2
Cunningham, F.G., 24
Cuomo-Dannenburg, G., 11
Cutrell, J.B., 92

D

da Costa, V.G., 5−6
Dao, B., 146
Dashraath, P., 41, 148−149
Davis, E.P., 24
Davis, S.S., 146
Davoudi-Monfared, E., 106

Debenham, L., 146−147, 147t, 149, 150t, 154−156
de Boer, M.G.J., 94, 95t
Debolt, C., 146
Deeks, J.J., 48−49
Delahoy, M.J., 146
Delaney, A., 146
Delgado-Rodríguez, M., 27
Dellicour, S., 10
Deloia, J.A., 24
Del Vecchio, C., 11
de Masson, A., 43
De Meyera, S., 93
De Moerloose, P., 146
Denison, M.R., 10
Der Nigoghossian, C., 97
De Sanctis, Y., 92
de Vos, P., 24
Diefenbach, A., 24−25
Díez, J.-M., 110
Ding, Q., 94
Di Nisio, M., 92
Di Paolo, M., 104
DI Santo, J.P., 24−25
Discher, D.E., 112
Divakar, H., 146
Dixit, A., 146−147, 147t, 149, 150t, 154−156
Dolk, H., 27
Dolladille, C., 96
Domínguez, A., 27
Donaldson, E.F., 10
Dong, L., 102
Dooling, K., 67
Dorsey, M., 110
Drici, M.D., 96
Driggin, E., 97
Duncan, A., 148−149
Duncan, B., 65
Duncan, D., 13
Duncan, H., 69
Du, R., 148−149

E

Eberl, G., 24−25
Eckerle, L.D., 10
Efrati, P., 24
Ekerfelt, C., 25−26
El-Gamal, Y., 110
Ellington, S., 146
Elliott, M.W., 70
El Masry, K.M., 13
Elrefaei, A., 146−147
Enninga, E.A.L., 146−147
Eppes, C.S., 148
Ernerudh, J., 25−26
Esser, E.S., 27
Ewer, K.J., 65

F

Faas, M.M., 24
Fan, G., 93, 102, 148−149
Fang, C., 28
Fan, Y., 94

Farley, M.M., 114
Fathman, C.G., 25−26
Favilli, A., 94
Fazzari, M., 146
Fedak, K.M., 5
Fell, D.B., 27−28
Felsen, C.B., 146
Feng, Y., 110−111
Fenioux, C., 96
Fernández-Hidalgo, N., 157
Ferrando, C., 103
Fineschi, V., 104
Fink, S.L., 40
Fish, E.N., 106
Folegatti, P.M., 65
Fouchier, R.A., 2
Fox, C., 146
Fox, K., 146
Franchin, E., 11
Frati, P., 104
Frigeni, M., 105
Fry, A.M., 146
Funck-Brentano, C., 96
Fung, G.P., 13

G

Gajardo, R., 110
Galang, R.R., 146
Gale, C., 146−147
Galeotti, C., 110
Galiano, M., 3
Gao, G., 65
Gao, G.F., 2, 65
Garcia-Manau, P., 157
Garcia-Ruiz, I., 157
Garg, S., 146
Garovic, V.D., 146−147
Gee, J., 146
Geldhof, V., 148−149
George, A., 146
Geraci, J., 27
Gerli, S., 94
Gershman, K., 114
Gessner, B.D., 27
GeurtsvanKessel, C., 109−110
Gharbharan, A., 109−110
Ghazanfari, T., 111
Giardina, I., 94
Gilbert, J.S., 24
Gilboa, S.M., 146
Gilchrist, M., 69
Gil, M.M., 147−148
Ginsberg, J.S., 24
Girod, J.P., 104
Glanz, J.M., 146
Glasziou, P., 4
Gleicher, N., 24
Godoy, P., 27
Goldsmith, J.A., 6−8
Goldstein, M.R., 110
Gong, J., 97, 101
Goodfellow, R.X., 24
Goodnight, W.H., 26, 40
Gorab, J., 13

Gosselink, R., 69
Gostin, L.O., 3
Graeber, C.W., 110
Graham, C., 69
Graham, R., 6, 9–10
Graham, R.L., 10
Granger, C.L., 69
Gran, J.M., 27
Grasselli, G., 148–149
Greenhalgh, T., 69
Greer, L.G., 24
Griffin, I., 146
Griffith, M., 4
Gritti, G., 105
Groen, B., 24
Gromiha, M.M., 2–3, 6–10
Gronski, J., 114
Groten, T., 29–30
Guan, L., 148–149
Guan, W.J., 10–11, 42
Guan, Y., 3
Guggemos, W., 4, 11, 28
Gui, M., 6–8
Gunaratne, M.D.S.K., 146–147
Guntupalli, K.K., 148
Guo, Y., 2–3, 6–8, 10, 15
Gupta, M., 104
Gu, X., 148–149

H

Haagmans, B.L., 3, 28
Haase, J., 13
Haines, D., 25–26
Hajiabdolbaghi, M., 106
Hakim, A., 148–149
Halberstadt, E., 26
Hall, A.J., 146
Hall, O.J., 40
Haluszczak, C., 24
Hambidge, S.J., 146
Han, H., 97
Hanna, M., 146
Hanna, N., 146
Hare, J.M., 112
Hashemi, A., 146
Hayden, F.G., 2
Ha, Y.E., 28
Head, K., 24
Heald-Sargent, T., 4
Heinen, E., 24
Heit, J.A., 101–102
He, J., 42
Henriksen, K.J., 24
Hessami, K., 146
Hess, D., 70
Hill, N.S., 70
Hodgson, C., 69
Hoelscher, M., 28
Hofmann, J., 4
Hollander, J.G, 109–110
Holmes, A., 69
Homayoon, N., 146
Hoversten, S., 114

Hsia, J., 146
Hsieh, C.L., 6–8
Hsueh, P.-R., 2
Huang, C., 28
Huang, C.L., 6
Hu, B., 6
Hughes, B., 43
Hui, D.S.C., 42
Hutchison, J., 69
Hu, Y., 10–11, 28, 42, 107

I

Iba, T., 92
Ibirogba, E.R., 146–147
Igout, A., 24
Ilias, I., 103
Iuliano, A.D., 27

J

Jackson, L.A., 146
Jacobsen, S.J., 146
James, A.H., 101–102
Jamieson, D.J., 27
Jani, J.T., 104
Jan Kullberg, B., 94, 95t
Jervis, R.H., 5
Jeslyn, W.J.L., 148–149
Ji, H.-L., 40–41
Jilma, B., 93
Jodlowski, T.Z., 92
Johnson, P.W., 108
Jo, I.J., 28
Jones, T.C., 4, 28
Jonkers, R.E., 94, 95t
Jordans, C.C.E., 109–110
Jørgensen, M., 146
Joyner, M.J., 108
Juan, J., 147–148

K

Kaiser, M., 28
Kakoulidis, I., 103
Kamimoto, L., 114
Kang, J.M., 28
Kapur, A., 146
Karim, F., 109–110
Karolyi, M., 93
Kasraeian, M., 146
Katz, R., 3
Kaufman, J.S., 27–28
Kaveri, S.V., 110
Kawasaki, B., 146
Kazemzadeh, H., 106
Kazmar, R.E., 25–26
Kazmi, A., 2–3
Kearney, M.P., 94
Kermali, M., 52
Kew, T., 146–147, 147t, 149, 150t,
    154–156
Keyaerts, E., 94
Khalili, H., 106
Khan, S., 2–3
Kharbanda, E.O., 146
Khaw, F.-M., 107

Khoury, R., 146
Kickler, T.S., 104
Kihara, A.B., 146
Kim, C.B., 5
Kim, D., 11
Kim, L., 146
Kim, S., 146
Kim, S.H., 28
Kim, Y.J., 28
Kim, Y.M., 5
Kinlaw, K., 67
Kirley, P.D., 146
Kitamura, N., 92
Klajnbard, A., 146
Klassen, S.A., 108
Klein, N.P., 146
Klok, P.A., 24
Klose, S.M., 3
Knight, M., 146–147
Kobbervig, C.E., 101–102
Kociolek, L.K., 4
Koh, G.C.H., 69
Koh, W.C., 4
Ko, J.H., 28
Kok, K.H., 2
Kollmann, T.R., 106
Kong, Y., 107
Koukkou, E., 103
Kourtis, A.P., 27
Kövér, Á., 24
Ko, W.-C., 2
Koyasu, S., 24–25
Krause, K., 146
Krause, T.G., 27
Kremsdorf, R.A., 24
Kühnert, M., 26
Kumar, G., 108
Kurinczuk, J.J., 146–147
Kwong, J.C., 27–28

L

Lai, C.-C., 2
Lam, C.S., 3
Lampé, R., 24
Lam, T.T., 3
Landi, F., 105
Landt, O., 28
Laracy, J., 97
La Russa, R., 104
Latha, B., 108
Lau, C.C., 3
Lau, J.H., 3
Lau, S.K., 3
Lavezzo, E., 11
Lavoie, P.M., 25–26
Layqah, L., 28
Lee, G.M., 67
Lee, J.Y., 28
Lee, K.H., 13
Lee, S., 5
Lee, S.I., 146–147, 147t, 149, 150t,
    154–156
Lei, C., 42
Lei, M., 110–111

Letko, M., 6
Levy, J.H., 92
Lewis, E., 146
Li, A.M., 13
Liang, W., 42
Liao, H., 146
Liao, Y.S., 3
Li, B., 6
Li, D., 97, 101
Li, D.-J., 25
Li, F., 2, 6, 9−10
Li, H., 148−149
Li, J., 97
Li, L., 107
Limaye, M.A., 146
Li, M.-Q., 25
Lim, W.S., 107
Lingappa, J.R., 13
Lipardi, C., 92
Lip, G.Y.H., 104
Lipkind, H.S., 146
Li, S., 94
Littauer, E.Q., 27
Liu, F., 97
Liu, J., 2
Liu, L., 42
Liu, M., 94
Liu, R., 97
Liu, W., 93, 102
Liu, X.H., 97
Liu, Y., 2−3, 6−8, 10, 15, 65, 148−149
Liu, Z., 110−111, 148−149
Li, X., 11, 28, 41
Llavall, A.C., 146−147, 147t, 149, 150t, 154−156
Locksley, R.M., 24−25
Loeb, M., 27
Lowther, S.A., 13
Lu, P., 94
Luppi, P., 24
Luteijn, J.M., 27
Lynfield, R., 146

M

MacIntyre, J., 94
Madhavan, M.V., 97
Maes, P., 94
Maeuerer, M., 93, 114−115
Magalis, B.R., 10
Magee, F.A., 94
Mahmoud, F., 25−26
Mahmoud Hashemi, S., 111
Maier, C.L., 148−149
Maiese, A., 104
Mair-Jenkins, J., 107
Maiz, N., 157
Makki, S., 107
Malhotra, P., 108
Malik, A., 13
Mammen, M.J., 102
Manetti, A.C., 104
Manna, S., 53

Manning, S., 146
Manuck, T.A., 148
Margalith, M., 24
Marini, S., 10
Marley, G., 146
Marmor, S., 114
Martínez, D., 103
Martín, V., 27
Marzi, A., 6
Mattei Gentili, M., 94
Mattern, J., 146−147, 147t, 149, 150t, 154−156
Matthew, D.B., 67
Matthiesen, L., 25−26
Mavian, C., 10
Maxwell, C., 12
Mayoral, J.M., 27
McCloskey, B., 93, 114−115
McClung, N., 67
McCullagh, B., 94
McDonald, C., 13
McElnay, J.C., 94
McGeer, A., 12
McGoogan, J.M., 41
McIntosh, K., 2
McIntyre, H.D., 146
Mckenzie, A.N.J., 24−25
McKinney, J.R., 148
McLaws, M.-L., 4
McMurray, J.J.V., 116
McNeil, C., 114
Meaney-Delman, D., 146
Mebius, R. E., 24−25
Meek, J., 114, 146
Mehdi Naghizadeh, M., 111
Mehra, M.R., 92
Meijer, A., 28
Melamed, N., 146
Melgert, B.N., 24
Memish, Z.A., 2
Mendes, J., 25
Mertz, D., 27
Messerole, E., 70
Messick, J., 13
Michalski, C., 25−26
Michel, T., 116
Mills, J.R., 108
Mills, L.K., 27
Mira, J.P., 96
Mishra, G.P., 104
Mohamoud, Y., 146
Molenkamp, R., 28
Mollema, F.P.N., 109−110
Monogue, M.L., 92
Monroe, M.L., 146
Moreli, M.L., 5−6
Mor, G., 26−27
Mori, M., 24
Morris, E., 146−147
Mortensen, L.H., 27
Mosaffa, N., 111
Moslehi, J.J., 96
Mota-Pinto, A., 25

Mühlemann, B., 4
Muigai, D., 148
Mukherjee, A., 108
Mulani, J., 104
Mulders, D.G., 28
Muller, M.A., 4, 11, 28
Muñoz, T., 103
Munster, V., 6
Murray, T., 114
Muscatello, D.J., 27
Muse, A., 146
Myers, R., 3
Myers, T.R., 146

N

Na, B.J., 5
Nahabedian, J.F., 146
Naing, L., 4
Naleway, A.L., 146
Narang, K., 146−147
Nates, J.L., 148
Navalesi, P., 70
Nava, S., 70
Neelam, V., 146
Nehdi, A., 28
Nelson-Piercy, C., 41
Nester, C.M., 24
Newton, S., 146
Ng, P.C., 13
Nguyen, L.S., 96
Nguyen-Van-Tam, J.S., 107
Niemeyer, D., 28
Ni, X.B., 3
Ni, Z.Y., 10−11, 42
Nkrumah, E.E., 3
Novembrino, C., 148−149
Nur, E., 148−149
Nybo Andersen, A.M., 27

O

O'Brien, P., 146−147
Oduyebo, T., 146
O'Halloran, A., 146
Okba, N.M.A., 48−49
Olsen, E.O., 146
Omu, A., 25−26
Openo, K.P., 146
Orange, J.S., 110
Ortiz, J.R., 27
Osterhaus, A.D., 2
Ou, C., 42
Owusu, M., 3

P

Palekar, R., 27
Pál, L., 24
Panagiotakopoulos, L., 146
Panigada, M., 148−149
Parashar, U.D., 13
Parazzini, F., 94
Park, G.E., 28
Park, S.Y., 5
Parshuram, C.S., 69

Pasut, A., 148–149
Patel, A.B., 4
Patel, J.V., 104
Péault, B.M., 112
Pecks, U., 30
Peiris, M.J., 13
Peluso, C., 92
Penfield, C.A., 146
Peng, S., 28
Pereira, G., 28
Pereira-Leite, L., 24
Perez, E.E., 110
Perez, M., 146
Perlman, S., 28
Perrier, A., 146
Perry, R., 13
Pesenti, A., 148–149
Petersen, E., 93, 114–115
Petterson, T.M., 101–102
Peyvandi, F., 148–149
Phelan, A.L., 3
Phinney, D.G., 112
Piasecki, A., 146
Pinsky, B., 11
Pitcher, G.R., 24
Pittenger, M.F., 112
Platt, R.W., 27–28
Póka, R., 24
Poland, G.A., 110
Polly, D.M., 148–149
Pond, S.K., 10
Poon, L.C., 146–148
Poon, R.W.S., 2
Posch, A., 104
Powrie, F., 24–25
Presentey, B., 24
Pruessner, J.C., 24
Pumarola, T., 27

Q

Qin, C., 49–50
Quigley, M., 146–147
Quintana, J.M., 27
Quintanilha, A., 24

R

Rahmani, H., 106
Raimondi, F., 105
Raj, V.S., 3
Ranganathan, N., 69
Rao, M., 93, 114–115
Raskob, G.E., 92
Rasmussen, I.S., 27
Raspa, F., 94
Ravi, M., 13
Rawson, T.M., 69
Ray, J.G., 24
Read, J.S., 27
Rebelo, I., 24
Reddy, S., 104
Regal, J.F., 24
Regan, A.K., 28
Reiken, M., 13

Reingold, A., 114
Ren, L., 28
Ren, W., 65
Reusken, C.B., 3
Reynolds, M., 146
Richard, C.A., 24
Righini, M., 146
Ripa, M., 49
Ripamonti, D., 105
Rippe, J., 4
Riser, A., 146
Riva, I., 105
Rivera, M., 146
Robertson, C.A., 13
Rocha, S., 24
Rochwerg, B., 70
Roden, D.M., 96
Rodo, C., 157
Rodrigues-Santos, P., 25
Roguski, K.M., 27
Roman, A.S., 146
Romero, C., 110
Rong, Z., 147–148
Rooney, K.D., 107
Rose, C.E., 5
Rosledzana, M.A., 4
Rothe, C., 28
Roth, N.M., 146
Rozenszajn, L., 24
Ruan, L., 93, 102
Ruano, R., 146–147
Ryan, D.S., 146
Ryan, P.A., 146
Rysman, E., 94

S

Saavedra-Campos, M., 107
Sadarangani, M., 25–26
Sadeghi, S., 111
Saivish, M.V., 5–6
Salehi, M., 106
Salem, J.E., 96
Samman, N., 28
Sancken, C., 146
Sanders, J.M., 92
Sandman, C.A., 24
Santos-Rosa, M., 25
Santos-Silva, A., 24
Satta, G., 69
Savoy, N., 109
Saw, J., 13
Sayed, F., 13
Schaffner, W., 146
Schanzer, D., 27
Scherag, A., 29–30
Schleussne, E., 29–30
Schmidt, M.L., 28
Schneider, E., 13
Schneider, J., 28
Schoergenhofer, C., 93
Scibona, P., 109
Seilmaier, M., 4, 11, 28
Senefeld, J.W., 108

Seok, H., 28
Sermer, M., 12
Shahbaba, B., 24
Shah, N.H., 11
Shamshirsaz, A., 148
Shang, J., 6, 9–10
Shan, H., 42
Shannon, C.P., 106
Shao, Z., 110–111
Sharma, S., 146
Sheedy, C., 13
Shek, C.C., 13
Shereen, M.A., 2–3
Shih, T.-P., 2
Shiltz, J., 146
Shimabukuro, T.T., 146
Shinde, N., 146
Shi, Z.-L., 2
Shum, M.H., 3
Siddiqi, H.K., 92
Siddique, R., 2–3
Siebman, S., 146
Siegel, I., 24
Sieswerda, E., 94, 95t
Si, H.R., 6
Silverfield, J.C., 25–26
Simonovich, V.A., 109
Simpson, L.L., 146
Simpson, N., 146–147
Skountzou, I., 27
Smith, K.A., 148
Smith, R.J.H., 24
Smoots, A., 146
Snead, M., 146
Soler, J.A., 103
Somsen, G.A., 3
Song, B., 148–149
Song, W., 6–8
Song, Z.G., 2
Soper, D.E., 26, 40
Sorhage, F., 13
Sosin, D.M., 146
Spaans, F., 24
Speake, H.A., 28
Spencer, M., 146
Sperati, C.J., 24
Spiro, T.E., 92
Spits, H., 24–25
Spyropoulos, A.C., 92
Srivastava, A., 2–3, 6–10
Stalenhoef, J.E., 109–110
Stallings, E., 146–147, 147t, 149, 150t, 154–156
Stauffer, W.M., 103
Stegmüller, M., 26
Stein, A., 4
Stender, S., 146
Stephens, G.M., 2
Stimpfl, T., 93
Stockman, L.J., 13
Stone, J., 146
Strid, P., 146
Strohmeier, R., 26

Sulleiro, E., 157
Sun, H., 65
Sun, S., 65
Sun, X., 41
Sun, Z., 97, 101
Sutton, D.M., 146
Sutton, M., 146
Swieboda, D., 27
Szecsi, P., 146
Szekeres-Bartho, J., 26
Szymanski, L.M., 146–147

T

Tagliabue, P., 148–149
Tai, K.F.Y., 12
Talbot, H.K., 146
Tamames, S., 27
Tan, C., 13
Tang, H.-J., 2
Tang, J., 102
Tang, K., 2–3, 6–8, 10, 15
Tang, M.W., 148–149
Tang, N., 97, 101
Tang, W., 146
Tanksley, C.L., 148–149
Taylor, R.P., 24
Tebbutt, S.J., 106
Tempia, S., 27
Ten Eyck, P., 28
Teuwen, L.A., 148–149
Thachil, J., 92
Theel, E.S., 48–49, 108
Theiler, R.N., 146–147
Thellin, O., 24
Thomas, A.R., 114
Thomas, P., 69
To, K.K.W., 2
Tolcher, M.C., 148
Tong, V.T., 146
Tong, X., 107
Tong, Y.G., 3
Trad, A.T.A., 146–147
Tripodi, A., 148–149
Trucco, M., 24
Truong, A.D., 148–149
Tsang, A.K., 3
Tsoi, H.W., 2
Turan, O., 148–149
Turillazzi, E., 104

V

Vaduganathan, M., 116
Vafaei, H., 146
Vallone, M.G., 109
Van Boheemen, S., 2
Van Dammea, E., 93
Vandamme, A.M., 10
Vandermeer, M.L., 114
Vanders, R.L., 40
van der Veer, B., 28
VAN Goor, H., 24
Van Loocka, M., 93
van Pampus, M.G., 24

Van Ranst, M., 94
Vardeny, O., 116
Vázquez, C., 109
Veith, T., 4
Verbeeck, J., 94
Victoroff, T., 5
Vijgen, L., 94
Villar, J., 103
Vincenzo Berghella, 43
Vitagliano, A., 94
Vivier, E., 24–25
Vollmar, P., 28
Vousden, N., 146–147

W

Walker, P.F., 103
Wallace, B., 146
Wallace, M., 67
Waltenburg, M.A., 5
Wang, C., 2, 110–111
Wang, J., 93, 102
Wang, L., 28
Wang, N., 6–8
Wang, W., 2, 46
Wang, X.G., 6–8, 106
Wang, Y., 28, 65, 93, 102, 148–149
Wang, Z., 146
Wang, Z.-H., 106
Wan, Y., 6, 9–10
Warrington, T.P., 104
Webster, C.M., 148
Wegmann, T.G., 26
Weintraub, E.S., 146
Wei, X.-S., 106
Wei, Y., 148–149
Wen, D., 93, 102
West, N., 146
Whaley, K., 25–26
Whitaker, M., 146
Whitehill, F., 146
Whitehouse, E., 146
Wiggins, C.C., 108
Williams, D.T., 27
Williams, J.T.B., 146
Wilson, K., 27–28
Winkup, J., 27
Wölfel, R., 4, 11, 28
Wong, K.H., 3
Wong, K.L., 3
Wong, S.F., 13
Wong, T.Y., 24
Woodfallc, B., 93
Woodworth, K., 146
Woodworth, K.R., 146
Woo, P.C., 3
Wortham, J.M., 146
Wrapp, D., 6–8
Wright, A., 148–149
Wu, C., 93
Wu, D., 146
Wu, F., 2
Wu, K.L., 97

Wu, X., 148–149
Wu, Z., 41
Wyllie, A.L., 46–47

X

Xiang, J., 148–149
Xiang, X., 106
Xiang, Y., 6–8
Xia, S., 28
Xia, X., 65
Xia, Y., 94
Xie, Q., 110–111
Xie, Y., 146
Xing, F., 2
Xu, J., 6–8, 92, 148–149
Xu, Y., 11

Y

Yang, H., 146–148
Yang, J., 2, 107
Yang, L., 97
Yang, X.L., 6
Yap, M., 146–147, 147t, 149, 150t, 154–156
Ye, C., 102
Yekaninejad, M.S., 106
Yesudhas, D., 2–3, 6–10
Yip, C.C.Y., 2
Yi, S., 5
Young, B.C., 40
Young, B.W.Y., 3
Yousey-Hindes, K., 146
Yuan, P., 65
Yuan, S., 2
Yu, B., 2
Yu, T., 148–149

Z

Zaki, A.M., 2
Zambrano, L.D., 43, 146
Zange, S., 4, 11, 28
Zapata, L., 146
Zhang, L., 6, 28
Zhang, W., 6, 107
Zhang, Y., 24, 147–148
Zhao, J., 28
Zhao, R., 65
Zhao, S., 2
Zheng, B.J., 3
Zheng, J., 28
Zheng, J.C., 65
Zheng, S., 107
Zheng, X., 4
Zhong, J., 102
Zhong, L., 110–111
Zhou, D., 146–147, 147t, 149, 150t, 154–156
Zhou, F., 52, 148–149
Zhou, H., 6–8
Zhou, J., 3
Zhou, P., 6
Zhou, Q., 106

Zhou, W., 28
Zhou, W.-J., 25
Zhou, Y., 40, 46–47, 51–52
Zhu, B., 11
Zhu, C.L., 97

Zhu, H., 28
Zhu, H.C., 3
Zhu, N., 69
Zhu, Y., 6
Zöllkau, J., 29–30

Zorzi, W., 24
Zoufaly, A., 93
Zou, L., 45
Zuchowski, M., 4
Zumla, A.I., 2, 93, 114–115

Subject Index

A

Acute hypoxemic respiratory failure, 121
Acute kidney injury (AKI), 149
Acute respiratory failure, 148
Adaptive immunity
  B lymphocytes, 26
  immunoglobulins, 26
  T lymphocytes, 25–26
Aerosol-generating procedures, 82–83
American College of Obstetricians and Gynecologists' (ACOG's), 66
Anesthesia, 78–84
Angiotensin-converting enzyme inhibitors, 116
Angiotensin-converting enzyme 2 (ACE2) receptors, 146
Angiotensin receptor blockers, 116
Asymptomatic mothers, 80–81
Asymptomatic/nonsevere COVID-19 infections, 76

B

B lymphocytes, 26
Breastfeeding, 81

C

Cesarean delivery management, 78
Chest computed tomography scan
  absorption-stage manifestations, 54
  advanced stage manifestations, 54
  early-stage manifestations, 54
  preparation for patient, 54
  preparation prior to admission, 53–54
  scanning parameters, 54
  scope and direction of scanning, 54
  severe-stage manifestations, 54
Complement system, 24
Contact personnel precautions, 82

D

Decontamination, 83
Delivery room management, 77–78

E

Echo heart, 56–57
Extracorporeal membrane oxygenation (ECMO), 57
  rationale, 126
  refractory hypoxemia, 126

F

Feeding pumped breast milk, 81
Fetal prognosis, 150–157
Formula feeding, 81

H

Healthcare personnel, 84–85
Hematological conditions, 149
High-flow nasal cannula, 121
HMG-CoA reductase inhibitors, 114–115
  pioglitazone, 115
  pregnancy, 115
  recommendations, 115
Host-directed therapy (HDT), 114

I

Immune system
  adaptive immunity
    B lymphocytes, 26
    immunoglobulins, 26
    T lymphocytes, 25–26
  COVID-19, 28
    course of the disease, 30–31
    fetal outcome, 31–32
    pregnancy issues, 28–32
    susceptibility to infection, 29–30
  infections, 26–27, 27t
  influenza, 27–28
  innate immunity, 24–25
    complement system, 24
    innate lymphoid cells (ILCs), 24–25
    leukocytes, 24, 25f
    monocytes, 24, 25f
  respiratory adaptations, 26
Immunoglobulins, 26
  COVID-19 clinical data, 110
  non-SARS-CoV-2, 110
  pregnancy, 110–111
  rationale for recommendation, 110
  SARS-CoV-2 specific, 110
Indeterminate NAAT, 47
Individuals contemplating pregnancy, 66–67
Influenza, 27–28
Innate immunity, 24–25
  complement system, 24
  innate lymphoid cells (ILCs), 24–25
  leukocytes, 24, 25f
  monocytes, 24, 25f
Innate lymphoid cells (ILCs), 24–25

Interleukin-6, 51
Intrinsically disordered regions (IDRs), 10

L

Lactate dehydrogenase (LDH) secretion, 51
Lactating individuals, 66
Lactoferrin (LF), 119–120
Leukocytes, 24, 25f
Lines of treatment
  adverse effects, 106–107
  antibacterial drugs, 94, 95t
  anticoagulants, 97–98
    thromboembolism, 97–98
  anti-interleukin-6 monoclonal antibody
    adverse effects, 104–105
    clinical data for, 105
    pregnancy, 105
    sarilumab, 104
    siltuximab, 105
  antimalarial drugs, 94–96
  antiparasitics, 97
  antithrombotic therapy, 101
  antiviral drugs, 92–93
  3-chymotrypsin-like protease (3CLpro), 93
  clinical data, 106
  coagulation markers, monitoring, 98–101, 99f
  convalescent plasma, 107
  corticosteroids
    adverse effects, 103
    dexamethasone, 103
    drug-drug interactions, 103
    interleukin (IL)-6 inhibitors, 104
    monitoring, 103
    patients, 102–103
  drug-drug interactions, 94, 96, 106
  immune-based therapy, 102
  interferon, 105–106
  interferon-alpha-2b, 106
  intravenous immunoglobulins (IVIGs), 110
  ivermectin, 97
  lactation, 100–102
  limitations, 109
  lopinavir/ritonavir (LPV/r), 93
  MERS, 106
  pregnancy, 100–102, 106–109

Note: Page numbers followed by "f" indicate figures and "t" indicate tables.

Lines of treatment (*Continued*)
   randomized controlled trial (RCT),
      92–93
   SARS, 106
   steroids, 102
   tocilizumab
      adverse effects, 105
      clinical data, 105
      pregnancy, 105
   venous thromboembolism
      treatment, NIH recommendations
      for, 100

M
Maternal death, 149
Maternal monitoring, 79
Mechanically ventilated adults
   ARDS, 122
   conventional mechanical ventilation,
      125
   helmet ventilation, 124
   high-flow nasal oxygenation (HFNO)
      therapy, 124
   high-frequency jet ventilation (HFJV),
      125
   high-frequency two-way jet
      ventilation (HFTJV), 125
   hypoxemia despite optimized
      ventilation, 123
   moderate-to-severe ARDS, 122–123
   noninvasive ventilation, 124
   oxygen face mask, 124
   oxygen nasal cannula, 123
   supraglottic jet oxygenation and
      ventilation (SJOV), 125–126
Mechanical ventilation, 147–148
Melatonin, 120–121
Mesenchymal stem cells (MSCs)
   clinical data, 112
   human umbilical cord
      mesenchymalstem cell (hUC-MSC)
      infusion, 112
   pregnancy, 112
Metformin, 114
Monocytes, 24, 25f
Mortality, 153–156
Mothernewborn contact, hospital,
   80

N
Negative NAAT result, 47
Neonatal COVID-19 infection,
   152–153, 154t
Neonatal prognosis, 150–157
Newborn evaluation, 80
Nonintubated patients, prone
   positioning for, 121–122
Noninvasive positive-pressure
   ventilation, 121
Nonmechanically ventilated adults
   with hypoxemic respiratory failure,
   121–122
Nonsteroidal antiinflammatory drugs
   (NSAIDs), 116

O
Obstertic outcomes, 150–157
Oxygenation, 121

P
Pathogen detection, secondary
   infection, 49–50
Pathophysiology
   acute kidney injury (AKI), 149
   acute respiratory failure, 148
   angiotensin-converting enzyme2
      (ACE2) receptors, 146
   cardiac complications, 149
   fetal prognosis, 150–157
   hematological conditions, 149
   ICU admission, 147–148
   maternal death, 149
   maternal prognosis, 150–157
   mechanical ventilation, 147–148
   mortality, 153–156
   neonatal COVID-19 infection,
      152–153, 154t
   neonatal prognosis, 150–157
   obstertic outcomes, 150–157
   perinatal morbidity, 153–156
   preeclampsia-like syndrome, 156
   severe disease and hospitalization,
      146
   susceptibility to infection, 146
   thromboembolic complications,
      148–149
   transient fetal skin edema, 157
   vertical transmission, 152–153
Patients' rooms, 83–84
Perinatal morbidity, 153–156
Permanent contraception, 81–82
Personal protective equipment,
   84–85
Platelet count, 50
Positive NAAT result, 47
Postpartum analgesia, 79
Preeclampsia-like syndrome, 156
Pregnancy
   aerosol-generating procedures,
      82–83
   American College of Obstetricians
      and Gynecologists' (ACOG's), 66
   anesthesia, 78–84
   antibody testing, 48–49
   asymptomatic individuals, 44
   asymptomatic mothers, 80–81
   asymptomatic/nonsevere COVID-19
      infections, 76
   at-risk populations, 4–6
   3-2 bedside X-ray, 54–57
   blood gas analysis, 53
   breastfeeding, 81
   cardiac troponin, 51–52
   chest computed tomography scan
      absorption-stage manifestations,
         54
      advanced stage manifestations, 54
      early-stage manifestations, 54
      preparation for patient, 54

Pregnancy (*Continued*)
   preparation prior to admission,
      53–54
   scanning parameters, 54
   scope and direction of scanning, 54
   severe-stage manifestations, 54
   chest imaging, 53
   clinical diagnosis
      coagulation response, 40–41
      endothelial cell function, 41
      immunological response, 40
      physiological changes, 40–41
      respiratory response, 40
   clinical findings, 41–57
      flu *vs.*, 43
      incubation period, 41
      signs, 41–43
      symptoms, 41–43
   contact personnel precautions, 82
   C-reactive protein (CRP), 49–50
   D-dimer, 51
   decontamination, 83
   delivery room management, 77–78
   discharge from hospital
      patients with COVID-19, 82
      patients without COVID-19, 82
      postpartum office visit, 82
   disseminated intravascular
      coagulation, 52
   elevated C-reactive protein, 50–51
   epidemiology, 3–6
   equipment, 83
   feeding pumped breast milk, 81
   formula feeding, 81
   gastrointestinal (GI) symptoms, 43
   genomic sequencing, 49
   healthcare personnel, 84–85
   before hospital admission, 72
   imaging of, 43
   imaging techniques, 54–56
      areas of the lesions, 54
      examination status, 54
      pulmonary lesions, 54–56
   inactive states, 8–9
   individuals contemplating
      pregnancy, 66–67
   infection history, 2
   interleukin-6, 51
   intrinsically disordered regions
      (IDRs), 10
   laboratory testing and, 43
   lactate dehydrogenase (LDH)
      secretion, 51
   lactating individuals, 66
   management during Cesarean
      delivery, 78
   modes of transmission, 3–4
   mothernewborn contact, hospital, 80
   newborn evaluation, 80
   pathogen detection, secondary
      infection, 49–50
   pathogenesis, 6–8
   patients' rooms, 83–84
   permanent contraception, 81–82

Pregnancy (*Continued*)
personal protective equipment, 84—85
general considerations, 84
individuals with potential or confirmed COVID-19, 84—85
platelet count, 50
postpartum care, 79
infection control measures, 79
maternal monitoring, 79
postpartum analgesia, 79
venous thromboembolism prophylaxis, 79
rapid diagnostic tests (RDTs), 47—48
remark, 69—70
renal markers, 52
reservoirs and hosts, coronaviruses, 3
reversible contraception, 81—82
route of delivery, 76—78
safety information, 66
SARS-CoV, 9—10
SARS-CoV-2
alpha coronaviruses, 14
beta coronaviruses, 14
clinical features, 10—11
COVID-19 Clinical course, 16
cytokines storm, 15—16
evolution of, 10
immunological response to, 15
middle east respiratory syndrome (MERS), 13—14
pathological changes, placenta, 13
pneumonia, 12
severe acute respiratory syndrome, 12—13
spike protein, 9—10
vertical transmission, 15
viral shedding, 11—16
symptomatic individuals, 44—45
fecal specimens, 45
nucleic acid amplification test, 46—47
respiratory specimen, 45
serum specimens, 45—46
symptomatic mothers, 80
test performance
assay type, 47
cycle threshold, 47
illness duration, 47
indeterminate NAAT, 47
negative NAAT result, 47
positive NAAT result, 47

Pregnancy (*Continued*)
specimen type, 46—47
test interpretation and additional testing, 47
timing of delivery, 72—76
total leukocytic count, 50
transmission of infection, 82
treatment plan, 67—72
ultrasonography, 56—57
echo heart, 56—57
extracorporeal membrane oxygenation, 57
fetal assessment, 57
intensive care unit, 57
peripheral vascular thrombosis, 57
pleural effusion, diagnosis and localization of, 56
pneumonia, additional diagnosis of, 56
pneumothorax sign, 56
pulmonary artery pressure, 57
thoracic examination, 56—57
vaccination, 66
viral isolation, 49
viral life cycle, 6—8
viral spike protein active, 8—9
viral structure, 6
viral taxonomy, 6
virology, 6—16
World Health Organization (WHO), 70—72, 71t, 73t—75t

R
Rapid diagnostic tests (RDTs), 47—48
Renal markers, 52
Respiratory adaptations, 26
Reversible contraception, 81—82
Route of delivery, 76—78

S
SARS-CoV-2, pregnancy
alpha coronaviruses, 14
beta coronaviruses, 14
clinical features, 10—11
COVID-19 Clinical course, 16
cytokines storm, 15—16
evolution of, 10
immunological response to, 15
middle east respiratory syndrome (MERS), 13—14
pathological changes, placenta, 13
pneumonia, 12

SARS-CoV-2, pregnancy (*Continued*)
severe acute respiratory syndrome, 12—13
spike protein, 9—10
vertical transmission, 15
viral shedding, 11—16
Symptomatic individuals, 44—45
fecal specimens, 45
nucleic acid amplification test, 46—47
respiratory specimen, 45
serum specimens, 45—46
Symptomatic mothers, 80

T
Test interpretation and additional testing, 47
Test performance
assay type, 47
cycle threshold, 47
illness duration, 47
Thromboembolic complications, 148—149
T lymphocytes, 25—26
Transient fetal skin edema, 157

V
Vaccination, 66
Venous thromboembolism prophylaxis, 79
Ventilation, 121
Vertical transmission, 152—153
Vitamin C, 117
critically ill patients without COVID-19, 117
noncritically ill patients with COVID-19, 117
Vitamin D
adverse effects, 118
drug interactions, 118
pregnancy, 118

W
World Health Organization (WHO), 70—72, 71t, 73t—75t

Z
Zinc
clinical data, 119
COVID-19, 118—119
pregnancy, 119
side effects, 119

Printed in the United States
by Baker & Taylor Publisher Services